African Edible Insects As Alternative Source of Food, Oil, Protein and Bioactive Components

Abdalbasit Adam Mariod
Editor

African Edible Insects As Alternative Source of Food, Oil, Protein and Bioactive Components

 Springer

Editor
Abdalbasit Adam Mariod
Indigenous Knowledge and Heritage Centre
Ghibaish College of Science and Technology
Ghibaish, Sudan

ISBN 978-3-030-32951-8   ISBN 978-3-030-32952-5   (eBook)
https://doi.org/10.1007/978-3-030-32952-5

This Springer imprint is published by the registered company Springer Nature Switzerland AG
The registered company address is: Gewerbestrasse 11, 6330 Cham, Switzerland

Contents

1   **Importance of Insects as Food in Africa**   1
Arnold van Huis

2   **African Edible Insect Consumption Market** 19
Cordelia Ifeyinwa Ebenebe, Oluwatosin Samuel Ibitoye,
Inwele Maduabuchi Amobi, and Valentine Obinna Okpoko

3   **Entomophagy in Africa** 53
Karanjit Das

4   **Microbiology of African Edible Insects** 59
Nils Th. Grabowski

5   **Food Safety of Edible Insects** 83
Miklós Mézes and Márta Erdélyi

6   **Interdisciplinary Uses of Some Edible Species** 95
Beatrice Mofoluwaso Fasogbon

7   **Sensory Quality of Edible Insects** 115
Marwa Yagoub Farag Koko and Abdalbasit Adam Mariod

8   **Automation of Insect Mass Rearing and Processing
Technologies of Mealworms (*Tenebrio molitor*)** 123
Nina Kröncke, Andreas Baur, Verena Böschen,
Sebastian Demtröder, Rainer Benning, and Antonio Delgado

9   **The Legislative Status of Edible Insects in the World** 141
Abdalbasit Adam Mariod

10   **Sorghum Bug (*Agonoscelis pubescens*) as a Source
of Edible Oil, Protein, and Gelatin** 149
Abdalbasit Adam Mariod

11   **Watermelon Bug (*Aspongopus viduatus*) as a Source
     of Edible Oil, Protein, and Gelatin**   159
     Abdalbasit Adam Mariod

12   **Nutritional Composition of African Edible Acridians**   169
     Sévilor Kekeunou, Alain Simeu-Noutchom,
     Marcelle Mbadjoun-Nziké, Mercy Bih Achu-Loh,
     Patrick Akono-Ntonga, Alain Christel Wandji,
     and Joseph Lebel Tamesse

13   **Nutrient Composition of Black Soldier Fly (*Hermetia illucens*)**   195
     Matan Shelomi

14   **Production, Nutrient Composition, and Bioactive
     Components of Crickets (Gryllidae) for Human Nutrition**   213
     Monica A. Ayieko and Mary A. Orinda

15   **Nutrient Composition and Bioactive Components of
     Ants (*Oecophylla smaragdina* Fabricius)**   225
     Abdalbasit Adam Mariod

16   **Nutrient Composition and Bioactive Components
     of the Migratory Locust (*Locusta migratoria*)**   231
     Suzy Munir Salama

17   **Nutrient Composition and Bioactive Components
     of Mopane Worm (*Gonimbrasia belina*)**   241
     Raphael Kwiri, Felix M. Mujuru, and Wishmore Gwala

18   **Nutrient Composition of Desert Locust (*Schistocerca gregaria*)**   257
     Abdalbasit Adam Mariod

19   **Nutritional Value of Brood and Adult Workers
     of the Asia Honeybee Species *Apis cerana* and *Apis dorsata***   265
     Sampat Ghosh, Bajaree Chuttong, Michael Burgett,
     Victor Benno Meyer-Rochow, and Chuleui Jung

20   **Nutrient Composition of Mealworm (*Tenebrio molitor*)**   275
     Abdalbasit Adam Mariod

21   **Nutrient Composition of Termites**   281
     Oladejo Thomas Adepoju

22   **Termites in the Human Diet: An Investigation
     into Their Nutritional Profile**   293
     Sampat Ghosh, Daniel Getahun Debelo, Wonhoon Lee,
     V. Benno Meyer-Rochow, Chuleui Jung, and Aman Dekebo

Index   307

Chapter 1
Importance of Insects as Food in Africa

Arnold van Huis

Abstract In Africa, about 470 insect species are recorded as edible, of which caterpillars are most consumed followed by grasshoppers, beetles, and termites. Most of those are collected from nature. There are several insect species, such as locusts and grasshoppers, that are pests of crops but which can be eaten at the same time. There are some edible insect species which are harvested in large number contributing to food security. Three of those species are discussed: the mopane caterpillar, the African bush cricket, and the shea caterpillar. However, when we would like to promote insects as food then harvesting from nature is not an option anymore, as overexploitation already occurs. Then we need to rear the insects. That can be done in semi-domesticated systems such as for the palm weevil or by farming insects as mini-livestock such as for crickets. We discuss the nutritional value of edible insects, and how they can contribute to food security. We also give examples of how insects can be processed and marketed. We conclude with the prospects of how edible insects can assure food security and improve the livelihood of the African people.

Keywords Sub-Saharan Africa · Edible insects · Insects as food · Nutrition · Food security · Harvesting insects · Farming insects

1.1 Introduction

The level of prevalence of undernourishment in sub-Saharan Africa aggravated over the last few years. It went first down from 24.3% of the population in 2005 to 20.7% in 2014 but then went up again to 23.2% in 2017, affecting 237 million people, while the number of people experiencing severe food insecurity in 2017 was 346

A. van Huis (✉)
Laboratory of Entomology, Wageningen University & Research,
Wageningen, The Netherlands
e-mail: arnold.vanhuis@wur.nl

© Springer Nature Switzerland AG 2020
A. Adam Mariod (ed.), *African Edible Insects As Alternative Source of Food, Oil, Protein and Bioactive Components*, https://doi.org/10.1007/978-3-030-32952-5_1

million (34%) (FAO et al. 2018). In Africa in 2017, 59 million (30%) of the children under five were affected by stunting (chronic malnutrition) and 14 million (7%) by wasting (acute malnutrition). The influence of climate change on production and livelihoods is strongest in Africa as dryland farming and pastoral rangeland systems dominate livelihood systems for 70–80% of the continent's rural population (Neely et al. 2009). Conditions of desertification and drought are aggravated by the impacts of human activities. Changes in climate impact heavily on nutrition through (1) impaired nutrient quality and dietary diversity of foods produced and consumed; (2) effects on water and sanitation, with their implications for patterns of health risks and disease; and (3) changes in maternal and child care and breast feeding (FAO et al. 2018).

Meat production in sub-Saharan Africa has been estimated to be 7334 thousand tons in 2005/2007 (only 2.6% of global meat production) and is expected to grow by 2.9% per annum up to the year 2050 (Alexandratos and Bruinsma 2012). Because of environmental reasons, dietary changes are urgently needed (Springmann et al. 2018), and insects, being already an important food source, should be considered.

The recorded number of insect species that are eaten in Africa is 472 (Fig. 1.1) (Jongema 2017). The most abundant group of species eaten are caterpillars (Lepidoptera) (31%), followed by grasshoppers, crickets, and locusts (Orthoptera) (23%). The Coleoptera, in particular the larvae follow (19%), and then termites (Isoptera) (7%), bees, wasps and ants (Hymenoptera) (7%), true bugs (Heteroptera) (6%), aphids, scale insects, cicadas, and leafhoppers (Homoptera) (4%), and flies (Diptera) (1%). The rest (3%) consists of cockroaches (Dictyoptera), mayflies

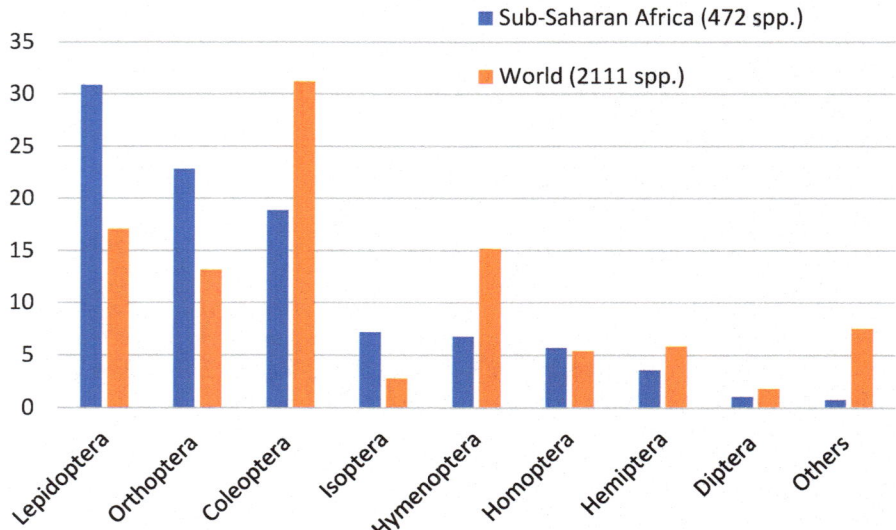

Fig. 1.1 Percentage of recorded edible insect species per insect order in sub-Saharan Africa and the world (Jongema 2017)

(Ephemeroptera), and other arthropod species such as hard ticks (Ixodidae) and spiders (Araneae).

When comparing these figures with those of the world (Fig. 1.1), then it seems that in Africa there are more recorded edible insect species in the Lepidoptera (caterpillars), Orthoptera (grasshoppers, crickets, and locusts), and Isoptera (termites) and less in the Coleoptera (beetles) and Hymenoptera than in the rest of world. There may be several reasons. They are recorded species which means that they have only been included in the database when known from literature. Taking the Lepidoptera, considerable effort has been made by several authors (Malaisse et al. 2017) to categorize those. Another reason is the continental diversity of an insect group; for example, termites have the largest diversity in Africa (38% of all species) (Lewis 2003).

Several review articles have been written about edible insects in Africa: Kelemu et al. (2015); Malaisse (2005) and van Huis (2003). We will deal with recent developments and cover traditional harvesting, semi-domestication, and farming of insects. We mention some economic important (pest) species and briefly deal with insects as feed for animals. Then we discuss nutritional issues, processing and marketing and how insects can contribute to food security. We finish with giving an outlook on insects as food in sub-Saharan Africa.

1.2   Pest Insects as Human Food

Certain insects can be a pest and at the same time are consumed by humans. The most well-known pests are the four locust species in Africa: the desert locust (*Schistocerca gregaria*), the migratory locust (*Locusta migratoria*), the brown locust (*Locusta napardalina*), and the red locust (*Nomadacris septemfasciata*) (Orthoptera: Acrididae). They may cause considerable damage to crops in Africa. When they occur during upsurges and plagues, they are eaten. However, it is a curse and a blessing at the same time. Sometimes it has been suggested that eating them could be used a control method. However, the extent of bands and swarms is so large and often in remote areas that this is not an option. The consumption of those locust may be a health risk as they are often treated with pesticides (Saeed et al. 1993).

The fall army worm *Spodoptera frugiperda* (Lepidoptera: Noctuidae), a serious pest of maize on the American continents feeding on young leaf whorls, ears, and tassels, has been reported for the first time in early 2016 in West and Central Africa (Goergen et al. 2016), and has spread to other parts of Africa (FAO 2017). Would it be an option to eat the caterpillar? Another indigenous pest, the African armyworm, *Spodoptera exempta* is capable of destroying entire crops by feeding on early stages of cereal crops. Mbata (1995) mentioned that this species is eaten by the local population in Zambia. The question is whether *S. frugiperda* can be eaten.

The caterpillar *Cirina forda* (Lepidoptera: Saturniidae) is found in western Africa, including Ghana and Nigeria, but also in Zimbabwe, the Democratic

Republic of the Congo, and South Africa. In Ghana and Nigeria, the larvae may cause heavy defoliation of the African shea tree (*Vitellaria paradoxa*). In southwestern Nigeria, it is widely marketed, and commonly consumed as a cheap delicacy being served as snacks or as an essential ingredient in vegetable soups along with carbohydrate food (Adepoju and Daboh 2013; Paiko et al. 2014). Its nutritional value is high (Akinnawo and Ketiku 2000) and it can be used as human food or animal feed (Omotoso 2006) replacing fishmeal in poultry diets (Amao et al. 2010; Oyegoke et al. 2006). Artificial rearing on the leaves of the African shea tree is possible (Ande and Fasoranti 1998).

Another pest, the variegated grasshopper *Zonocerus variegatus* (Orthoptera: Pyrgomorphidae) attacks food and cash crops such as banana, cassava, cocoa, citrus, cowpea, maize, soybeans, and yam in central and West Africa (Kekeunou et al. 2006). In Nigeria, the control of the variegated grasshopper has been reported as a control method (Iduwu and Modder 1996). In this country the grasshoppers are caught at night during the cold weather (November–February); at low temperatures because inactivity of the cold-blooded animals enables easy catching. After being boiled and roasted, they are displayed in markets and sold like meat (Solomon et al. 2008). Considering the high nutritional content, the grasshopper can both be used as food for humans and feed for animals and fish (Ademolu et al. 2010, 2017; Alegbeleye et al. 2012; Olaofe et al. 1998; Sani et al. 2014; Solomon et al. 2008). This is despite the fact that when molested, both sexes and instars expel an odorous, milky secretion of which the odour is repulsive or unpleasant to human beings (Idowu and Idowu 1999). This is the reason that the insect species is called "criquet puant" in French which means "stinking locust."

1.3   Economic Important of Edible Insect Species

Although there are quite some edible insect species that are economically important we only discuss the mopane caterpillar, the African edible bush cricket and the shea caterpillar. Some other species were already mentioned in Sect. 1.2.

1.3.1   Mopane Caterpillar

The mopane caterpillar *Imbrasia belina* (Lepidoptera: Saturniidae) (Fig. 1.2) feeds on the mopane tree *Colophospermum mopane*. The larvae are not only popular food in many cultures in southern Africa, they also are often an important source of income for rural households. In South Africa in 1983, it was estimated that 1.6 million kg of traditionally prepared dried caterpillars were traded (Dreyer and Wehmeyer 1982). In Stack et al. (2003) it was mentioned that Styles (1994) estimated that yearly 9500 million caterpillars were harvested in an area of 20,000 km²,

Fig. 1.2 Dried mopane caterpillar *Imbrasia belina* (Lepidoptera: Saturniidae). Photo credits and copyright: Hans Smid—www.bugsinthepicture.com

only in South Africa, for the value of about US\$ 85 million at that time. Approximately 40% goes to producers, who are primarily poor rural women.

The protein content of the larvae is high with large amounts of all of the essential amino acids, essential, fatty acids (linoleic acid, α-linolenic acid), and many minerals (such as iron and zinc) critical to normal growth, development, and health maintenance (Glew et al. 1999). In the Limpopo Province in South Africa, the trade and consumption of mopane caterpillars contributed to rural household food security (Baiyegunhi et al. 2016). However, overexploitation and commercialisation threatens the long-term management of the mopane woodlands, and a balance need to be found between sustainable harvesting of mopane caterpillar and improving the livelihoods of the rural poor (Baiyegunhi and Oppong 2016). Strategies proposed are delaying the supply of the stock to the market and practices to maintain a sufficient number of fifth-instar mopane caterpillars, as well as safeguard the host tree against exploitation and ways to preserve the pupae (Gondo et al. 2010). Also restrictive harvesting periods have been proposed but there are doubts about its effectiveness (Akpalu et al. 2007). As the occurrence of mopane caterpillar is erratic and periodically fails to produce caterpillars of harvestable size, there is now an increased interest in developing domestic farming techniques of the caterpillars at the household level. However, this depends on the technical feasibility as parasitism, predation and diseases proved to be challenge besides that it was costly (Ghazoul 2006). It depends also on a number of other issues, such as farming only becoming interesting when levels of wild populations that can be harvested are low (Hope et al. 2009).

Cereals have low iron and zinc bioavailability, and it has been attempted to enrich cereals with mopane caterpillar in Zimbabwe (Gabaza et al. 2018). However, the bio-accessibility of iron and zinc was not improved, it only increased the iron and zinc content of the enriched fermented cereals. Also the nutritional potential of the mopane caterpillar has been studied in diets through its use in fortified blended foods formulations (Kwiri et al. 2014). Allergic reactions to the consumption of mopane caterpillar are possible (Kung et al. 2011; Okezie et al. 2010).

1.3.2   The African Edible Bush Cricket

The African edible bush cricket *Ruspolia differens* (Orthoptera: Tettigoniidae) is considered a delicacy in Uganda and is called by the Lugandan name "nsenene." It also occurs in Kenya, Rwanda, Tanzania, and Madagascar. The insect species appears in nocturnal flying swarms of a high density, during May–April and November–December and are gathered as a highly prized item of human food (McCrae 1982). *Ruspolia nitidula* locally may be boiled or eaten raw, or sun-dried, fried, and flavored with onions, or used to make a soup (Agea et al. 2008). Sun-dried insects may be kept for several months.

The grasshopper oviposits on the leafage of grasses on which they develop (Bailey and McCrae 1978). Due to their inactivity during cold weather they are collected in the morning by hand from the grasses. As the grasshoppers during their nocturnal flights are attracted by light, women and children often engage in collecting from street lights; however, with the traffic that is a dangerous undertaking. There is homestead collection and harvesting in a more commercial way. Light traps are made by folding corrugated iron sheets into a cone shape. The lights attract the insects, which hit the sheets and then fall in large buckets with a hole on the lid (Mmari et al. 2017). It is also attempted to mass-rear the insects (Lehtovaara et al. 2018; Malinga et al. 2018a, b; Rutaro et al. 2018a, b).

1.3.3   Shea Caterpillar

The shea caterpillar *Cirina butyrospermi* (Lepidoptera: Saturniidae) only has the African shea tree as a host. It is highly valued as a human food item in Burkina Faso and Mali (Séré et al. 2018). Caterpillars can be dried, fried or boiled and used in various meals. The insect has exceptional nutritional characteristics, with 63% protein, 15% fat, as well as vitamins and minerals (Anvo et al. 2016b). It has also potential as fish feed (Anvo et al. 2016a, 2017). However, one of the main constraints on the consumption of this insect is its seasonal availability, due to its univoltine cycle. Therefore, it has been studied whether it would be possible to rear the insects by breaking the diapause (Bama et al. 2018; Rémy et al. 2018).

1.4   Insect as Feed

Small farmers in Africa feed chickens and guinea fowls with termites. They do so by breaking open small termite nests such as those of *Microtermes* spp. (van Huis 2017). However, in Togo they also used a method to lure termites to clay containers in which they have placed for example dry stems of sorghum or other cereals. They add some water and then put the pot upside down with the opening on a termite gallery. Then they wait until there are enough termites (3–4 weeks) and then they empty the pot for the chicks (Farina et al. 1991).

The costs of feeding fish and poultry is often 60–80% of the total production costs, and this is due to the relative expensive protein sources such as fish meal and soy meal (Ssepuuya et al. 2017). These conventional protein sources could be replaced by 10–100% with insects (grasshoppers, house fly maggots, *Cirina forda* larvae, termites) without affecting the growth performance of fish and poultry. Moreover, insect-based feed in some cases performed better than conventional feed.

Worldwide there is a lot of interest to develop insect-based food and in particular the black soldier fly *Hermetia illucens* (Diptera: Stratiomyidae) and the housefly *Musca domestica* (Diptera: Muscidae) are candidates and were also considered of interest for West Africa (Kenis et al. 2014). Roffeis et al. (2018) looked at the economic implications of implementation in this part of Africa. Considering the prices of conventional feeds, there seems to be potential to substitute imported fishmeal with insect-based feeds, but there were no economic advantages over plant-based feeds.

1.5   Farming Insects

Simple rearing methods are sometimes carried out by the local population. It is also called semi-domestication in which the captive state of wild insects in which the living conditions are controlled by humans. Van Itterbeeck and Van Huis (2012) gave a number of examples of how to increase the predictability and availability of edible insects such as manipulating shifting cultivation and fire regimes in Africa to improve forest caterpillar exploitation. Below we give an example of the palm weevil. However, when insects are produced as mini-livestock, we can call it farming. We give an example of crickets which can be farmed by households.

1.5.1   Palm Weevil

Palm weevil larvae are one of the edible insects that are popularly eaten in different parts of the world. In Africa, the African palm weevil *Rhynchophorus phoenicis* (Coleoptera: Curculionidae) is considered a delicacy. The larvae (Fig. 1.3) are widely consumed raw, boiled, fried, smoked, and sometimes used in the preparation of stews and soups, as part of a meal or as a complete meal (Nrior et al. 2018). The African palm weevil is common in the humid lowland forest and savannah areas of Africa. It feeds mainly on oil palm, date palm, raffia palm, and coconut palm. The larvae are important pests of these plant species, due to their boring action into plant stems. Oil palms cut down for palm wine production, but also trunks of dead or wounded palms attract adult weevils. The females after mating lay their eggs in the

Fig. 1.3 African palm
weevil larvae
(*Rhynchophorus phoenicis*;
Coleoptera: Curculionidae)
sold at a local market in
Yaoundé, Cameroon.
(Photo by author)

trunks and when the larvae are full grown in about 4 weeks, they are harvested
(Muafor et al. 2015). Traditional harvesting methods of African palm weevil larvae
are very destructive to the ecosystem, as a single collector can cut down more than
1100 raffia trunks (Ayemele et al. 2016). For this reason it is studied whether the
larvae can be reared. Quayea et al. (2018) showed that agro-waste materials from
fruits, banana, pineapple, and millet waste can be used as alternative feed resources.
Muafor et al. (2017) developed simple rearing techniques using as substrate Raphia
palms. They were able to produce the insect year round and increase the production
at least four times in comparison to the harvest obtained in the wild.

The palm weevil larvae have a high nutritional value (Cito et al. 2017; Edijala
et al. 2009; Ekpo 2010; Ekpo and Onigbinde 2005, 2007; Mba et al. 2017; KOFFI
et al. 2017; Lenga et al. 2012; Nzikou 2010; Okunowo et al. 2017; Omotoso and
Adedire 2007; Quaye et al. 2018). Processing also influences the nutritional value;
for example, grilling, roasting, or boiling increased the biological value (Ekpo
2011) and solubility of proteins (Womeni et al. 2012). Adeboye et al. (2016)
developed cookies of good quality and acceptability by supplementing wheat

flour with 10% palm weevil larvae flour. Due to the high microbial (bacteria and fungi) load, adequate hygienic practices and proper processing are needed before they can be consumed (Nrior et al. 2018).

1.5.2 Crickets

Among the different species of edible insects, crickets are one of the most interesting as they have shown particular promise in addressing food and nutrition insecurity. Their high nutritional quality makes a variety of cricket species attractive for rearing. Besides their life cycle is considerably shorter than that of traditional livestock species. They also require a reduced-size rearing space and are able to recycle agricultural by-products, while having a high feed conversion ratio (Caparros Megido et al. 2017). Cricket rearing is popular in Thailand where 20,000 smallholder farms produce more than 7500 tonnes a year (Hanboonsong et al. 2013). The Flying Food project, a consortium between partners from Kenya, Uganda and the Netherlands, produced a manual for trainers who will teach small holder farmers how to setup and maintain a cricket rearing production system (Beckers et al. 2019). Also in Kenya, farming of crickets (*Acheta domesticus* and *Gryllus bimiculatus*) have been introduced for households (Ayieko et al. 2016).

In Kenya, a new cricket of the genus *Scapsipedus* (Orthoptera: Gryllidae) has been described (Tanga et al. 2018). This species has been reared for 3 years in the research facility at the International Centre of Insect Physiology and Ecology (icipe), Duduville Campus, Nairobi, Kenya. It has been demonstrated through several research activities that it is a very promising species for mass rearing for food and feed. They named the species *Scapsipedus icipe* n. sp.

1.6 Nutrition

Edible insect species in Cameroon (R. *phoenicis*, R. *differens*, Z. *variegatus*, *Macrotermes* sp., *Imbrasia* sp.) are considerable sources of fat (Womeni et al. 2009). Their oils are rich in polyunsaturated fatty acids, of which essential fatty acids are linoleic and linolenic acids. The "polyunsaturated fatty acid–saturated fatty acid" ratio is in the majority of cases higher than 0.8. The major fatty acids (occurring at more than 10%) of R. *phoenicis* are palmitic acid, oleic acid and linoleic acid while those of G. *belina* and C. *forda* are palmitic, oleic, linoleic, and stearic acids (Amadi and Kiin-Kabari 2016).

Two caterpillar species sold in South Africa and Zimbabwe had high iron and zinc, in particular iron is an important nutrient for combating anaemia (Payne et al. 2015). In the Democratic Republic of Congo the efficacy of a cereal made from caterpillars was assessed on reducing stunting and anaemia in infants at 18 month of age (Bauserman et al. 2015). It did not reduce the prevalence of stunting, but those infants had higher Hb concentration and fewer were anaemic.

Payne et al. (2016) compared energy and 12 relevant nutrients for three commonly consumed meats (beef, pork and chicken), and six commercially available insect species (the cricket *A. domesticus*, the honeybee *Apis mellifera*, the domesticated silkworm *Bombyx mori*, the mopane caterpillar *I. belina*, the African palm weevil larva *R. phoenicis* and the yellow mealworm *Tenebrio molitor*). They used two models. According to the Ofcom model, no insects were significantly "healthier" than meat products. According to the Nutrient Value Score, crickets, palm weevil larvae and mealworm had a significantly healthier score than beef and chicken. It was difficult to draw general conclusions as there was a large variation between edible insect species in nutrient content. It seems that in the context of overnutrition meat product may be preferred over certain insect species, while in the context of undernutrition certain insects species are a better choice.

1.7 Contribution to Food Security

Niassy et al. (2016) showed a gender bias toward women and children in the collection and consumption of edible insects. Women and children play an active role in the whole value chain of edible insects including collection, processing, and packaging. By collecting they secure stocks of proteins for household consumption and generate income as has been shown for the mopane caterpillar in South Africa (Baiyegunhi et al. 2016). In Zimbabwe, the harvesting and processing of the mopane caterpillar is carried out by women and children while men in general dominate the more lucrative long-distance and large-volume trading chains, but in general the collection and marketing is carried out by relatively poor people (Kozanayi and Frost 2002). In South Africa, where the edible stink bug *Encosternum delegorguei* (Hemiptera: Tessaratomidae) is a traditional delicacy of some ethnic groups in South Africa and Zimbabwe, women control 72% of the market (Dzerefos et al. 2014).

1.8 Processing and Marketing Insects

In Kenya, termites and lake flies were collected and processed in the laboratory under different types of cooking methods such as baking, boiling, and steam cooking under pressure. The processed products, such as crackers, muffins, sausages, and meat loaf, were readily accepted (Ayieko et al. 2010). In the same country a number of processed products with crickets were developed and presented to consumers. Children were particularly attracted to biscuits and the fried foods such as fritters, samosa, and pancakes (Ayieko et al. 2016). Processing can have an effect on the digestibility and vitamin content of edible insects as was shown by Kinyuru et al. (2010) with the termite *Macrotermes subhyalinus* and the grasshopper *R. differens*. In Kenya, it was also investigated how consumers evaluated the sensory attributes of a common bakery product (buns) that was blended with cricket flour. Providing information on the product could enhance consumer acceptance of the insect-based food (Pambo et al. 2018). However, Kinyuru et al. (2009) also showed that it depends on the concentration of insect flour in the product.

1.9 Conclusions

Although insect consumption is common in Africa, the danger exists that western food habits are copied and that insects as traditional food are being considered as a poor man's diet (Niassy et al. 2016; van Huis 2013). This seems strange as in the western world insects are now increasingly being appreciated as nutritious food and more sustainable than the common meat products (Van Huis and Oonincx 2017). Kelemu et al. (2015) called it for Africa *"an overlooked food source"* and she mentions with others (Kelemu et al. 2015) that *"within the context of sustainable diet, the use of insects as food and feed has a significant role to play in assuring food security and improving the livelihood of the African people."* However, traditionally insects have been harvested from nature and when this food source is going to be promoted than this is not an option anymore. Insects need to be reared and urgently rearing methods need to be developed for a number of species which are highly popular in Africa. This would change the food source from being seasonal to a continuous available food item. Besides also processing methods need to be developed to make them into readily available insect-based products with a long shelf life.

References

Adeboye AO, Bolaji TA, Fatola OL (2016) Nutritional composition and sensory evaluation of cookies made from wheat and palm weevil larvae flour blends. Ann Food Sci Technol 17:543–547

Ademolu KO, Idowu AB, Olatunde GO (2010) Nutritional value assessment of variegated grasshopper, *Zonocerus variegatus* (L.) (Acridoidea: Pygomorphidae), during post-embryonic development. Afr Entomol 18:360–364. https://doi.org/10.4001/003.018.0201.

Ademolu KO, Simbiat ES, Concilia I, Adeyinka AA, Abiodun OJ, Adebola AO (2017) Gender variations in nutritive value of adult variegated grasshopper, *Zonocerus variegatus* (L) (Orthoptera:Pygomorphidae). J Kansas Entomol Soc 90:117–121. https://doi.org/10.2317/170325.1.

Adepoju OT, Daboh OO (2013) Nutrient composition of *Cirina forda* (Westwood) enriched complementary foods. Ann Nutr Metab 63:139–144

Agea JG, Biryomumaisho D, Buyinza M, Nabanoga GN (2008) Commercialization of *Ruspolia nitidula* (Nsenene grasshoppers) in central Uganda. Afr J Food Agricult Develop 8:319–332

Akinnawo O, Ketiku AO (2000) Chemical composition and fatty acid profile of edible larva of *Cirina forda* (Westwood). Afr J Biomed Res 3:93–96

Akpalu W, Muchapondwa E, Zikhali P (2007) Can the restrictive harvest period policy conserve mopane worms in Southern Africa? A bio-economic modelling approach. Working paper number 65, University of Pretoria.

Alegbeleye WO, Obasa SO, Olude OO, Otubu K, Jimoh W (2012) Preliminary evaluation of the nutritive value of the variegated grasshopper (*Zonocerus variegatus* L.) for African catfish *Clarias gariepinus* (Burchell. 1822) fingerlings. Aquac Res 43:412–420. https://doi.org/10.1111/j.1365-2109.2011.02844.x.

Alexandratos N, Bruinsma J (2012) World agriculture towards 2030/2050: the 2012 revision global perspective studies team. ESA working paper no 12-03, Agricultural Development Economics Division. Food and Agriculture Organization of the United Nations, Rome. www.fao.org/economic/esa

Amadi E, Kiin-Kabari D (2016) Nutritional composition and microbiology of some edible insects commonly eaten in africa, hurdles and future prospects: a critical review. J Food Microbiol Safety Hygiene 1:107. https://doi.org/10.4172/2476-2059.1000107

Amao OA, Oladunjoye IO, Togun VA, Olubajo K, Oyaniyi O (2010) Effect of Westwood (*Cirina forda*) larva meal on the laying performance and egg characteristics of laying hen in a tropical environment. Int J Poult Sci 9:450–454

Ande AT, Fasoranti JO (1998) Some aspects of the biology, foraging and defensive behaviour of the emperor moth caterpillar, *Cirina forda* (Westwood). Int J Tropic Insect Sci 18:177–181. https://doi.org/10.1017/S1742758400023377.

Anvo MPM, Toguyéni A, Otchoumou AK, Zoungrana-Kaboré CY, Kouamelan EP (2016a) Evaluation of *Cirina butyrospermi* caterpillar's meal as an alternative protein source in *Clarias gariepinus* (Burchell, 1822) larvae feeding. Int J Fish Aquatic Stud 4:88–94

Anvo MPM, Toguyéni A, Otchoumou AK, Zoungrana-Kaboré CY, Kouamelan EP (2016b) Nutritional qualities of edible caterpillars *Cirina butyrospermi* in southwestern of Burkina Faso. Int J Innov Appl Stud 18:639–645

Anvo MPM, Aboua BRD, Compaoré I, Sissao R, Zoungrana-Kaboré CY, Kouamelan EP, Toguyéni A (2017) Fish meal replacement by *Cirina butyrospermi* caterpillar's meal in practical diets for *Clarias gariepinus* fingerlings. Aquac Res 48:5243–5250. https://doi.org/10.1111/are.13337.

Ayemele AG, Muafor FJ, Levang P (2016) Indigenous management of palm weevil grubs (*Rhynchophorus phoenicis*) for rural livelihoods in Cameroon. J Insects Food Feed 31:43–50. https://doi.org/10.3920/JIFF2016.0002

Ayieko MA, Oriamo V, Nyambuga IA (2010) Processed products of termites and lake flies: improving entomophagy for food security within the Lake Victoria region. Afr J Food Agric Nutr Dev 10:2085–2098

Ayieko MA, Ogola HJ, Ayieko IA (2016) Introducing rearing crickets (gryllids) at household levels: adoption, processing and nutritional values. J Insects Food Feed 2:203–211. https://doi.org/10.3920/JIFF2015.0080

Bailey WJ, McCrae AWR (1978) The general biology and phenology of swarming in the East African tettigoniid *Ruspolia differens* (Serville) (Orthoptera). J Nat Hist 12:259–228. https://doi.org/10.1080/00222937800770151.

Baiyegunhi LJS, Oppong BB (2016) Commercialisation of mopane worm (*Imbrasia belina*) in rural households in Limpopo Province. South Afr Forest Policy Econom 62:141–148. https://doi.org/10.1016/j.forpol.2015.08.012.

Baiyegunhi LJS, Oppong BB, Senyolo GM (2016) Mopane worm (*Imbrasia belina*) and rural household food security in Limpopo province. South Afr Food Security 8:153–165. https://doi.org/10.1007/s12571-015-0536-8.

Bama HB, Dabire RA, Ouattara D, Niassy S, Ba MN, Dakouo D (2018) Diapause disruption in *Cirina butyrospermi* Vuillet (Lepidoptera, Attacidae), the shea caterpillar, in Burkina Faso. J Insects Food Feed. https://doi.org/10.3920/JIFF2017.0068

Bauserman M et al (2015) A cluster-randomized trial determining the efficacy of caterpillar cereal as a locally available and sustainable complementary food to prevent stunting and anaemia. Public Health Nutr 18:1785–1792. https://doi.org/10.1017/S1368980014003334.

Beckers E, et al (2019) Training manual: cricket rearing for small holder farmers using the 30/3 crate system. Flying Food project, Wageningen

Caparros Megido R, Haubruge É, Francis F (2017) Chapter 5. Small-scale production of crickets and impact on rural livelihoods. In: Van Huis A, Tomberlin JK (eds) Insects as food and feed: from production to consumption. Wageningen Academic Publishers, Wageningen, pp 101–111

Cito A, Longo S, Mazza G, Dreassi E, Francardi V (2017) Chemical evaluation of the *Rhynchophorus ferrugineus* larvae fed on different substrates as human food source. Food Sci Technol Int 23:529–539. https://doi.org/10.1177/1082013217705718.

Dreyer JJ, Wehmeyer AS (1982) On the nutritive value of mopanie worms. S Afr J Sci 78:33–35

Dzerefos CM, Witkowski ETF, Toms R (2014) Use of the stinkbug, *Encosternum delegorguei* (Hemiptera, Tessaratomidae), for food and income in South Africa. Soc Nat Resour 27:882–897. https://doi.org/10.1080/08941920.2014.915368.

Edijala JK, Egbogbo O, Anigboro AA (2009) Proximate composition and cholesterol concentrations of *Rhynchophorus phoenicis* and *Oryctes monoceros* larvae subjected to different heat treatments. Afr J Biotechnol 8:2346–2348

Ekpo KE (2010) Nutrient composition, functional properties and anti-nutrient content of *Rhynchophorus pheonicis* (F) larva. Ann Biol Res 1:178–190

Ekpo KE (2011) Effect of processing on the protein quality of four popular insects consumed in Southern Nigeria. Arch Appl Sci Res 3:307–326

Ekpo KE, Onigbinde AO (2005) Nutritional potentials of the larva of *Rhynchophorus phoenicis* (F). Pak J Nutr 4:287–290

Ekpo KE, Onigbinde AO (2007) Characterization of lipids in *Rhynchophorus pheonicis* larval oil. Pak J Sci Ind Res 50:75–79

FAO (2017) FAO advisory note on fall armyworm (FAW) in Africa. Advisory note 5. June 2017. Food and Agriculture Organization of the United Nations, Rome. http://www.fao.org/3/a-bs914e.pdf

FAO, IFAD, UNICEF, WFP, WHO (2018) The State of Food Security and Nutrition in the World 2018. Building climate resilience for food security and nutrition. FAO Licence: CC BY-NC-SA 30 IGO, Rome

Farina L, Demey F, Hardouin J (1991) Production de termites pour l'aviculture villageoise au Togo. Tropicultura 9:181–187

Gabaza M, Shumoy H, Muchuweti M, Vandamme P, Raes K (2018) Baobab fruit pulp and mopane worm as potential functional ingredients to improve the iron and zinc content and bioaccessibility of fermented cereals. Innovative Food Sci Emerg Technol 47:390–398. https://doi.org/10.1016/j.ifset.2018.04.005.

Ghazoul J (2006) Mopani woodlands and the mopane worm: enhancing rural livelihoods and resource sustainability. Final technical report. DFID, London

Glew RH, Jackson D, Sena L, VanderJagt DJ, Pastuszyn A, Millson M (1999) *Gonimbrasia belina* (Lepidoptera: Saturniidae), a nutritional food source rich in protein, fatty acids and minerals. Am Entomol 45:250–253

Goergen G, Kumar PL, Sankung SB, Togola A, Tamò M (2016) First report of outbreaks of the fall armyworm *Spodoptera frugiperda* (J E Smith) (Lepidoptera, Noctuidae), a new alien invasive pest in West and Central Africa. PLoS One 11:e0165632. https://doi.org/10.1371/journal.pone.0165632.

Gondo T, Frost P, Kozanayi W, Stack J, Mushongahand M (2010) Linking knowledge and practice: assessing options for sustainable use of mopane worms (*Imbasia belina*) in southern Zimbabwe. J Sustain Develop Africa 12:281–305

Hanboonsong Y, Jamjanya T, Durst PB (2013) Six-legged livestock: edible insect farming, collection and marketing in Thailand. Food and Agriculture Organization of the United Nations. Regional Office for Asia and the Pacific, Bangkok

Hope RA, Frost PGH, Gardiner A, Ghazoul J (2009) Experimental analysis of adoption of domestic mopane worm farming technology in Zimbabwe. Dev South Afr 26:29–46

Idowu AB, Idowu OA (1999) Pharmacological properties of the repellent secretion of *Zonocerus variegatus* (Orthoptera: Prygomorphidae). Rev Biol Trop 47:1015–1020

Iduwu A, Modder W (1996) Possible control of the stinking grasshopper *Zonocerus variegatus* (L) (Ortohptera: Pyrgomorphidae) in ondo state, through human consumption. The Nigerian Field 61:7–14

Jongema Y (2017) List of edible insect species of the world. Laboratory of Entomology, Wageningen University, Wageningen. https://www.wur.nl/en/Research-Results/Chair-groups/Plant-Sciences/Laboratory-of-Entomology/Edible-insects/Worldwide-species-list.htm

Kekeunou S, Weise S, Messi J, Tamo M (2006) Farmers' perception on the importance of variegated grasshopper (*Zonocerus variegatus* (L.)) in the agricultural production systems of the humid forest zone of Southern Cameroon. J Ethnobiol Ethnomed 2:17. https://doi.org/10.1186/1746-4269-2-17.

Kelemu S (2015) Insects: an overlooked food source. Int J Tropic Insect Sci 35:1–2. https://doi.org/10.1017/S174275841500003X.

Kelemu S et al (2015) African edible insects for food and feed: inventory, diversity, commonalities and contribution to food security. J Insects Food Feed 1:103–119. https://doi.org/10.3920/JIFF2014.0016

Kenis M, Koné N, Chrysostome CAAM, Devic E, Koko GKD, Clottey VA, Nacambo S, Mensah GA (2014) Insects used for animal feed in West Africa. Entomologia 2:107–114

Kinyuru JN, Kenji GM, Njoroge MS (2009) Process development, nutrition and sensory qualities of wheat buns enriched with edible termites (*Macrotermes subhylanus*) from Lake Victoria region, Kenya. Afr J Food Agricult Nutr Develop 9:1739–1750

Kinyuru JN, Kenji GM, Njoroge SM, Ayieko M (2010) Effect of processing methods on the in vitro protein digestibility and vitamin content of edible winged termite (*Macrotermes subhylanus*) and grasshopper (*Ruspolia differens*). Food Bioprocess Technol 3:778–782. https://doi.org/10.1007/s11947-009-0264-1.

Koffi DM, Cisse M, Koua GA, Niamke SI (2017) Nutritional and functional properties of flour from the palm (*Elaeis guineensis*) weevil *Rhynchophorus phoenicis* larvae consumed as protein source in south Côte d'Ivoire. Ann Univ Dunarea de Jos of Galati Fascicle VI – Food Technol 41:9–19

Kozanayi W, Frost P (2002) Marketing of mopane worm in southern Zimbabwe. Institute of Environmental Studies, Harare, p 31

Kung SJ, Fenemore B, Potter PC (2011) Anaphylaxis to mopane worm (*Imbrasia belina*). Ann Allergy Asthma Immunol 106:538–539. https://doi.org/10.1016/j.anai.2011.02.003.

Kwiri R, Winini C, Muredzi P, Tongonya J, Gwala W, Mujuru F, Gwala ST (2014) Mopane worm (*Gonimbrasia belina*) utilisation, a potential source of protein in fortified blended foods in zimbabwe: a review. Global J Sci Front Res 14:55–67

Lehtovaara VJ, Roininen H, Valtonen A (2018) Optimal temperature for rearing the edible*Ruspolia differens* (Orthoptera: Tettigoniidae). J Econom Entomol 111:234. https://doi.org/10.1093/jee/toy234

Lenga A, Kezetah C, Kinkela T (2012) Conservation et étude de la valeur nutritive des larves de *Rhynchophorus phoenicis* (Curculionidae) et *Oryctes rhinoceros* (Scarabeidae), deux coléoptères d'intérêt alimentaire au Congo-Brazzaville. Int J Biol Chem Sci 6:1718–1728. https://doi.org/10.4314/ijbcs.v6i4.28

Lewis VR (2003) Isoptera (Termites). In: Resh VH, Cardé RT (eds) Encyclopedia of insects. Academic, Amsterdam, pp 604–608

Malaisse F (2005) Human consumption of Lepidoptera, termites, Orthoptera, and ants in Africa. In: Paoletti MG (ed) Ecological implications of minilivestock. Science Publishers, Inc., Enfield, pp 175–230

Malaisse F, Mabossy-Mobouna G, Latham P (2017) Un atlas des chenilles et chrysalides consommées en Afrique par l'homme (An Atlas of caterpillars and chrysalises consumed by man in Africa). Geo-Eco-Trop 41:55–66

Malinga GM, Valtonen A, Lehtovaara VJ, Rutaro K, Opoke R, Nyeko P, Roininen H (2018a) Diet acceptance and preference of the edible grasshopper *Ruspolia differens* (Orthoptera: Tettigoniidae). Appl Entomol Zool 53:229–236. https://doi.org/10.1007/s13355-018-0550-3.

Malinga GM, Valtonen A, Lehtovaara VJ, Rutaro K, Opoke R, Nyeko P, Roininen H (2018b) Mixed artificial diets enhance the developmental and reproductive performance of the edible grasshopper, *Ruspolia differens* (Orthoptera: Tettigoniidae). Appl Entomol Zool 53:237–242. https://doi.org/10.1007/s13355-018-0548-x

Mba FAR, Kansci G, Viau M, Hafnaoui N, Meynier A, Demmano G, Genot C (2017) Lipid and amino acid profiles support the potential of *Rhynchophorus phoenicis* larvae for human nutrition. J Food Compos Anal 60:64–73. https://doi.org/10.1016/j.jfca.2017.03.016.

Mbata KJ (1995) Traditional uses of arthropods in Zambia. In: DeFoliart G, Dunkel FV, Gracer D (eds) Food insect newsletter volumes 1013; 1988 through 2000. Aardvark Global Publishing Company, Salt Lake City, pp 235–237

McCrae AWR (1982) Characteristics of swarming in the African Edible Bush-Cricket *Ruspolia differens* (Serville) (Orthoptera,Tettigonioidea). J East Afr Nat History Soc National Museum 178:1–5

Mmari MW, Kinyuru JN, Laswai HS, Okoth JK (2017) Traditions, beliefs and indigenous technologies in connection with the edible longhorn grasshopper *Ruspolia differens* (Serville 1838) in Tanzania. J Ethnobiol Ethnomed 13:60. https://doi.org/10.1186/s13002-017-0191-6.

Muafor FJ, Gnetegha AA, Gall PL, Levang P (2015) Exploitation, trade and farming of palm weevil grubs in Cameroon. Center for International Forestry Research (CIFOR), working paper 178, Bogor, Indonesia. https://doi.org/10.17528/cifor/005626

Muafor FJ, Gnetegha AA, Dounias E, Le Gall P, Levang P (2017) Chapter 6. African Palm Weevil farming: a novel technique contributing to food security and poverty alleviation in rural sub-Saharan Africa. In: van Huis A, Tomberlin JK (eds) Insects as food and feed: from production to consumption. Wageningen Academic Publishers, Wageningen, pp 113–125

Neely C, Bunning S, Wilkes A (eds) (2009) Review of evidence on drylands pastoral systems and climate change: implications and opportunities for mitigation and adaptation Land Tenure and Management Unit (NRLA), Land and Water Division, land and water discussion paper 8. Food and Agriculture Organization of the United Nations, Rome

Niassy S, Affognon HD, Fiaboe KKM, Akutse KS, Tanga CM, Ekesi S (2016) Some key elements on entomophagy in Africa: culture, gender and belief. J Insects Food Feed 2:139–144. https://doi.org/10.3920/JIFF2015.0084

Nrior RR, Beredugo EY, Wariso CA (2018) Dual purpose edible insect larva (*Rhynchophorus phoenicis*) in south south Nigeria—microbiological assessment of body parts, IOSR. J Environ Sci Toxicol Food Technol 12:59–68. https://doi.org/10.9790/2402-1209035968

Nzikou JM (2010) Characterisation and nutritional potentials of *Rhynchophorus phoenic* larva consumed in Congo-Brazzaville. Marien Ngouabi University, Brazzaville

Okezie OA, Kgomotso KK, Letswiti MM (2010) Mopane worm allergy in a 36-year-old woman: a case report. J Med Case Rep 4:42. https://doi.org/10.1186/1752-1947-4-42.

Okunowo WO, Olagboye AM, Afolabi LO, Oyedeji AO (2017) Nutritional value of *Rhynchophorus phoenicis* (F.) larvae, an edible insect in Nigeria. Afr Entomol 25:156–163. https://doi.org/10.4001/003.025.0156.

Olaofe O, Arogundade LA, Adeyeye EI, Falusi OM (1998) Composition and food properties of the variegated grasshopper, *Zonocerus variegatus*. Trop Sci 38:233–237

Omotoso OT (2006) Nutritional quality, functional properties and anti-nutrient compositions of the larva of *Cirina forda* (Westwood) (Lepidoptera: Saturniidae). J Zhejiang Univ Sci 7:51–55

Omotoso OT, Adedire CO (2007) Nutrient composition, mineral content and the solubility of the proteins of palm weevil, *Rhynchophorus phoenicis* f. (Coleoptera: Curculionidae). J Zhejiang Univ Sci B 8:318–322

Oyegoke OO (n.d.). H-index : 1Aj AkintolaH-index: 1Jo FasorantiH-index: 1

Oyegoke OO, Akintola AJ, Fasoranti JO (2006) Dietary potentials of the edible larvae of *Cirina forda* (westwood) as a poultry feed. Afr J Biotechnol 5:1799–1802. https://doi.org/10.5897/AJB06.189.

Paiko YB, Jacob JO, Salihu SO, Dauda BEN, Suleiman MAT, Akanya HO (2014) Fatty acid and amino acid profile of emperor moth caterpillar (*Cirina forda*) in Paikoro local government area of Niger State, Nigeria. Am J Biochem 4:29–34. https://doi.org/10.5923/j.ajb.20140402.03.

Pambo KO, Okello JJ, Mbeche RM, Kinyuru JN, Alemu MH (2018) The role of product information on consumer sensory evaluation, expectations, experiences and emotions of cricket-flour buns. Food Res Int 106:532–541. https://doi.org/10.1016/j.foodres.2018.01.011.

Payne CLR, Umemura M, Dube S, Azuma A, Takenaka C, Nonaka K (2015) The mineral composition of five insects as sold for human consumption in Southern Africa. Afr J Biotechnol 14:2443–2448. https://doi.org/10.5897/AJB2015.14807.

Payne CLR, Scarborough P, Rayner M, Nonaka K (2016) Are edible insects more or less 'healthy' than commonly consumed meats? A comparison using two nutrient profiling models developed to combat over- and undernutrition. Eur J Clin Nutr 70:285–291. https://doi.org/10.1038/ejcn.2015.149.

Quaye B, Atuahene CC, Donkoh A, Adjei BM, Opoku O, Amankrah MA (2018) Nutritional potential and microbial status of african palm weevil (*Rhynchophorus phoenicis*) larvae raised on alternative feed resources. Am Sci Res J Eng Technol Sci 48:45–52

Quayea B, Atuahene CC, Donkoh A, Adjei BM, Opoku O, Amankrah MA (2018) Alternative feed resource for growing african palm weevil (*Rhynchophorus phoenicis*) larvae in commercial production. Am Sci Res J Eng Technol Sci 48:36–44

Rémy DA, Hervé BB, Sylvain ON (2018) Study of some biological parameters of *Cirina butyrospermi* Vuillet (Lepidoptera, Attacidae), an edible insect and shea caterpillar (*Butyrospermum paradoxum* Gaertn. F.) in a context of climate change in Burkina Faso. Adv Entomol 6:81510. https://doi.org/10.4236/ae.2018.61001

Roffeis M et al (2018) Life cycle cost assessment of insect based feed production in West Africa. J Clean Prod 199:792–806. https://doi.org/10.1016/j.jclepro.2018.07.179.

Rutaro K, Malinga GM, Lehtovaara VJ, Opoke R, Nyeko P, Roininen H, Valtonen A (2018a) Fatty acid content and composition in edible *Ruspolia differens* feeding on mixtures of natural food plants. BMC Res Notes 11:687. https://doi.org/10.1186/s13104-018-3792-9.

Rutaro K et al (2018b) Artificial diets determine fatty acid composition in edible *Ruspolia differens* (Orthoptera: Tettigoniidae). J Asia Pac Entomol 21:1342–1349. https://doi.org/10.1016/j.aspen.2018.10.011.

Saeed T, Dagga FA, Saraf M (1993) Analysis of residual pesticides present in edible locusts captured in Kuwait. Arab Gulf J Sci Res 11:1–5

Sani I, Haruna M, Abdulhamid A, Warra A, Bello F, Fakai I (2014) Assessment of nutritional quality and mineral composition of dried edible *Zonocerus variegatus* (grasshopper). J Food Dairy Technol 2:1–6

Séré A et al (2018) Traditional knowledge regarding edible insects in Burkina Faso. J Ethnobiol Ethnomed 14:59. https://doi.org/10.1186/s13002-018-0258-z.

Solomon M, Ladeji O, Umoru H (2008) Nutritional evaluation of the giant grasshopper (*Zonocerus variegatus*) protein and the possible effects of its high dietary fibre on amino acids and mineral bioavailability. Afr J Food Agric Nutr Dev 8:238–248

Springmann M, Clark M, Mason-D'Croz D, Wiebe K, Bodirsky BL, Lassaletta L, de Vries W, Vermeulen SJ, Herrero M, Carlson KM, Jonell M, Troell M, DeClerck F, Gordon LJ, Zurayk R, Scarborough P, Rayner M, Loken B, Fanzo J, Godfray HCJ, Tilman D, Rockström J, Willett W (2018) Options for keeping the food system within environmental limits. Nature 662:519–525

Ssepuuya G et al (2017) Use of insects for fish and poultry compound feed in sub-Saharan Africa—a systematic review. J Insects Food Feed 3:289–302. https://doi.org/10.3920/jiff2017.0007

Stack J, Dorward A, Gondo T, Frost P, Taylor F, Kurebgaseka N (2003) Presentation title: mopane worm utilisation and rural livelihoods in southern Africa. Paper presented at the international conference on rural livelihoods, forests and biodiversity, Bonn, Germany.

Styles CV (1994) The big value in mopane worms. Farmer's Weekly 22:20–22

Tanga CM, Magara HJO, Ayieko MA, Copeland RS, Khamis FM, Mohamed SA, Ombura FLO, Niassy S, Subramania S, Fiaboe KKM, Roos N, Ekesi S, Hugel S (2018) A new edible cricket species from Africa of the genus *Scapsipedus*. Zootaxa. https://doi.org/10.11646/zootaxa.0000.0.0

Van Huis A (2003) Insects as food in sub-Saharan Africa. Insect Sci Appl 23:163–185

Van Huis A (2013) Potential of insects as food and feed in assuring food security. Annu Rev Entomol 58:563–583. https://doi.org/10.1146/annurev-ento-120811-153704.
Van Huis A (2017) Cultural significance of termites in sub-Saharan Africa. J Ethnobiol Ethnomed 13(8). https://doi.org/10.1186/s13002-017-0137-z.
Van Huis A, Oonincx DGAB (2017) The environmental sustainability of insects as food and feed. A review. Agron Sustain Dev 37:43. https://doi.org/10.1007/s13593-017-0452-8.
Van Itterbeeck J, Van Huis A (2012) Environmental manipulation for edible insect procurement: a historical perspective. J Ethnobiol Ethnomed 8:1–19. https://doi.org/10.1186/1746-4269-8-3.
Womeni HM, Linder M, Tiencheu B, Mbiapo FT, Villeneuve P, Fanni J, Parmentier M (2009) Oils of insects and larvae consumed in Africa: potential sources of polyunsaturated fatty acids. J Oleo Sci 16:230–235
Womeni HM, Tiencheu B, Linder M, Nabayo EMC, Tenyang N, Mbiapo FT, Villeneuve P, Fanni J, Parmentier M (2012) Nutritional value and effect of cooking, drying and storage process on some functional properties of *Rhynchophorus phoenicis*. Int J Life Sci Pharma Res 2:203–219

Chapter 2
African Edible Insect Consumption Market

Cordelia Ifeyinwa Ebenebe, Oluwatosin Samuel Ibitoye, Inwele Maduabuchi Amobi, and Valentine Obinna Okpoko

Abstract Consumption is the utilization of economic goods to satisfy needs. Africa is home to the rich diversity of insects with over 1500 species of insects. Several reports highlighted the nutritional, medicinal values and industrial uses of some edible insects. The global edible insects market is mainly segmented by insect type, product type, application, and geography. Insects can be grown on organic waste. The potential of edible insects in curbing the menace of malnutrition and ensuring food security has necessitated so much interest in the production, marketing, and utilization of edible insects.

Keywords Edible insect · Consumption · Commercialization · Market

2.1 Introduction

Consumption is often defined as the utilization of economic goods to satisfy needs. The dictionary of marketing terms defined consumption as the process of using consumer products in order to satisfy desires, real or imaginable needs so that the products are used up, transformed or deteriorated in such a manner as not to be either reusable or recognizable in their original form (https://www.allbusiness.com/

C. I. Ebenebe (✉)
Microlivestock Unit, Department of Animal Science and Technology, Nnamdi Azikiwe University, Awka, Nigeria
e-mail: ci.ebenebe@unizik.edu.ng

O. S. Ibitoye
Forestry Research Institute of Nigeria, Onigambari Forest Reserve, Ibadan, Oyo State, Nigeria

I. M. Amobi
Federal University of Kashere, Gombe, Nigeria

V. O. Okpoko
Bioconservation Unit, Department of Zoology, Nnamdi Azikiwe University, Awka, Nigeria

© Springer Nature Switzerland AG 2020
A. Adam Mariod (ed.), *African Edible Insects As Alternative Source of Food, Oil, Protein and Bioactive Components*, https://doi.org/10.1007/978-3-030-32952-5_2

barrons_dictionary/dictionary-consumption-4965423-1.html). African edible insect consumption market therefore refers to the business of promoting production, buying, and selling of edible insects and products derived from insects as well as other beneficial services from insects including entotherapy/zootherapy. Promotion of edible insect production requires preservation of forests where the insects thrive or provision of simulated environment or other devices for mass production of insects. In the wild, the volume of edible insects and by-products from it correlates with the wealth of forest resources. The African continent is endowed with rich forest resources, especially tropical rain forests (Bernard and Womeni 2017).

According to FAO (2005), forests and woodlands in Africa occupy an estimated 650 million hectares (21.8%) of the land area of this continent and account for 16.8% of the global forest cover. On average, forests account for 6% of gross domestic product (GDP) in Africa, which is the highest in the world. In Uganda, for example, forests and woodlands are now recognized as an important component of the nation's stock of economic assets and they contribute in excess of US$546.6 million to the economy. Leal et al. (2016) highlighted the relationship between the forest ecosystem and insect biodiversity. Insects according to Alaloiun (2014) play an essential role in forest ecosystems by affecting the primary production and evolution of plants. They form a critical link between plants and higher trophic levels (Crawley 1997, cited in Alaloiun 2014). There are over 1500 species of edible insects across Africa, this is closely related to its tropical climate and magnitude of tropical rainforest that is majorly home for an enormous species of insects. Today, these vast species of insects have become a singular resource in sustaining food security and curbing the menace of malnutrition in the continent.

Sidiki Sow (2016) in his report on "Insect Protein for Food Security in West Africa" stated that by 2050 the world population will be 9.4 billion, with 2.4 billion people in Africa. This teeming population will demand double amount of food and animal feed production to meet nutritional needs of man and animal. Van Huis (2015) and Alexandratos and Bruinsma (2012) reported that meat consumption in sub-Saharan Africa is expected to double from 202% from 2010 to 2050. It will therefore be a herculean task to feed such a huge population especially in Africa where the World Food Program (2016) already reported that one out of every four Africans is undernourished, 1.2 M people are in urgent need of food assistance in Mali, Niger, and Burkina Faso, and USS 11.5 M is needed to offset food deficit challenges in Mali and Niger. In Kenya USAID in 2014 reported that 1.5 million people needed food assistance, while cases of kwashiorkor is presently reported in northern part of Nigeria (Hamidu et al. 2003). If the extrapolation on population growth for 2050 becomes real, feeding of such a massive population will mean increase in agricultural activities and its concomitant environmental degradation. Of all food nutrients, animal protein deficiency is more pronounced, with records as low as 7–10 g/person/day in many African countries against the 35 g/person/day recommended by FAO (1991) for normal growth and development.

Besides, massive production of livestock to increase animal protein supply and consumption will lead to environmental issues especially higher levels of greenhouse gases (GHG), pollution of water resources and making land unavailable for other uses. According to Food and Agricultural Organization (FAO), livestock is the world's largest user of land resources, with grazing land and cropland dedicated to the production of feed representing almost 80% of all agricultural land. Judging by the hunger rate in Africa, if 50% of this land is used in producing food for human consumption, hunger rate would have drastically reduced. Livestock also requires large amount of water, grains used in feed formulation required about 1000–5000 kg of water to grow depending on the region (Chapagain and Hoekstra 2003). Livestock itself contains between 5 and 20 times more water per kg product (Chapagain and Hoekstra 2003). In most African countries, there is little or no factual record on the level of environmental degradation associated with livestock activities.

However, given the triad doom-spelling factors in Africa (fast growing population, natural resource degradation by slash-and-burn agriculture, and high level of malnutrition), edible insect cultivation and utilization as food and feed appear very promising in ensuring food security in the continent. FAO (2008) (cited in Van Huis 2013 recommended insects as alternative source of food, capable of meeting the animal protein demands of a growing population while preserving the environment. As a follow-up to the FAO workshop in Chiang Mai in 2008, the Non-Wood Forest Products Programme of the FAO Forestry Department and the Wageningen University and Research Centre (WUR) (Laboratory of Entomology) initiated a collaborative effort to promote entomophagy; thus FAO (2013) outlined common insects consumed globally, including beetle, grasshoppers, locusts, and crickets. Consumption of larva of many insects has also been documented. In an earlier work FAO (2013) reported consumption of mopane caterpillar (*Imbrasia belina*) in Angola, Botswana, Mozambique, Namibia, South Africa, Zambia, and Zimbabwe. Malaise (1997) identified 38 different species of caterpillar consumed across the Democratic Republic of Congo, Zambia, and Zimbabwe. Evidently, rearing insects requires remarkably less land than farming other categories of livestock (Conincx and de Boer 2012).

Several reports in literature highlighted the nutritional, medicinal values and industrial uses of some edible insects (Ekpo and Onigbinde 2004, 2005; Banjo et al. 2006; Edijala et al. 2009; Alamu 2014; Mbah and Elekima 2007; Ebenebe and Okpoko 2014; Schabel 2010). According to Braide et al. (2010) protein content of edible insects ranged from 21 to 65% and therefore compares favorably with meat and fish proteins. Igwe et al. (2011) also reported that insect larvae are rich in essential amino acids like lysine and threonine which are deficient in grain and cereals. Similarly, Ekpo and Onigbinde (2004) had earlier reported that edible insect larvae are rich in essential fatty acids like linoleic and linolenic acids while Igwe et al. (2011) reported on vitamin content of edible insects. Apart from the nutritional and medicinal benefits, there are other ecological, magical, and spiritual benefits of insects; therefore edible insect consumption market in Africa will address all of these aspects of insect benefits and the level of commercialization in Africa. According to FAO (2012), edible insects contain high-quality protein, vitamins, and amino acids for humans. Insects have a high food conversion rate; for example, crickets need six times less feed than cattle, four times less than sheep, and twice less than pigs and broiler chickens to produce the same amount of protein. Besides, they emit less greenhouse gases and ammonia than conventional livestock (Oonincx et al. 2010). Insects can be grown on organic waste. The potential of edible insects in curbing the menace of malnutrition and ensuring food security has necessitated so much interest in the production, marketing, and utilization of edible insects.

The global edible insects market is mainly segmented by insect type (crickets, mealworms, black soldier flies, buffalos, grasshoppers, ants, silkworms, cicadas, and others), product type (whole insects, insect meal, insect powder, insect protein bars and protein shakes, insect baked product and snacks, insect confectionaries, insect beverages, insect oil, and others), application (human consumption, animal nutrition, and cosmetics and pharmaceutical), and geography.

2.2   Edible Insects as Food in Africa

Edible insect consumption in Africa is as old as the continent. The continent is home to the richest diversity of insects with over 1500 species of insects (Raheem et al. 2019) ranging from caterpillars (Lepidoptera) to termites (Isoptera), locust, grasshoppers, cricket (Orthoptera), ants and bees (Hymenoptera), bugs (Heteroptera and Homoptera), and beetles (Coleoptera) (Saliou and Ekesi 2017). Although almost all African countries practice entomophagy, there is considerable variation in the most consumed insect order in the continent.

Saliou and Ekesi (2017) listed the most dominant insect eating countries to include Democratic Republic of Congo, Congo, The Central African Republic, Cameroun, Uganda, Zambia, Zimbabwe, Nigeria, and South Africa. Kelemu et al. (2015) and Kelemu (Kelemu 2016) gave a country-by-country run down of common species/orders of insects consumed in each of the African countries (Fig. 2.1 and Table 2.1), while Adeoye et al. (2014) showed diversity of edible insects in the African subregion.

In the Central African Republic, 96 species of insects are consumed. Of these 96 species, Roulon-Doko (1998) cited in Raheem et al. (2019) stated that insects of the order Orthoptera (locust and grasshopper) were the most consumed (40% consumption level), followed by Lepidoptera (36%), Isoptera (termites 10%), Coleoptera (beetles 6%), and others such as cicadas and crickets (8%) (Table 2.2).

In Kenya, Kinyuru et al. (2012, 2013) showed that six species of edible insects are consumed in the western part of Kenya. Of the six species four are termites of four different subspecies (*Pseudocanthotermes militaris* H., *Macrotermes bellicosus* S., *Macrotermes subhyalinus* R., and *Pseudocanthotermes spiniger* S.), the rest are black ant and long-horned grasshopper. Ayieko et al. (2012) posited that insect species like "agoro" termites, black ants, crickets, and grasshoppers form part of traditional menu in the western part of Kenya (Table 2.3).

In Ghana, the report of Anankware et al. (2016) showed that nine edible insects are consumed in Ghana, with scarab beetle (2%), field cricket (5%), shea butter tree

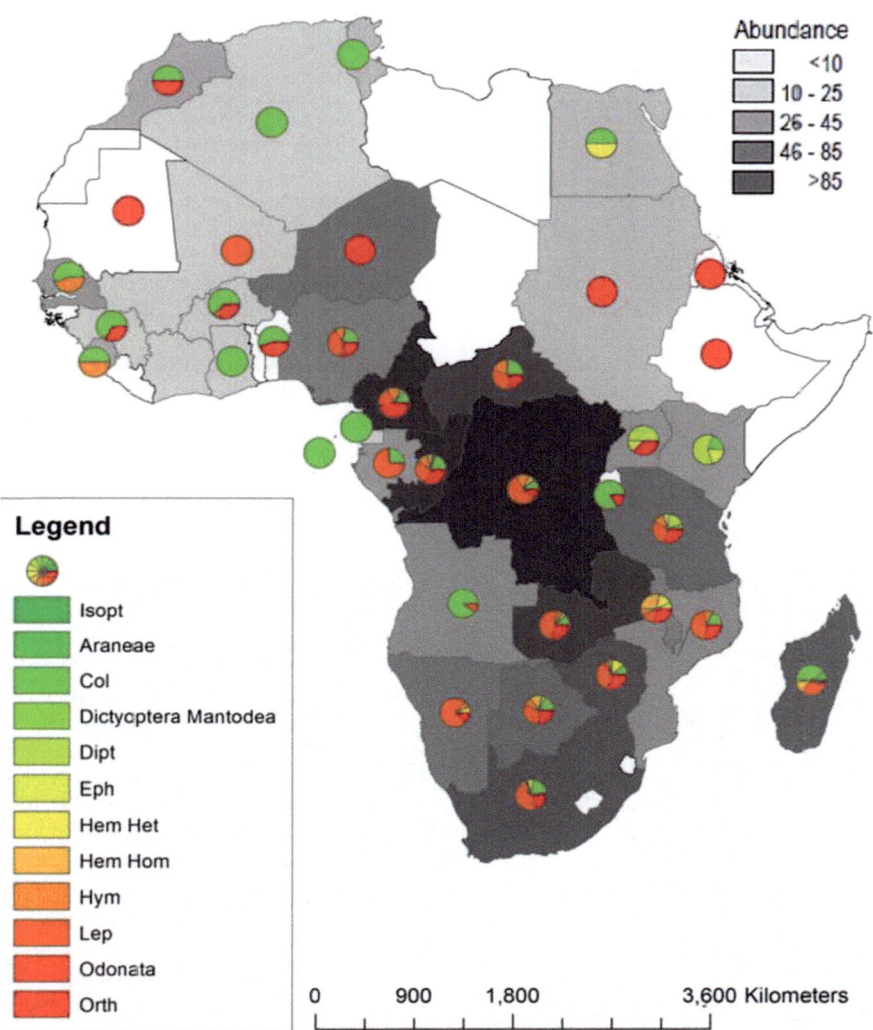

Fig. 2.1 Diversity and abundance of main groups of edible insects in Africa. (Source: Kelemu 2016 in Raheem et al. 2019)

caterpillar (8.7%), house cricket (9.5%), and locust (10%), African palm weevil larvae (47.2%), termites (45.9%), ground cricket (*Scapteriscus vianus*, 33.3%), and grasshoppers (30.5%). His report showed that Northern Ghana dominates in entomophagy as eight out of the nine edible insects consumed are mostly eaten in that region (Table 2.4). In Uganda, Raheem et al. (2019) reported that termites (*Macrotermes* spp.) and grasshopper (*Ruspolia nitidula*) are the most consumed edible insects.

Table 2.1 The most consumed insect species in Africa. Countries and regions of Africa where species are mostly consumed

Order	Scientific and common names	Countries
Coleoptera	*Oryctes owariensis* (Palisot de Beauvois) (Rhinoceros beetle)	DRC, South Africa, Ivory Coast, Sierra Leone Guinea, Ghana, Equatorial Guinea, Guinea Bissau
	Rhynchophorus phoenicis (Fabricius) (African palm weevil)	DRC, Cameroon, Congo, CA Republic, Nigeria, Angola, Ivory Coast, Niger, *São Tomé and Príncipe*, Guinea, Togo
	Oryctes boas (Fabricius) (Boas rhinoceros beetle)	Nigeria, Ivory Coast, Sierra Leone, Guinea, Liberia, Guinea Bissau DRC, Congo, South Africa, Botswana, Namibia
Hemiptera	*Encosternum delegorguei* (Spinola) (Stinkbug)	South Africa, Swaziland, Mozambique, Malawi, Zimbabwe, Botswana, Namibia
Hymenoptera	*Apis mellifera mellifera* Linnaeus (European dark bee)	DRC, Zambia, Botswana, Nigeria, Tanzania, Senegal, Sierra Leone, Ghana, South Sudan, Togo, Lesotho, Benin
	Apis mellifera adansoni (Latreille) (Africanized honey bee)	DRC, Zambia, CA Republic, Nigeria, Tanzania, Sierra Leone, Ghana, Benin
	Carebara vidua (Smith)	DRC, Zambia, South Africa, Zimbabwe, Botswana, Malawi, Sudan, Kenya, South Sudan
	Carebara lignata (Westwood)	Zambia, South Africa, Zimbabwe, Botswana, Sudan, Mozambique, Namibia, South Sudan
	Macrotermes spp. (African mound-building termites)	DRC, Zambia, Zimbabwe, Nigeria, Tanzania, Malawi, Senegal, Uganda, Cote d'Ivoire, Guinea, Ghana, Togo, Burundi
	Macrotermes bellicosus (Smeathman) (Termites)	DRC, Cameroon, Congo, CA Republic Nigeria, Cote d'Ivoire, Kenya, *São Tomé and Príncipe*, Guinea, Togo Liberia, Guinea Bissau, Burundi
	Macrotermes subhyalinus (Rambur) (Mendi Termite)	Zambia, Angola, Kenya, Togo, Burundi
	Macrotermes falciger (Gerstäcker)	Zambia, Zimbabwe, Burkina Faso, Burundi, Benin
	Macrotermes natalensis (Haviland)	DRC, Cameroon, Congo, CA Republic, Nigeria, Burundi, South Africa, Zimbabwe, Nigeria, Malawi
Lepidoptera	*Bunaea alcinoe* (Stoll) (African moth)	DRC, Zambia, South Africa, Cameroon, Congo, Central African Republic, Zimbabwe, Nigeria, Tanzania
	Anaphe panda (Boisduval) (Silk moth)	DRC, Zambia, Cameroon, Congo, CA, Republic, Zimbabwe, Nigeria, Tanzania
	Cirinaforda (Westwood) (Emperor moth)	DRC, Zambia, South Africa, Botswana, Burkina Faso, Nigeria, Mozambique, Namibia, Ghana, Togo, Chad
	Dactyloceraslucina (Drury) (Drury's Owl Moth)	DRC, Zambia, South Africa, Cameroon, Congo, Angola, Gabon, Sierra Leone, São Tomé and Príncipe, Equatorial Guinea

(continued)

Table 2.1 (continued)

Order	Scientific and common names	Countries
	Platysphinx stigmatica (Mabille) (Red spot moth)	DRC, Zambia, Congo, CA Republic, Sierra Leone, São Tomé and Príncipe, Equatorial Guinea, Rwanda, Burundi
	Cirina butyrospermi (Vuillot) (Shea tree caterpillar)	DRC, Zambia, South Africa, Zimbabwe, Burkina Faso, Nigeria, Mali, Ghana
	Epanaphe carteri (Walsingham)	DRC, Zambia, Angola, Gabon, Sierra Leone, São Tomé and Príncipe, Equatorial Guinea
	Imbrasiabelina (Westwood) (Mopane caterpillar, mopane worm, emperor moth)	DRC, Zambia, South Africa, Zimbabwe, Botswana, Malawi
	Gynanisaata (Strand) (African moth)	DRC, Zambia, Malawi, South Sudan
	Eumeta cervina (Druce) (Bagworm)	DRC, Cameroon, Congo, CA Republic, Angola, Gabon, Sierra Leone, Sao Tome
	Imbrasia ertli (Rebel) (Confused Emperor)	Zambia, South Africa, Cameroon, Congo
	Anaphe venata (Butler) (African silkworm)	Zambia, South Africa, Cameroon, Congo, CA Republic, Zimbabwe, Botswana, Angola
	Imbrasia epimethea (Drury) (African moth)	DRC, Zambia, South Africa, Cameroon, Congo, CA Republic, Zimbabwe
	Urota sinope (Westwood) (Tailed Emperor)	DRC, South Africa, Zimbabwe, Botswana, Gabon, Mozambique, Namibia
Orthoptera	*Schistocerca gregaria* (Forskål) (Desert locust)	Zambia, South Africa, Cameroon, Congo, Botswana, Tanzania, Sudan, Uganda, Ethiopia, Kenya, Sierra Leone, Morocco, Guinea, Lesotho, Mauritania, Somalia, Eritrea, Guinea Bissau
	Acanthacris ruficornis (Fabricius) (Garden locust)	DRC, Zambia, South Africa, Cameroon, Congo, CA Republic, Zimbabwe, Burkina
	Brachytrupes membranaceus (Drury) (Tobacco cricket)	Zambia, Cameroon, Congo, CA Republic, Zimbabwe, Burkina Faso, Nigeria, Tanzania, Angola, Togo, Benin
	Nomadacris septemfasciata (Serville) (Red locust)	Zambia, South Africa, Congo, Zimbabwe, Uganda, Mozambique
	Ruspolia differens (Serville) (Longhorn grasshopper)	DRC, Zambia, South Africa, Cameroon, Zimbabwe, Kenya, Uganda, Tanzania
	Zonocerus variegatus (Linnaeus) (Variegated grasshopper)	DRC, Cameroon, Congo, CA Republic, Nigeria, Côte d'Ivoire, São Tomé and Príncipe, Guinea, Ghana, Liberia, Guinea Bissau
	Locusta migratoria migratorioides (Reich & Fairmaire) (Migratory locust)	Zambia, Cameroon, Congo, Zimbabwe, Sudan, South Sudan
	Locusta napardalina (Walker) (Brown locust)	Zambia, South Africa, Zimbabwe, Botswana, Malawi, Libya

(continued)

Table 2.1 (continued)

Order	Scientific and common names	Countries
	Gastrimargus africanus (Saussure) (African grasshopper)	Cameroon, Congo, Niger, Lesotho, Liberia
	Phymateus viridipes brunneri (Bolivar) (Gaudy grasshopper)	Zambia, South Africa, Congo, Zimbabwe, Botswana, Mozambique, Namibia
	Gryllus bimaculatus (De Geer)	Guinea Bissau, Sierra Leone, Guinea, Liberia, Benin, Togo, Nigeria, DRC, Kenya, South Sudan, Zambia
	Anacridium melanorhodon melanorhodon (Walker) (Sahelian tree locust)	Cameroon, Sudan, Niger
	Paracinema tricolor (Thunberg)	Cameroon, Malawi, Lesotho
	Acheta spp. (Crickets)	Zambia, Zimbabwe, Kenya

Source: Kelemu et al. (2015) cited in Raheem et al. (2019)

According to Silow (1976) cited in Kelemu et al. (2015), a single community in Africa consume different kinds of insect species. In the report of Kelemu et al. (2015) communities like Mbunda people in Angola, Zambia, and Namibia 31 species of edible insects, 21 species are consumed by Ngandu people in Democratic Republic of Congo (DRC), 96 species by Gbaya people (Takeda 1990), 30 species consumed among the Bemba people in northern Zambia, southern DRC, and northeastern Zimbabwe (Malaisse 2005) and 27 species in Botswana (Obopile and Seeletso 2013). Van Huis (2003) earlier stated that 246 species of edible insects are consumed in 27 African countries.

Nigeria appear to be taking the lead in edible insect consumption in Africa, probably due to her very large population (over 180 million) even though the number of species is not as much compared to other African countries but the quantity consumed is by far more appreciable. Alamu et al. (2013) listed 22 insect species commonly consumed in Nigeria: 27.3% Lepidoptera (moths), 27.3% Coleoptera (beetles), 22.7% Orthoptera (grasshoppers and crickets), 13.6% Isoptera (termites), and 9.0% Hemiptera and Hymenoptera (bees). Besides, there are numbers of regional report on edible insect consumption. Ebenebe and Okpoko (2016) reported that even though the cricket (*Brachytrupes* spp.) is the most preferred edible insect in the south east Nigeria, termite is the most consumed, due to unavailability of cricket in the local markets. However, termites are seasonal, while African palm weevil larvae, the third most preferred edible insect, have the advantage of year round availability. Insect consumption in the six geopolitical zones in Nigeria is summarized in Amobi and Ebenebe (2018) (Table 2.5). Ebenebe et al. (2017a) also reported on the survey of edible insects consumed in the south eastern Nigeria and the local dishes cooked with edible insects (Table 2.6).

Table 2.2 Diversity of edible insects in Nigeria and their geographical distribution

Scientific name	English name	Local name	Order	Family	Location	Consumption stage
Cirina forda Westwood	Pallid Emperor moth	Yoruba: Kanni, Munimuni	Lepidoptera	Saturniidae	North Central, North East, South-West	Larvae
Bunaea alcinoe Cram	Emperor moth	–	Lepidoptera	Saturniidae	North Central, North-East, North-West	Larvae
Rhynchophorus Phoenics	Palm weevil	Yoruba: Ipe, Itun	Coleoptera	Curculionidae	South-South, South-West	Larvae
Oryctes boas Oryctes monocerus Oliver	Snout beetle	Yoruba: Ogongo	Coleoptera	Scarabaeidae	South-South, South-West	Larvae
Analeptes trifasciata	Rhinocerus beetle	Ibo: Ebe	Coleoptera	Scarabaeidae	South-South, South-West	Larvae
Anaphe venata	Caterpillar	Yoruba: Ekuku	Lepidoptera	Notodontidae	South-West	Larvae
Heteroligus meles Billberger	Yam beetle	–	Coleoptera	Scarabaeidae	South-West South-South South-East	Adults
Zonocerus	Grasshopper	Yoruba:	Orthoptera	Pyrgomorphidae	North Central	Adults

Source: Adeoye et al. (2014)

Table 2.3 Major edible insects of Ghana

Common name	Scientific name	Local name in Ghana	Stage consumed
Palm weevil larva	*Rhynchophorus phoenicis*	Akokono in Twi/Akan	Larva and adult
Beetle larva	*Phyllophaga nebulosa*	Chibionabra in Kasem	Larvae
Termite	*Macrotermes bellicosus* S	Kwena in Kasem	Adult
Shea tree caterpillar	*Cirina butyrospermi* V.	Kan Tuli in Frafra Dagari	Larva
Locust	*Locusta migratoria* L	Gbameda in Ewe	Adult
Grasshopper	*Zonocerus variegatus* L.	Manchogo in Kasem	Adult
House cricket	*Acheta domesticus* L.	Cheri in Kasem	Adult
Field cricket	*Gryllus similis* C.	Paan-terkyiirae in Dagoan	Adult
Ground cricket	*Scapteriscus vicinus* (Scudder)	Tigachari in Kasem	Adult

Source: Anankware et al. (2015)

Table 2.4 Traditional animal source foodstuffs and their edible parts consumed in Western Kenya

Common name	Scientific name	Local name (in Kenya)	Stage consumed
Winged termite	*Pseudacanthotermes militaris*	Hagen	Whole: dewinged
Winged termite	*Macrotermes bellicosus*	Smeathmen	Whole: dewinged
Winged termite	*Macrotermes subhyalinus*	Rambier	Whole: dewinged
Winged termite	*Pseudacanthotermes spiniger*	Oyala	Whole: dewinged
Black ant	*Carebara vidua*	Onyoso	Abdomen: dewinged
Long-horned grasshopper	*Ruspolia differens*	Senesence	Whole: dewinged

Sources: Kinyuru et al. (2012, 2013)

Table 2.5 Edible insects in each of the six geopolitical zones in Nigeria

Zone	Insects
North-East	Grasshopper, locust, cricket, pallid emperor moth larva, emperor moth (Amobi and Ebenebe 2018; Adeoye et al. 2014)
North-West	Grasshopper, locust, cricket, and emperor moth (Amobi and Ebenebe 2018; Adeoye et al. 2014)
North-Central	Grasshopper, locust and cricket, pallid emperor moth larva, emperor moth, mole cricket, green stink bug, termite (*Macrotermes natalensis*), rhinoceros beetle (*Bunaea alcinoe* Cram) (Amobi and Ebenebe 2018; Adeoye et al. 2014).
South-East	Winged termite, raffia palm grub/African palm weevil, house cricket, mole cricket, termite, grasshopper, locust, caterpillar of butterfly and moth, yam beetle, praying mantis, rhinoceros beetle, greenish beetle, two unidentified species (Ebenebe et al. 2017a)
South-West	Raffia palm grub/African palm weevil, snout beetle, rhinoceros beetle, local silkworm, yam beetle, grasshopper, honey bee, termite, cricket, mole cricket (Banjo et al. 2006; Adeoye et al. 2014)

(continued)

Table 2.5 (continued)

Zone	Insects
South-South	Raffia palm grub/African palm weevil, snout beetle, rhinoceros beetle, yam beetle, grasshopper, bug beetle, termite, honey bee and caterpillar of butterfly, cricket, mopane worms (Okore et al. 2014; Okweche and Abanyam 2017)
Niger Delta[a]	Termite, crickets, locust, grasshopper, African palm weevil, rhinoceros beetle, praying mantis, yam beetle, rice weevil, bean beetle, egg fruit borers, mopane worms, bees, house fly, cotton stainer (Okore et al. 2014)

[a]Niger Delta is a part of South-South Zone. (Source: Amobi and Ebenebe 2018)

Table 2.6 Season of harvest/method of harvest and preservation

Edible insect	Season of harvest/ month of the year	Site of harvest	Method of collection/ harvesting	Method of preparation
Winged termite	May–July	Farmlands and open fields	Light trapping	Drying/ frying
Raffia palm grub	All year round	Raffia palm	Nibbling sound of larvae in the rotting tree/ handpicking from rotting palm tree	Roasting/ Frying
Cricket	October–May	Home stead/farmland and small sand dunes on the bank of water bodies or sandy areas	Light trapping during the rains/digging out from the tunnels during the dry season and handpicking	Frying
Rhinoceros beetle	All year round	Oil palm plantation	Handpicking	Drying

Of all the 22 species of edible insects in Nigeria, four (African palm weevil, termite, locust, and grasshopper) are harvested in large quantities and sold in urban markets, while the rest are consumed locally.

2.3 Market Values of Edible Insects Used as Food

According to the report by Persistence Market Research (2018), the global edible insects market is expected to grow tremendously between 2017 and 2024. The global market for edible insects is estimated to reach US$ 722.9 million revenue. Global Opportunity Analysis And Industry Forecast (2018–2023) stated that the edible insect market is expected to reach US$ 1181.6 million by 2023, supported by a CAGR of 23.8% during the forecast period of 2018–2023. Owing to the short generation interval of insects and the high nutritional benefits, it is proposed worldwide as a tool to reduce hunger and poverty especially in developing countries such as Africa. DeFoliart (2002), reported that scores of species of edible insects are prominent items of commerce in the town and village markets of Africa and tropical and semitropical regions of the world. The growing trend on insect as food and feed has necessitated the establishment of large scale insect farms especially in the developed countries (Europe and America) and also in Asia; African countries known to consume one-fourth of the global edible insect inventory (Van Huis 2013) and to have favorable climate for insect growth and multiplication is yet to develop large scale insect farms except for agri protein in South Africa, experimental work of Aspire in Ghana and INSFEED project in Kenya and Uganda. Production therefore is on low scale and mostly on wild collection and many of the edible insects are seasonal.

The value of edible insects in many African countries therefore fluctuates with location, season of harvest, community's attitude to insect consumption, type of insects, and time spent in search for insects. In Nigeria, for instance, insects are consumed in all the six geopolitical zones, and each region has one or two insects that are most cherished. Apart from domestic consumption by the gatherers and their families, the excesses are sold for cash at local markets or by the street corners. Sales to urban markets is rare among some insects but huge in some others, where such sales occur, the insects are often sold to middlemen who sell to retailers. The interaction and number of middlemen involved usually set the final price of the insects for the end consumer. In Gombe state located in north eastern part of Nigeria for example, a measure of grasshoppers, crickets, winged termites and locusts sells for 500–800 and above depending on bargaining power of the customer in various local markets: Babbarkasuwa, Kumo; Tashargwari, Kumo; Gombe old market, Gombe; Saturday village market, Billiri; Wednesday village market, Dukku; Friday village market, Kaltingo; Thursday village market, Dadinkowa; Saturday village market, Malancidi (Amobi M, Personal observation, 2017). In the south eastern Nigeria, winged termites of similar measure sells for $8.22 to $10.96, while in the south southern Nigeria, a similar measure of African palm weevil larva sells for $4.11 and above (Amobi M, Personal Observation, 2017). In the southeast Nigeria, 3–4 roasted African palm weevil larvae sells for $0.37 in the cities while in the villages, the price can be as low as 7–8 larvae at the same $0.37 about 100 g of termite for $0.74.

Despite these challenges, edible insect market continues to thrive and grow tremendously. Kelemu et al. (2015) noted that in southern Africa alone, the trade value of mopane worm is over $85 million. The market for insect usually have a value chain that includes the following:

Input supply	Production	Processing	Marketing	Consumer
⇨ Human resources	⇨ Management of resources	⇨ Livestock Feed production	⇨ Production type	⇨ Willingness to buy or eat insect
⇨ Money Tools Parent Stock Housing	⇨ Prior knowledge of the insect biology Knowing the target market	⇨ Insect Powder Insect Manure Other products	⇨ Insect Type ⇨ Product type ⇨ Application type	⇨ Monitory capabilities ⇨ Cultural preference

2.3.1   Input Supply in Edible Insect Value Chain

Edible insects in Africa is at subsistence level and mainly gathered from the wild in many African countries. The insects are handpicked or harvested through indigenous techniques for capturing. Tools such as brooms, water, sound, lights traps, and other forms of traps are used for catching insect. Adeoye et al. (2014) reported a case in Lagos state Nigeria where women pay money to palm wine tappers to collect African palm weevil larvae from felled raffia palms especially in swampy areas that are inaccessible. Ebenebe and Okpoko (2016) also reported similar intervention by palm wine tappers at Ngbo swamp, Ebenebe town in Anambra State, Nigeria. The women involved in the APW larvae business purchase directly from wine tappers and sell to consumers. The women are often owners of restaurants, hotels and other forms of eateries where the African palm weevil larvae are sold to men who use it together with beer to cool off after a hectic day job. Young people who do not have localized business centers are often seen hawking the roasted insects in open or glass sealed containers. Insects like grasshopper and locusts are gathered largely from the northern part of Nigeria especially by women and children. The preparation forms of insect intended for sale include dried, cooked or boiled, and fried (Fig. 2.2). The present subsistence level of insect business is not yet specialized and therefore does not attract funding either in form of loan from banking industries or grant from other funding agencies.

2.3.2   Production Factor in Edible Insect Value Chain

Large-scale production in any business endeavor is the standpoint on which profitability of the business is hinged. Mass production is the only means of ensuring low cost of production and profitability of the insect business. High productivity

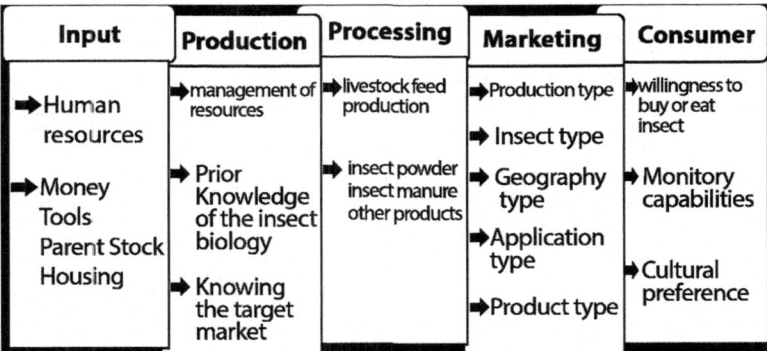

Fig. 2.2 Value Chain of Edible Insects

and profitability requires a good knowledge of the insect biology, disease problems of such insects, and health risks associated with edible insect consumption (Chai et al. 2009; Graczyk et al. 2005). Many rural dwellers who are in the business of gathering edible insects from the wild do not have knowledge of the insect, technique of captive rearing, and pathogens and parasites that can hamper their productivity or usability. Disease outbreak can marked the entire insect rearing business. Production in this line of business also requires the market for the produce. Braide (2012) cited in Kelemu et al. (2015) noted that leaves of host plant of emperor moth (*Bunaea alcinoe* S.) are primary source(s) of microorganisms associated with foodborne illnesses in the guts and skins of the larvae. Mastery of the biology of each of the edible insect species is sine qua non in sustaining mass production of the insect.

As important as the knowledge of the biology of the animal, is the knowledge of the target market. Camilleri (2017) opined that knowing the target market is a prerequisite for the development of customer-centric strategy which specify the target markets. Large-scale profit making edible insect business is still at experimental stage in most parts of Africa. At the subsistence level found in most parts of Africa, edible insect market is mainly consumer market either on the roadside or designated market centers or hotels and restaurants. Large scale export market is monopolized by few players such as AgriProtein established in Elsenburg, South Africa by Jason and David Drew. AgriProtein feeds black soldier flies with municipal waste and sells them as feed for the livestock industry. The largest insect farm in Africa sited near Cape Town (an extension of AgriProtein), is a farm conceived by a group of scientists and environmentalists eager to find protein alternatives owing to high production cost of livestock. AgriProtein plans to roll out 100 fly factories by 2024, and a further 100 by 2027. Another large scale farm is established by the Aspire group in Ghana. Aspire operates in Ghana and the USA; in Ghana they commercially farm palm weevil larvae and run a program which empowers peri-rural farmers to raise palm weevils locally. In Kenya and Uganda, a large-scale insect farm is coordinated by Insect for Feed for Poultry and Fish Production (INSFEED) funded by International Development Research Centre, Canada and the Australian Centre for International Agricultural Research (ACIAR).

In Nigeria, USAID Market II (United States Agency for International Development Project on Aquaculture Development In Nigeria), Next Generation Nutrition (NGN), Netherlands and the Netherland government sponsored the training of 14 Nigerians on rearing of black soldier fly larva for replacement of fish meal in livestock and fish feed in the Netherlands and Belgium. Between second to 11th October, 2017. During the training the group were trained in three Universities in the Netherlands (Wageningen University of Research (WUR), Thomas Moore University, University of Leuven and Hagues School and many large-scale insect farms (Next Generation Nutrition, Hertogenbosch, Protix, Millibeter, Proti-farm) and subsidiary industries (Feed Design Lab, Christiaens, Bonda, Blue Acres Aquaponics, Furen & Nooijen, and Skreeton). The group is already championing the developing the black soldier fly larva production under the name Quality Insect for Feed and Food in Nigeria (QUIFFAN). Mass production of edible insects could help solve food insecurity problems in the continent.

2.4 Marketing by Production Type/Processing

Processing of edible insects into more durable, more attractive forms is important strategy for improving acceptability. As earlier stated, this is already a common place in America and Asia. Globally, the edible insect market demand from flour application was valued over USD 19.5 million in 2017. Omotoso (2006) investigated the functional properties of *Cirina forda* larvae powder and particularly obtained high oil and water absorption capacities. Osasona and Olaofe (2010) also observed a high water absorption capacity and oil absorption capacity, respectively, of *C. forda* larvae powder and concluded that the larvae powder could be applied especially in baked products due to its high water absorption capacity, as a flavor retainer and to improve the mouth feel of food products because of its high oil absorption capacity. Omotoso (2006) also observed relatively high emulsion stability and emulsion capacity sand therefore suggested that the larvae powder could function as a texturizing agent in food products. Similar reports have also been recorded for many other insect flour especially cricket. Food products that use grasshoppers and cricket powder owing to its essential minerals, amino acids, and pastry-improving qualities will further increase edible insect market demand. Cricket flour is reported to possess gluten-free properties and so is used in food products which propel the market demand.

In Africa, insect consumption is either raw, cooked, boiled, stewed or even dried (Ebenebe et al., 2017b), processing of insects into other product types: chocolate bars, beer, ice cream, flakes, and noodles as found in Europe and America is still at experimental stage in Africa. Thus seasonal fluctuations affect supply negatively as insect to sell or eat may not be available in certain seasons of the year. Given the

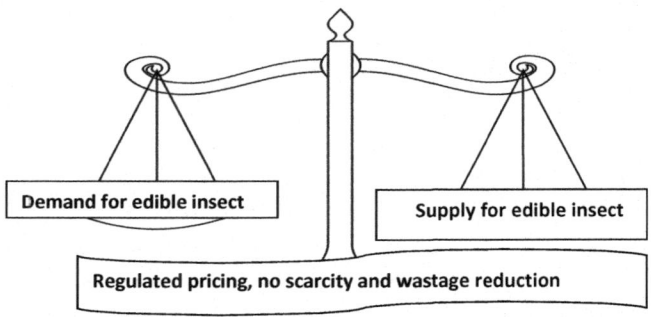

Fig. 2.3 Edible insect demand and supply at equilibrium

Table 2.7 Reasons why some people do not practice entomophagy

Insects	Reasons
Termites	Associated with breakdown of coffin and corpse buried underground
Variegated grasshopper	1. Some communities associate the obnoxious odor in the insect with evil spirits. 2. Unpleasant spots on the insect.
African palm weevil	Appears like a bloated maggot (housefly larvae)
Rhinoceros beetle	Some grow in goat manure and so they are unhygienic
Cricket	1. Unavailability 2. Regarded as food for children

[a]Source: Ebenebe and Okpoko (2014)

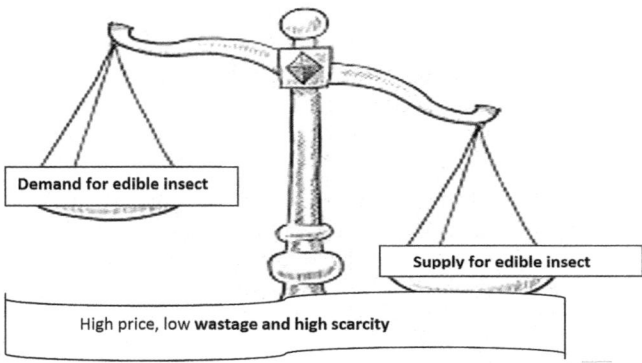

Fig. 2.4 High demand and low supply

issues raised in Ebenebe and Okpoko (2016) especially the fact that many Nigerians associate insect consumption with poverty, the demand and supply in the 1980s to earlier part of 2000 was at equilibrium (Fig. 2.3) (Table 2.7).

When the supply for edible insects is equal to the demand for them, it resulted into a stable market known as equilibrium point (Fig. 2.4). This stage is what every market is trying to attain. It is characterized by regulated pricing, no scarcity, and reduced wastage. However, the point at which the supply and demand is at equilibrium is low, there is need for upward shift of this point to reflect the true realities on ground.

2.5 Edible Insect Supply and Demand Gap

The increasing demand of edible insects in developing countries is a result of rapid rate of population growth and the malnutrition problems. Globally, a high demand for edible insects is driven by socioeconomic changes such as rising incomes, increased urbanization, and aging populations as the contribution of protein to

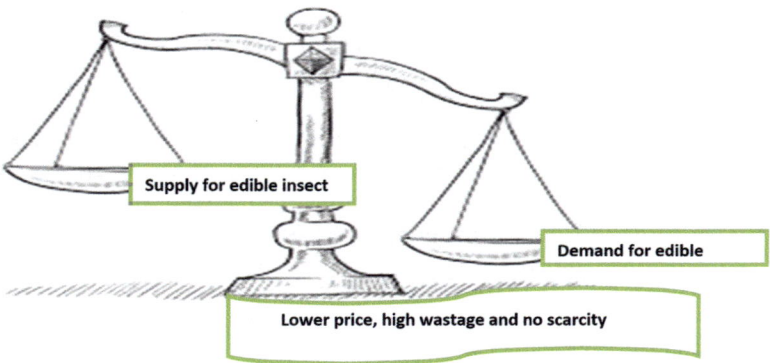

Fig. 2.5 Supply–demand gap

healthy aging is increasingly recognized (Henchion et al. 2017). In Africa, increases in demand for edible insects is presently driven by awareness of the nutritional values of edible insects, impact of consumption on edible insects on the environment and human health. FAO has aroused people's interest on the nutritional importance of edible insects and its environmental health potentials especially in the face of human induced climate change. Meludu and Onoja (2018) reported that edible insects are abundantly eaten in Nigeria and other Africa countries however the impending decline in population of edible insects reported by (Akunne et al. 2013) with the rate of deforestation. Gathering from the wild and subsistence scale of production can no longer meet the demand. Like in any other market system, the supply–demand gap for edible insects is in a situation when the supply of edible insects is higher than the demand that is lower price, high wastage, and low scarcity (Fig. 2.3). The condition when the demand is higher than supply of edible insects referred to as edible insect demand–supply gap. This is characterized by high price, low wastage, and high scarcity of insect products. Presently in Africa edible insect market is at this stage.

While the farmer aim at equilibrium, the consumer needs a condition where the supply is more than demand, so priced can be forced down (Fig. 2.5). The only way this can be possible is by establishing large scale insect farms.

2.6 Reasons Why Demand for Insect Protein Is Growing in Africa

2.6.1 Shortage of Animal Protein/Malnutrition in Africa

FAO (1991) recommended that animal protein requirement for normal growth and development is 34 g/person/day but in most African countries. In Nigeria animal protein consumption level is 7–10 g/Person/day while her counterparts like Somalia and Mauritania were getting 32–34 g (Okoro 2000). In America and some other developed countries animal protein consumption is 54 g/person/day. The poor state of animal protein consumption is responsible for Kwashiorkor reported in some African countries (Hamidu et al. 2003), Geoffery Njoku the communication expert of UNICEF, in 2008 reported on the infant mortality, low intelligent quotient, short stature, and sometimes maternal mortality in Nigeria. In these African nations where meat and fish are expensive, edible insects can provide an alternative source of animal protein and help to avoid protein deficiency problems.

2.6.2   *Alternative to Meat Protein*

Meat consumption especially red meat has been associated with many chronic diseases like sclerosis, hypertension, diabetes, and other heart conditions; the need to stay healthy has elicited a new interest in alternative protein resource. Essential fatty acids (EFA) like linoleic and linolenic acids are present in sufficient quantity in many insect larvae (Ekpo and Onigbinde 2004). According to Van Huis (2013), beef contains more palmitoleic, palmitic, and stearic acid than mealworms, while mealworms contain higher values of essential linoleic acids than beef. Ekpo and Onigbinde (2004) reported that APW larval oil has high medicinal value in terms of its high iodine value which according to them is an index of the degree of unsaturation of the oil. Edible insects can conveniently be used as an alternative meat protein.

2.6.3   *Insects Have a Huge Potential in Animal and Fish Feed Production*

The quantity of fishmeal required for production of fish and livestock feed is quite astronomical with the high level of human population that require meat and fish protein for survival. The high cost of fishmeal for livestock and fish feed, necessitates the use of alternative cheaper animal protein ingredient. At present, around 10% of global fish production goes to fishmeal (i.e., either whole fish or fish remains resulting from processing) and is used mainly in aquaculture (FAO 2012b). The animal feed supply business stated that the international trade in animal feed has an estimated turnover of nearly $400 billion every year. Van Huis (2013) noted that in 2008, aquaculture used 61% of world fishmeal production and 74% of fish-oil production. Insect appears to be perfect replacement for fishmeal in livestock feed and fish feed production.

Rumpold and Schlutter (2013) noted that Acridids (grasshoppers) have been identified as one potential feed component for poultry feed since they have higher protein contents than other protein sources such as soybean meal and fish meal and are rich in the micronutrients Ca, Mg, Zn, Fe, and Cu (Anand et al. 2008). According to him, in the Philippines, grasshoppers are fed to chickens raised on pasture. The chicken fed on pasture (and grasshoppers) are to have a delicious taste and are sold for a much higher price than chickens reared on commercial chicken feed (Litton 1993). Oyegoke et al. (2006) also investigated the dietary potential of the moth larvae *Cirina forda* as a poultry feed component rich in proteins in comparison to fishmeal, a complete replacement of the fish meal by larvae powder resulted in no significant differences in growth rate and weight gain of broiler chicks. Rumpold and Schlutter (2013) therefore opined that the larvae powder represents a potential alternative for the highly nutritious and rather expensive fish meal.

2.7 Examples of Large Scale Insect Farm and Other Insect Projects in Africa

2.7.1 AgriProtein

AgriProtein is the first and foremost large insect farm in Africa. The farm is established in Elsenburg, South Africa by Jason and David Drew. AgriProtein rears fly larvae (on an industrial scale) on organic waste and harvests the larvae to make natural, high-protein animal feed products The company focuses on nutrient recycling (i.e., recycling of organic waste) by insects to develop insect meal/insect larval meal used to replace expensive fishmeal in fish and livestock feed. By using common house fly larvae fed or black soldier fly larvae fed on abundant waste nutrient sources, AgriProtein has developed and tested a new large-scale and potentially sustainable source of protein. The bioconversion process therefore uses low-cost waste materials and generates a valuable commodity (insect larva for feed production). According to Van Huis (2013), the production process starts with rearing stock flies in sterile cages, each holding over 750,000 flies. Various types of waste are used, including human waste (faces), abattoir blood, and spent food. Depending on the species, a single female fly can lay up to 1000 eggs over a 7-day period, which then hatch into larvae. Housefly larvae go through three life stages in a 72-h period and are harvested just before becoming pupae. The harvested larvae are dried on a fluidized bed dryer, milled into flake form and packed according to customer preferences. Van Huis (2013) further stated that the product contains nine essential amino acids, with high levels of cysteine and similar levels of lysine, methionine, threonine, and tryptophan—similar to marine fishmeal. The company aims at production 100 tonnes of larvae per day. At present an Austria-based engineering group Christ of Industries has partnered with waste-to-nutrient company AgriProtein to build up to 25 fly farms a year. The maggot-based animal feed meal developed by AgriProtein is more than 15% cheaper than other alternatives and has been proven to be highly nutritious for livestock, especially chickens (poultry), fish, and pigs.

2.7.2   Aspire, Ghana

Aspire Food Group is an international Social Enterprise dedicated to provision of sustainable means of farming edible insects. The company has offices both in Ghana and USA, In Ghana, Aspire farms African palm weevil larva which is a delicacy in most parts of Ghana. The company develops and teaches farmers in Southern and Central Ghana on the techniques of rearing Africa Palm weevil in their homes instead of felling the oil or raffia palm. They intend to scale up production and also develop packaged products so that they can export to other countries and at the same time advance further research in African palm weevil.

2.7.3   INSFEED: Insect Feed for Poultry and Fish Production in Kenya and Uganda

INSFEED is a project funded by International Development Research Centre, Canada (IDRC) and Australian Centre for International Agricultural Research (ACIAR) in collaboration with Egerton University, University of Nairobi, Sanergy Ltd., Kenya Bureau of Standards (KEBS), Unga Feed, Lasting Solution in Kenya and Makerere University; UNBS, NaLiRRI, NaFiRRI, UGACHICK in Uganda. The major thrust of research in the project is identification of suitable insect species, assessing the potential market and nutritional attributes of the products, and development and adaptation of cost-effective insect rearing, harvesting, and post-harvest techniques for smallholder producers. It will also establish the risk factors associated with the insect-based feeds along the food chain and their mitigation. Strategies as well as conduct research to inform policies for promotion of safe, sustainable and cost-effective use of insect in the feed sector. The project is coordinated by Dr. Komi Fiaboe, who is instrumental in getting regulatory changes needed to allow insect feed to be approved for use in all animals and fish feed in both countries (Bryne 2018).

2.7.4   FasoPro

FasoPro is not a large-scale insect farm, rather an initiative of a young engineer from the International Institute of Engineering of Water and the Environment (2 IE). FasoPro project born in the business incubator of 2 IE. It is a kind of off take, where the company buys the edible insects harvested by women gatherers in the locality for further processing and more attractive and healthy packaging. It is located a small town called Somousso, about 400 km from Ouagadougou, the capital of Burkina Faso. Over a hundred women are engaged in managing the shea trees. The women harvest the caterpillars from these centuries-old fruit trees that abound in the region.

The women coordinating the Somousso collection center were organized by the developer, Kahitouo Hien and the plan included five villages (Somousso, Bare, Piére, Yegueresso, and Sare). FasoPro group buy their harvest unpackaged in kilograms, a method which benefits the gatherers. At present, a standard box of about 3 kg is used for measuring the caterpillar at costs between 600 and 700 CFA in Bobo-Dioulasso, market. However, where the caterpillars are boiled or dried a measure with the standard box trades for around 2000 F CFA (Tao). According to Kahitouo Hien the developer of FasoPro, the trained farmers had the ingenious idea of preserving the caterpillars in cans, a method of transformation and conservation that extends their period of consumption, stabilizes prices, and increases production. His ultimate goal is to fight child malnutrition and poverty in rural areas. The engineer wishes to create a permanent nutrient resource containing 63% protein that lasts beyond the winter season and the labeling of the local product.

Ms. Ouattara, owner of the agroecologic farm Guiriko, coordinates the union of women's groups engaged in this activity. Their mission is to gather the 10 tonnes needed to start the pilot phase. To achieve this, they regrouped into associations to sort the correct species and to precondition them according to the technical standards provided by the developer.

2.8 Factors Affecting Supply of Edible Insects

1. *Storage*: Insect farming in a closed or indoor environment is an important means of making insect and insect products available throughout the year. As production increases there will be need to store the food to avoid wastage (Fig. 2.5). Processing insect into a more attractive and acceptable product will ensure reduction of wastage (Ebenebe 2016).
2. *Availability of organized market*: Unavailability of organized market where consumer can get or secure edible insects is affecting its marketability in Africa. Organized market that will stabilize price in all parts of each of the African country. Establishment of cooperative societies aimed at promoting edible insect conservation is required.
3. *Packaging*: It was clear that insect food alone would not be able to defend itself in a crowded market place of other proteins source. We needed to synthesize a culture around insects, to serve as a platform for their popularity. Aguirre-Joya et al. (2018) and Ebenebe (2016) recognized that fanciful packaging of edible insects will attract more consumers, reduce waste and to create novel applications for improving desired features of a product, such as stability, quality, safety, variety, and convenience for consumers.
4. *Technology and research*: Insect based technology is needed for mass production, processing, and marketing of edible insects. Data gathering surrounding insect as food feed should be encouraged to deepen the knowledge of people and further grow the sector in Africa.

5. *Consumer acceptance of edible insects*: Africa especially Nigeria is known to relish insect, but due to advancement in education, urbanization, and civilization, edible insect acceptance dropped. The perception that only poor people eat insects (Ebenebe 2016) started to thrive, especially among younger generation, which needs to be changed. Taboos and cultural influences affecting the urge to eat insect need to be visited; training and awareness campaigns showcasing the benefits of eating insect should be encouraged.

2.9   Marketing by Insect Type

The continent is home to the richest diversity of edible insects, an inventory of 250 species of edible insects were recorded. Among which 78% are Lepidoptera (30%), Orthoptera (29%), and Coleoptera (19%), and 22% Isoptera, Homoptera, Hymenoptera, Heteroptera, Diptera, and Odonota, respectively (Van Huis 2003). In Nigeria however,22 edible insect species from six orders were compiled. Of these, 77.3% were Lepidoptera (27.3%), Coleoptera (27.3%), Orthoptera (22.7%) and 22.7% Isoptera, Hemiptera and Hymenoptera (Alamu et al. 2013). Marketing edible insects according to their type requires knowledge of the cultural acceptability of the insect in the region before venturing into it. Tables 2.1, 2.2, 2.3, 2.4 and 2.5 showed edible insect types consumed in many parts of Africa.

2.10   Marketing by Product Type

Research and development strategies have improved the awareness of the enormous importance of edible insects as a global food. Edible insects have gone beyond basic processing methods other than eaten raw, roasted, sun dried, fried, and boiled in more developed countries increasing acceptability and usage. Edible insect is now incorporated into many food products sweet bars, flakes, noodles, ice cream, and beer, this owing to the fact that a food's preparation could strongly influence its desirability by changing its hedonic sensory qualities and perceived appropriateness. In Africa, there are few insect products. In Nigeria, an infant formula containing insect protein that goes with the name "Cerovil" was developed by Dr. C.I. Ebenebe. In Kenya, termites and lake flies were baked, boiled, and cooked to increase shelf life and processed into conventional consumer products such as crackers, muffins, sausages, and meat loaf to encourage entomophagy (Ayieko et al. 2012). In the western world, products like insect flour, burger, pasta, juice, smoothie, wine, and so on can be found sold online (Entomarket 2018). Despite the acceptability of edible insects in Africa, processing edible insects into acceptable products has not been fully explored. Therefore, there is a need for creating of more culturally acceptable products to increase insect marketability in Africa.

2.11 Marketing by Application

Edible insect can also be marketed based on what their intending use is. Edible insects have been used to formulate feed for livestock including poultry, pigs, and fish (Stamer 2015; Kenis et al. 2014) as well as dogs and other pets. Food product made from edible insect could also be used by zoos to feed their animals (Entomarket 2018; Van Huis 2013, 2015).

2.12 Marketing by Geography

Species dominance of edible insects varies from region to region. The species coverage of some insect may span throughout a country while some may be restricted. The more widespread a species is the more acceptable it is and vice versa. In Table 2.1, *Apis mellifera* is found in all parts of Nigeria, therefore its acceptance level is all over Nigeria but others are region specific. There is possibility for exporting edible insects within and beyond Africa.

2.13 Processing of Edible Insects

Some ethnic groups in Africa eats insect raw (Van Huis 2013), while in other tribes processed them into a more palatable product to eat. In Nigeria, *Rhynchophorus phoenicis* commonly known as palm weevil, is processed by frying before eaten (Opara 2003). *Oryctes monoceros*, a common pest of coconut tree, is eaten raw, boiled, smoked or fried (Ifie and Emeruwa 2011). *Microtermis nigeriensis*, a very common food found in almost all ecological zones in Nigeria is usually processed by washing, salting to taste and mild frying or roasting. This delicacy is usually rich in oil; therefore, there is no need for oil while frying (Igwe et al. 2011). It can also be consumed raw by some tribe in Yoruba land (Fasoranti and Ajiboye 1993). The larvae of *Bunaea alcinoe* a common pest of forest trees in the country is reported eaten boiled and sun dried (Fasoranti and Ajiboye 1993; Agbidye et al. 2009). The large African cricket, *Brachytrupes membranaceus*, is a pest of forest nurseries with severe defoliation. They are commonly processed roasted in mild fire and fried. Edible insects can also be processed into other products for value addition. Akullo et al. (2017) stated that termite flour has a high potential in fortifying food products and feed with acceptable sensory and nutritional qualities.

In Africa, use of insects to feed animals has been documented in Angola, Benin, Burkina Faso, Nigeria, Togo, Cameroon, Democratic Republic of Congo (DRC), and South Africa (Mutungi et al. 2017).The most widespread industrial method of harvesting insect species is by chilling them to freezing temperatures. This process causes the insects to enter a state of sleep much like a coma as their body temperature lowers. After an extended period of being frozen, which varies by species but is generally 2–3 days, the insects die without regaining consciousness. As compared with modern methods of slaughtering traditional livestock, the pain levels are believed to be drastically lower than those of cows, pigs, and chickens; however, we lack the full understanding of the ways insects experience pain or if they indeed do at all (Dossey et al. 2016). Insect are usually processed by drying before grinding into fine form, which is then used for feed formulation.

Milling edible insects into flour is reportedly a way to take away the icky factor from eating insects (Dossey et al. 2016). Most commercial food products are processed the same way they are processed at home. Commercial production might use motorized machine for processing, this makes marketing of edible insects easier and wider area is often covered. Saprophagous insects such as black soldier fly and house fly have been used to produce manure which is of value in biodegradation and aquaculture. Combination of waste treatment capacity together with generation of a valuable product makes the black soldier fly technology a highly promising tool for waste management in low- and middle-income countries. It offers small entrepreneurs the possibility of income generation without high investment costs, and concurrently reduces the environmental impact. Improving existing methods of processing of commonly consumed insects is important if edible insects are to meet the global market. Massive research needs be encouraged along the line of processing.

2.13.1 Market Values of Products from Edible Insects

Given that consumption in marketing terms involve processes of using consumer products in order to satisfy desires, real or imaginable needs so that the products are used up, transformed or deteriorated in such a manner as not to be either reusable or recognizable in their original form. Edible insect consumption market will therefore include whole insects, insect products, other uses especially services rendered by insect.

2.13.2 Market Values of Edible Insect Services

2.13.2.1 Medicinal Services

The name given to the medicinal usage of animals and its derived products is zootherapy (Marques and Costa-Neto, 1994). Traditionally, insects and insect products are used in preparing concoction for treatment of several ailments (e.g., *Camponotini brutus* for treatment of wounds, *Belanogaster* spp. for treatment of various diseases in children; Banjo et al. 2003). Termites and crickets apart from being used as food are also used for rituals. Lawal et al. (2003) reported the traditional utilization of honey bee among Ijebus which include rituals, forceful command, defense, and favor. Molan (2006) stated that the ancient Egyptians and Greeks used honey for wound care. Simon et al. (2009) posited that honey works differently from antibiotics, which attack the bacteria's cell wall or inhibit intracellular metabolic pathways. According to them, honey is hygroscopic, meaning it draws moisture out of the environment and thus dehydrates bacteria. Its sugar content is also high enough to hinder the growth of microbes, but the sugar content alone is not the sole reason for honey's antibacterial properties. When honey is diluted with water, reducing its high sugar content, it still inhibits the growth of many different bacterial species that cause wound infections. Today the first medically certified honey-based licensed product: Medihoney™ has been developed as a medical product for professional wound care in Europe and Australia. The global market price of honey stands at $3.63/kg and the two leading African countries in honey production. FAO estimated that Africa accounted for roughly 9% of global honey production (155,789 t), representing a 10% increase since 2000, which has since increased to 13% by 2016 and Ethiopia leading with 50,000 t, followed by Tanzania (30,000 t), Angola (23,300 t) and Central African Republic (16,200 t). Ethiopia is also the fourth largest beeswax producer in the world. Honey exports throughout the continent appear to have grown sharply by 613% from 2000 to 2013, representing an increase to 3195 t, worth €8.9 million. Nigeria only fulfills 10% of its total domestic consumption demand (380,000 t), while it annually imports €1.84 billion worth of honey, according to David Victor Musa, general manager of Barg Natural Honey (Nigeria) and US Agency for International Development consultant (Spore 2017). Srivastava et al. (2009) has made a detailed documentation of insect uses in traditional medicine; over ten insects are described as medicine in native medicine. According to him, trembling red ant locally known as "*L. Nkaam*" used for the treatment of muyeem (bronchitis). Again, grasshopper locally known as Mpaylaar in Zaire is used for the treatment of violent headaches. Another important insect in traditional medicine he documented is that of worker wasp locally known as "*Ngankoy*" in Zaire that strengthens a weak infant, the community believes that the nest of the worker wasp has a substance, which gives life strength to the weak. He also mentioned the use of cockroach locally known as "Kembaar" in Zaire to cure scabies/mange in animals. Praying mantis locally known as"*Kayakua*" in Zaire are also used for the treatment of epilepsy while builder/worker caterpillar locally known as "*KenbulMpiak*" in Zaire, cures hemorrhage during childbirth or during pregnancy.

2.13.2.2   Magical Insects

Cherry (2005) described the branch of magic associated with insect as imitative magic based on the assumption that "likes produces like" (Frazer 1998). By this principle individual can produce any effect they want by imitating it. The most common use of insects in imitative magic hinges on expression of behavior, special power or morphological trait associated with the insect (Cherry 2005). Van Huis (2003) noted that in sub-Saharan Africa, it is believed that the morphological trait or specific behavior of an arthropod can be acquired by humans when they treat themselves with the insect or a preparation made from it. The most commonly used insects in this regard are the sacred scarab of ancient Egypt, the cicada and butterfly. These three insects are associated with religious concept of rebirth. For the sacred scarab of Egypt (*Scarabaeus sacer* L.), it is reported that amulets to protect the dead were carved in the form of scarab and buried with the body with the intent of ensuring rebirth of such person. The nymph crawling of the as the insect sheds off the nymphal skin is symbolic of the spirit of the deceased leaving the body and emerging out of the dead body (Kristy and Cherry 2000).

Apart from regeneration and rebirth magical doctrines, insects use in imitative magic also includes the aspect of the impartation of ferocious tendencies to humans by some ferocious insects (Cherry 2005). Van Huis (2003) posited that in Africa, wasps are crushed and put into incisions made on the back of the hand of warriors/ wrestlers to give the person a punch like the sting of wasp.

Another aspect of imitative magic that involve the use of insect is based on the proposition of Berenbaum (1995) for the "doctrine of signatures" in which she opined that God provided all things for human use provided in signs such as shape or color to show their use. Van Huis (2003) reported that some of the best examples of this medical magic occur in sub-Saharan Africa. According to him, in Chad, children who are slow to walk can be stimulated by using fast running ants which are crashed into powder and rubbed into incisions made on the legs of the children. Similarly, Cherry (2005) reported that in several countries of Africa, cicadas, cricket and other singing insects are crushed, mixed with herbs and eaten to obtain a pleasantly high and clear voice among women.

Imitative magic is also extended to the aspect of manipulation of insects (Campbell 1988 cited in Cherry 2005) to behave in certain desired way or yield higher products. Cherry (2005), reported that the pygmies of Africa use imitative magic to enhance their honey harvest from bees. According to him, the natives first simulate unfruitful efforts in the hunt for honey in form of a dance facilitated by singing. The dance culminate into setting up of a great hone fire. The fire is accompanied by a song of magic that will travel with the smoke to call the bees to come back to make more honey. In the south east Nigeria, Ebenebe et al. (2017b) reported on the magical invocation of bees into an enemy's house with the intent of having the bees sting the enemy to death. Another insect reported by Ebenebe et al. (2017b) is the larva of butterfly locally called "Nwaigu" usually found on leaves of very tall trees, which freely fall to be harvested as the natives call/sing its name "Nwaigu, Nwaigu, Nwaigu" or "Wee, Wee, Wee." The more the song, the more the number of larvae that hit the ground to be caught (Figs. 2.6, 2.7, 2.8, 2.9, 2.10, 2.11, and 2.12).

Fig. 2.6 Locust (Fara) from village market, Kumo

Fig. 2.7 Grasshopper (Fara) with various measuring containers Gombe, Gombe State

Fig. 2.8 Boiled cricket (*Brachytrupes* spp.)

Fig. 2.9 Grasshopper sale in Nigeria

Fig. 2.10 African palm weevil larva displayed for sale with African chicken at Oba, Idemili South, LGA, Anambra State

Fig. 2.11 Winged termite collected

Fig. 2.12 Winged termite
roasted

2.14   Conclusion

There is high prospect for edible insects in Africa, but seeing it as a business opportunity is new. Edible insects are highly nutritious, environmentally friendly and it is tipped to end protein crisis in Africa. With the global awareness of edible insect at its peak. Africa must grab this opportunity and invest heavily in edible insects. The market for edible insects is large in Africa, but the supply is low, meeting this supply gap is the greatest challenges we face in Africa edible insect sector. Therefore, increase in awareness, research, mass production, packaging, and marketing should be encouraged.

References

Adeoye TO, Job OO, Abiodun AFT, Dare AO (2014) Eco-diversity of edible insects of Nigeria and its impact on food security. J Bio Life Sci 5(2):175. https://doi.org/10.5296/jbls.v5i2.6109

Agbidye FS, Ofuya TI, Akindele SO (2009) Marketability and nutritional qualities of some edible forest insects in Benue State, Nigeria. Pakistan J Nutri 8:917–922. https://doi.org/10.3923/pjn.2009.917.922

Aguirre-Joya JA, De Leon-Zapata MA, Alvarez-Perez OB, Torres-León C, Nieto Oropeza DE, Ventura-Sobrevilla JM, Aguilar CN (2018) Basic and applied concepts of edible packaging for foods. In: Grumezescu AM, Holban AM (eds) Food packaging and preservation. Academic, Cambridge, pp 1–61. https://doi.org/10.1016/b978-0-12-811516-9.00001-4

Akullo J, Agea JG, Obaa BB, Acai JO, Nakimbugwe D (2017) Process development, sensory and nutritional evaluation of honey spread enriched with edible insects flour. Afri J Food Sci 11(2):30–39

Akunne CE, Ononye BU, Mogbo TC (2013) Insects: friends or enemies? Global J Biol Agri Health Sci 2(3):134–140

Alaloiun U (2014) Insects in forests: assemblages, effects of tree diversity and population dynamic. Philipps-Universität Marburg 15(2014):685–692

Alamu OT, Amao AO, Nwokedi CI, Oke OA, Lawa IO (2013) Diversity and nutritional status of edible insects in Nigeria: a review. Inter J Biod Conserv 5(4):215–222. https://doi.org/10.5897/IJBC12.121

Alexandratos N, Bruinsma J (2012) World agriculture towards 2030/2050: the 2012 revision. ESA Working paper no. 12–03. FAO, Rome

Amobi MI, Ebenebe CI (2018) Performance of broiler chicks fed two insect based diets in South East, Nigeria. J Insect Food Feed 4(4):263–268. https://doi.org/10.3920/JIFF2017.0078

Anankware PJ, Fening KO, Osekre E, Obeng-Ofori D (2015) Insects as food and feed: a review. Int. J Agric Res Rev 3(1):143–151

Anankware JP, Osekre EA, Obeng-Ofori D, Khamala C (2016) Identification and inter. J Entomo Res 1(5):33–39

Anand P, Kunnumakkara AB, Sundaram C et al (2008) Cancer is a preventable disease that requires major lifestyle changes. Pharm Res 30(25):2200

Ayieko MA, Kinyuru JN, Ndong'a MF, Kenji GM (2012) Nutritional value and consumption of Black insects (*Carebara vidua* Smith) from the Lake Victoria Region in Kenya. Advan Food Sci Technol 4(1):39–45

Banjo AD, Hassan AT, Ekanyaka IJ, Dixon AGO, Jackal EN (2003) Developmental and behavioural study of spiralling whitefly (A. disperses) on three cassava (Manihot esculents Crantz). J Res Crops 5(2–3):252–260

Banjo AD, Lawal OA, Songonuga EA (2006) The nutritional value of fourteen species of edible insects in southwestern Nigeria. Afri J Biotech 5(3):298–301

Bernard T, Womeni HM (2017) Entomophagy: insects as food, insect physiology and ecology, Vonnie DC. Shields, IntechOpen. https://doi.org/10.5772/67384. https://www.intechopen.com/books/insect-physiology-and-ecology/entomophagy-insects-as-food

Berenbaum M (1995) Bugs in the system. Adison Wesley, New York

Braide W (2012) Perspectives in the microbiology of the leaves of three plant species as food for an edible caterpillar of an emperor moth. Inter J Res Pure Appl Microbio 2:1–6

Braide W, Sokari TG, Hart AD (2010) Nutrition quality of an edible caterpillar of lepidopteran (Bunea alcinoe). Adv Sci Technol 4:49–53

Bryne J (2018) MacDonald championing research into insect feed for chickens. Report from feed protein vision 2018. https://www.feednavigator.com

Camilleri MA (2017) Market segmentation, targeting and positioning. Travel marketing, tourism economics and the airline product (Chapter 4). Springer, Milan, pp 69–84

Campbell J (1988) Mythologies of Primitive Hunters and Gatherers. Harper and Row, New York

Chai J-Y, Shin E-H, Lee S-H, Rim H-J (2009) Food-borne intestinal flukes in Southeast Asia. Korean J Parasitol 47:S69–S102

Chapagain AK, Hoekstra AY (2003) Virtual water flows between nations in relation to trade in livestock and livestock products. UNESCO-IHE, Delft

Cherry R (2005) Magical insects. In: Insect Mythology. Writers Club Press, New York

DeFoliart GR (2002) The human use of insects as a food resource: a bibliographic account in progress. University of Wisconsin, Madison, WI

Dossey AT, Morales-Ramos JA, Rojas MG (eds) (2016) Insects as sustainable food ingredients: production, processing and food applications. Academic, London

Ebenebe CI, Okpoko VO (2014) Edible insect in Southeastern Nigeria. Conference Paper Presented at Eating Insect Conference, Future Food Salon, Montreal, Quebec, Canada, August 26th–28th, 2014

Ebenebe CI, Okpoko VO (2016) Preliminary studies on alternative substrate for multiplication of APW under captive management. J Insect Food Feed 2(3):171–177

Ebenebe CI, Amobi MI, Udeagbala C, Ufele AN, Nweze BO (2017a) Survey of edible insect consumption in Southeastern Nigeria. J Insect Food Feed 3(4):241–252

Ebenebe CI, Okpoko VO, Ufele AN, Amobi MI (2017b) Survivability, growth performance and nutrient composition of African Palm Weevil (*Rynchophorus phoenicis* Fabricius) reared on four different substrates. J BioSci Biotechnol Discov 2:1–9

Edijala TK, Eghogbo O, Anigboro AA (2009) Proximate composition and cholesterol concentrations of *Rhyncophorusphoenicis* and *Orytcesmonocerus* larvae subjected to Different Heat Treatments. Afri J Biotech 8(10):2346–2348

Ekpo KE, Onigbinde AO (2004) Pharmaceutical potentials of *Rhyncophorus phoenicis* larva oil Nigeria. Annal Nat Sci 5:28–36

Ekpo KE, Onigbinde AO (2005) Nutritional potentials of the larva of *Rhynchophorus*. Pak J Nut 4(5):287–290

Entomarket (2018) Edible Bugs as Snacks. https://www.edibleinsects.com

FAO (1991) State of forest and tree genetic resources in dry zone Southern African Development Community Countries. FAO, Rome

FAO (2005) State of the World's Forests 2005. Food and Agriculture Organization of the United Nations, Rome. ftp://ftp.fao.org/docrep/fao/007/y5574e/y5574e00.pdf

FAO (2008) The state of food insecurity in the world: high food prices and food security—threats and opportunities. www.fao.org/3/i029ie/

FAO (2012) The state of food and agriculture: investing in agriculture for a better future. www.fao.orga-13028epdf

FAO (2012a) FAOSTAT database collections. Food and agriculture organization of the United Nations. http://faostat.fao.org/default

FAO (2012b) The state of world fisheries and aquaculture 2012, Rome, Italy.

FAO (2013) Edible insects future prospects for food and feed security. http://www.fao.org/i3253e.pdf

Frazer J (1998) The Golden Bough. Oxford University Press, New York

Fasoranti JO, Ajiboye DO (1993) Some edible insects of Kwara State, Nigeria. Amer Entom 39:113–116

Graczyk TK, Knight R, Tamang L (2005) Mechanical transmission of human protozoan parasites by insects. Clin Microbiol Rev 18:128–132

Hamidu JL, Salami HA, andEkanem AU (2003) Prevalence of protein-energy mal-nutrition in Maiduguri, Nigeria. Afri J Biomed 6:123–127

Henchion M, Hayes M, Mullen A, Fenelon M, Tiwari B (2017) Future protein supply and demand: strategies and factors influencing a sustainable equilibrium. Foods 6(7):53. https://doi.org/10.3390/foods6070053

Ifie I, Emeruwa CH (2011) Nutritional and anti-nutritional characteristics of larva of *Oryctesmonoceros*. Agric Biol J N Am 2(1):42–46

Igwe CU, Ujowundu CO, Nwaogu LA, Okwu GN (2011) Chemical analysis of an edible African termite, *Macrotermesnigeriensis*; a potential antidote to food security problem. Biochem Anal Biochem 1:105. https://doi.org/10.4172/2161-1009.1000105

Kelemu S (2016) African edible-insects: diversity and pathway to food and nutritionalsecurity. https://es.slideshare.net/SIANIAgri/segenet-kelemu-african-edible-insects-diversity-and-pathway-to-foodand-nutritional-securit

Kelemu S, Niassy S, Torto B, Fiaboe K, Affognon H, Tonnang H, Maniania NK, Ekesi S (2015) African edible insects for food and feed: inventory, diversity, commonalities and contribution to food security. J Insects Food Feed 1(2):103–119. https://doi.org/10.3920/JIFF2014.0016

Kinyuru J, Konyole S, Kenji G, Onyango C, Owino V, Owuor B, Estambale B, Friis H, Roos N (2012) Identification of traditional foods with public health potential for complementary feeding in Western Kenya. J Food Res 1(2):148–158. https://doi.org/10.5539/jfr.v1n2p148

Kinyuru J, Konyole S, Roos N, Onyango C, Owino V, Owuor B, Estambale B, Friis H, Aagaard-Hansen J, Kenji G (2013) Nutrient composition of four species of winged termites consumed in Western Kenya. J Food Comp Analy 30(2):120–124. https://doi.org/10.1016/j.jfca.2013.02.008

Kristy G, Cherry R (2000) Insect mythology. Writers Club Press, New York

Leal CRO et al (2016) Vegetation structure determines insect herbivore diversity in seasonally dry tropical forests. J Insect Conserv 20:979–988

Lawal OA, Banjo AD, Junaid SO (2003) A Survey of the Ethnozoological Knowledge of Honey Bees (*Apis mellifera*) in Ijebu Division of South Western Nigeria. Indilinga African Journal of Indigenous Knowledge System 2:75–87

Litton E (1993) Grasshopper consumption by humans and free range chicken reduce pesticide use in the Phillipines. Food Insect Newslett 3:6

Malaisse F (2005) Human consumption of lepidoptera, termites, orthoptera, and ants in Africa. In: Paoletti MG (ed) Ecological implications of minilivestock: potential of insects, rodents, frogs and snails. Science Publishers, Enfield, MT, pp 175–230

Marques Jose GW, Costa-Neto EM (1994) Insects as Folk Medicine in the State of Alagoas, Brazil. In: Proceedings of the 8th international conference on traditional medicine and folklore, Vol. 4, Scrutinies from western medicine, Newfoundland, Canada, pp. 115–119

Mbah CE, Elekima GOV (2007) Nutrient composition of some terrestrial insects in Ahmadu Bello University, Samaru, Zaria Nigeria. Sci World J 2(2):17–20

Meludu NT, Onoja MN (2018) Determinants of edible insect consumption level in Kogi State, Nigeria. J Agri Exten 22(1):156. https://doi.org/10.4314/jae.v22i1.14

Molan PC (2006) Using honey in wound care. Int J Clin Aromatherap 3(2):21029

Mutungi C, Irungu FG, Nduko J, Mutua F, Affognon H, Nakimbugwe D, Ekesi S, Fiaboe KKM (2017) Postharvest processes of edible insects in Africa: a review of processing methods, and the implications for nutrition, safety and new products development. Criti Rev Food Sci Nut. https://doi.org/10.1080/10408398.2017.1365330

Obopile M, Seeletso TG (2013) Eat or not eat: an analysis of the status of entomophagy in Botswana. Food Sec 5:817–824

Okoro FU (2000) Adoption of Rabbitry Technologies Among Farmers in Okigwe Agricultural Zone of Imo State, Nigeria. In: Proceedings of 25th Conference of Nigerian Society of Animal Production 19th–23rd of March, 2000. Michael Okpara University of Agriculture Umudike, Abia State, Nigeria, pp 414–416

Okweche SI, Abanyam VA (2017) Perception of rural households on the consumption of edible insects in Cross River State, Nigeria, Book of Abstract, African Association of Insect Scientist (AAIS) Conference 23rd to 26th October, 2017. Agricultural Research Centre, Wad Medina, Sudan

Oonincx DG, de Boer IJ (2012) Environmental impact of the production of mealworms as a proteins source for humans—a life cycle assessment. PLoS One 7(12):e51145. https://doi.org/10.1371/journalpone.0051145

Oyegoke OO, Akintola A, Fasoranti JO (2006) Dietary potentials of the Edible Larva of Cirina forad (Westwood) as a poultry feed. Afr J Biotechnol 5(19):1799–1802

Okore O, Avaoja D, Nwana I (2014) Edible insects of the Niger delta area in Nigeria. J Nat Sci Res 4(5):1–9

Omotoso OT (2006) Nutritional quality, functional properties and anti-nutrientcompositions of the larva of Cirinaforda (Westwood) (*Lepidoptera: Saturniidae*). J Zhejiang Uni Sci B 7:51–55

Oonincx DGAB, Van Itterbeeck J, Heetkamp MJW, Van den Brand H, Van Loon JJA, Van Huis A (2010) An exploration on greenhouse gas and ammonia production by insect species suitable for animal or human consumption. PLoS One 5(12):e14445

Opara LU (2003) Traceability in agriculture and food supply chain: a review of basic concepts, technological implications, and future prospects. J Food Agri Enviro 1:101–106

Osasona AI, Olaofe O (2010) Nutritional and functional properties of *Cirinaforda* larva from Ado-Ekiti, Nigeria. Afri J Food Sci 4:775–777

Persistence Market Research (PMR) (2018) Increasing food demand to drive the consumption of edible insects. https://www.Persistencemarketresearch.com. Accessed 5 Sept 2018

Raheem D, Carrascosa C, Oluwole OB, Nieuwland M, Saraiva A, Millán R, Raposo A (2019) Traditional consumption of and rearing edible insects in Africa, Asia and Europe. Crit Rev Food Sci Nutr 59(14):2169–2188. https://doi.org/10.1080/10408398.2018.1440191

Roulon-Doko P (1998) Chasse, cueilletteet cultures chez les Gbaya de Centrafrique. L'Harmattan, Paris

Rumpold BA, Schlüter OK (2013) Potential challenges of insects as an innovative source for food and feed production. Innov Food Sci Emerg Technol 17:1–11

Saliou N, Ekesi S (2017) Eating insects has long made sense in Africa. The world must catch up. www.conversation.com/institution/University-of-Pretoria/1645

Schabel HG (2010) Forest insects as food: a global review. In: Durst PB, Johnson DV, Leslie RN, Shono K (eds) Food and Agriculture Organization of UN, Regional Office for Asia and Pacific, Bangkok, Thailand, pp 37–64

Sidiki S (2015) Cricket protein for food security in West Africa. http://www.kickstarter.com. Accessed 5 Sept 2018

Spore (2017) Honey exports take off in Africa, Spore Magazine, CTA. https://spore.cta.int

Silow CA (1976) Edible and other insects of mid-western Zambia; studies in Ethno-Entomology II. Antikvariat Thomas Andersson, Uppsala. 223 pp

Simon A, Traynor K, Santos K, Blaser G, Bode U, Molan P (2009) Medical honey for wound care—still the 'Latest Resort'? Evid Based Compl Alternat Med 6(2):165–173

Srivastava SK, Babu N, Pandey H (2009) Traditional insect bioprospecting—as human food and medicine. Indian J Trad Know 8:485–494

Takeda J (1990) The dietary repertory of the Ngandu people of the tropical rain forest: an ecological and anthropological study of the subsistence activities and food procurement technology of a slash-and burn agriculturist in the Zaire river basin. Afri Study Monographs Supplement 11:1–75

Van Huis A (2003) Insects as food in Sub-Saharan Africa. Insect Sci Appl 23:163–185

Van Huis A (2013) Potential of insects as food and feed in assuring food security. Annual Rev Entom 58:563–583. https://doi.org/10.1146/annurev-ento-120811-153704

Van Huis A (2015) Edible insects contributing to food security? Agric and Food Security 4:20. https://doi.org/10.1186/s40066-015-0041-5

World Food Hunger Statistics—Food Aid Foundation (2016). http://www.foodaidfoundation.org.

Chapter 3
Entomophagy in Africa

Karanjit Das

Abstract Edible insects are natural resources that serves as food to different ethnic groups worldwide. The present record of studies suggest that numerous species of insects belonging to orders such as Hymenoptera, Lepidoptera, Hemiptera, Coleoptera, Orthoptera and Isoptera of class "Insecta" are consumed by a large number of communities. The edible insects are consumed either raw or fried or roasted. Edible insects have prime nutritional value of the consumed it is recorded by various authors that contain high content of nutrients such as proteins and carbohydrates, advocating their utilization as quality nutritional supplements of balanced diet. Eating insects is a common practice among rural and urban people of Africa. Generally, the edible insects can be collected in large numbers and is cheaper option to the consumers in comparison to other sources of protein. The edible insects or their products are sold in most of African food markets. The main focus of this study is to understand the potential of edible insects in Africa, its conservation strategy and the importance for documentation of traditional rearing and cultivation of edible insects.

Keywords Edible insects · Entomophagy · Insect · Nutritional value · Proteins

Abbreviations

ADG   Average daily gain
FAO   Food and Agriculture Organization of the United Nations

3.1   Introduction

The word entomophagy is derived from the Greek word "éntomon", meaning "insect", and "phagein", meaning "to eat" which describes the practice of eating insects by humans (as well as by non-human species). Since the dawn of mankind

K. Das (✉)
College of Health and Life Sciences, Department of Life Sciences, Biosciences,
Brunel University London, London, UK

© Springer Nature Switzerland AG 2020

A. Adam Mariod (ed.), *African Edible Insects As Alternative Source of Food, Oil, Protein and Bioactive Components*, https://doi.org/10.1007/978-3-030-32952-5_3

insects have played an important role in human nutrition (Bodenheimer 1951). The history of consumption of insects as food is as old as the history of mankind. However, consumption of insects form a small diet, they are important for compensation deficiency in proteins, fats and calories which occurs among the poorer section of societies (Narzari and Jatin 2015). Biochemical analysis of the edible insects exhibited rich sources of many essential nutrients (Das et al. 2019). Edible insects contain very high amount of proteins, carbohydrates, fats, minerals and vitamins (Ene 1963; Ashiru 1988; DeFoliart 1989, 1992) and they serve as an alternative food source in many developing countries.

The most common edible insects are beetles (Coleoptera) constituting 31%, followed by caterpillars (Lepidoptera) (18%), bees, wasps and ants (Hymenoptera) (14%). Other edible insects are grasshoppers, locusts and crickets (Orthoptera) (13%), cicadas, leafhoppers, plant-hoppers, scale insects and true bugs (Hemiptera) (10%), termites (Isoptera) (3%), dragonflies (Odonata) (3%), flies (Diptera) (2%) and other orders (5%) (Thakur et al. 2017). There is also a strong social and religious influence on entomophagy in many cultures and communities and insects are usually consumed in various parts of the world during different festivities (Cerritos 2009).

The population of the earth will reach 9.6 billion in 2050 (United Nations, Department of Economic and Social Affairs, Population Division 2013) which will demand increased food and feed outputs. Traditionally edible insects are consumed in many parts of the world (DeFoliart 1999) and are considered to be prospective alternative food source to contribute to the world's food security (van Huis 2013).

Insects are food with towering levels of energy and protein, good amino acid and fatty acid content and possess a variety of micronutrients such as iron, magnesium, copper, manganese, phosphorous, zinc, selenium and vitamins riboflavin, pantothenic acid, biotin and, in few cases, folic acid (Rumpold and Schlüter 2013). Besides its role as food, edible insects are presumed to have an impact on livelihood and social conditions of many rural people. Gathering and farming of edible insects can be performed with a minimal input of technical or capital resources which provides the poorest section of society a likelihood to generate income (van Huis 2013). Therefore, entomophagy is growing popular globally, mostly among the poor masses as a supplement of protein as well as source for income.

3.2   History

The eggs, larvae, pupae and adults of numerous insect species have been consumed by humans since prehistoric times. Around 3000 ethnic groups practice entomophagy. Entomophagy is common to cultures and communities in most parts of the world, including Africa, Australia, Central America and Southeast Asia. A vast variety of wild foods and edible insects is composed of many African diets. Eighty percent of the world's nations eat insects of 1000–2000 species. However, in some societies or cultures eating insects is not practiced or taboo. On this present day, entomophagy is taboo in North America and Europe, but insects remain a popular foodie other parts of the world, and different companies are looking into the possibility to introduce insects as food into Western diets. Around 1900 edible insect species has been registered by FAO and estimates that there were some two billion insect consumers worldwide in 2005.

Entomophagy, despite seeming odd, eating insects had an important influence for human diet throughout history. In fact, there are several accounts of entomophagy in different religious documents from Christianity, Islam and Judaism. In Europe, the first references of eating insects come from Ancient Greece, where eating Cicadas was considered a delicacy. Aristotle left proof of this practice at Historia Animalium (384–322 BC), that female cicadas tasted better after mating because they are full of eggs. The people from Ethiopia were called "Acridophagi" by Diodorus (200 BC), from Sicilia as their diet was based on grasshoppers and locusts (family Acrididade). The author of Historia Naturalis, Pliny the Elder from the Ancient Rome, refers in his work to a dish loved by romans called "cossus", according to Bodenheimer (1951), the dish was prepared with beetle larvae of the species *Cerambyx cerdo*. Ancient Chinese literature usually refers to the practice of eating insects and the use of insects in traditional medicine. In the Compendium of Materia Medica [Li Shizhen, Ming Dynasty, (1368–1644)], numerous amount of recipes are listed which was based on the use of insects along with their medicinal attributes.

Despite insects being an essential element in human diet since the dawn of time, different countries around the world started to see "entomophagy" as a taboo as societies became modern (specially on Europe and the USA).

The main reason behind the change of perspective is due to development of agriculture and livestock. The Fertile Crescent, a historical region containing western territories of Asia, the Nile Valley and the Nile Delta, is considered the birth place of agriculture and, secondarily, livestock (Western Neolithic Revolution). As the practice of agriculture started to spread towards Europe, it replaced the practice of hunting and gathering for food sources.

3.3 Benefits of Entomophagy

The protein content of the many edible insects is equivalent with that meat or fish and can be used as a cheaper and readily available alternative food resource to combat protein malnutrition among the poor (Kariuki and White 1991). The major role entomophagy will play in future, is to provide food security, serve as an alternative food source to meet the demands of ever increasing population. More attention is required to fully understand the potential of edible insects. As a natural source, they provide of essential carbohydrates, proteins, fats, minerals and vitamins, also lighten the ecological footprint (Gahukar 2011). Various edible insects contain large amount of lysine, an amino acid deficient in the diets of large number people that depend mainly on grain (Gordon 1998). The combination of increasing land use demand, climate change, and food grain shortages due will cause serious challenges in the future to meet protein demand (Premalatha et al. 2011). The suggestion that edible insects may solve the problems of global food shortages was first published by Meyer-Rochow in 1975.

In the twenty-first century insects as food and feed have emerged as a potential source of food due to the rising cost of animal protein, food and feed insecurity, climate change, alarming population growth and increasing demand for protein. The cultivation of insects and edible arthropods for human consumption is termed as "minilivestock", which is emerging as an ecologically sound concept in animal husbandry. Studies have shown that rearing insect, is a more environmentally sustainable alternative to animal live-stocking (Paoletti 2005).

Two species of edible insects (cricket and palm weevil larvae) are commonly reared in Northern and Southern Thailand, respectively (FAO 2013). The technique of cricket farming and breeding has not changed much since the technology was introduced 15 years ago. Cooperatives of cricket farmers have been established to provide information on technical farming, marketing and combating business issues, throughout Thailand. Cricket farming has evolved into a significantly important animal husbandry sector and main income source for a large number of farmers. In 2013, approximately 20,000 farms were operating 217,529 rearing pens (FAO 2013). The total production during the last 6 years (1996–2011) was found to average around 7500 tonnes per year.

Many authors have proposed entomophagy as a solution to control "pests". Elimination of the grasshopper *Sphenarium purpurascens* in different parts of Mexico is performed by its capture and use as food. This strategy lowers the use of pesticide and generates an alternative source of income for farmers contributing to nearly $3000 per family. Besides environmental impact, many argue that the use of pesticides is economically due to its destruction of insects which may contain up to 75% animal protein in order to save crops containing no more than 14% protein (Premalatha et al. 2011).

The process of matter assimilation and nutrient transport performed by insects for converting plant material into biomass is a more efficient method than traditional livestock rearing. To produce 1 kg of meat, ten times more plant material is required than to produce 1 kg of insect (Premalatha et al. 2011). Also the spatial use and requirements for water is only a fraction of that required to produce the same mass of food for livestock rearing. For production of 150 g of grasshopper meat requires minimal amount of water, while cattle rearing requires 3290 L of water to produce the equivalent amount of beef (Walsh 2008). This shows that use of lower natural resource and lesser strains ecosystem can be expected from insects and also edible insects have a much faster life cycle than traditional livestock. According to analysis conducted by the University of Wageningen, Netherlands on the carbon intensity of the studied five edible insect species found that "the average daily gain (ADG) of the five insect species was 4.0–19.6%, the minimum value of this range being close to the 3.2% reported for pigs, whereas the maximum value was found to be six times higher. When compared to cattle (0.3%), insect ADG values were much higher." Moreover, all insect species studied produced very low amounts of ammonia than traditional livestock, however further research is required to fully understand the long-term impact. The authors concluded that edible insects could potentially serve as a more environmentally sustainable source of dietary protein (Oonincx et al. 2010).

3.4 Disadvantages

3.4.1 Spoilage/Wastage

Bacteria growing on both raw and cooked insect can spoil the protein, threatening to cause food poisoning. All edible insects must be processed with care by various methods to prevent spoilage. The insects should be boiled before refrigeration, drying, acidification, or use in fermented foods.

3.4.2 Toxicity

Cooking of edible insects is advisable, as parasites may be present which may be harmful. Pesticide and herbicides can accumulate in insects through bioaccumulation which can make the insects unsuitable for human consumption. For example, when combating locust outbreaks, spraying of chemicals is used, the insects become unsuitable for human consumption.

In many cases, some insects are eaten despite of their toxicity. In Carnia region, Italy, moths of Zygaenidae family are eaten. Both larvae and adults are known to produce hydrogen cyanide precursors. But the crops of adult moths carry cyanogenic chemicals in exceedingly low quantities along with high concentrations of sugar, making Zygaena a favourable supplementary alternative of sugar during early summer. The moths are commonly found, and they are easy to catch by hand; minimal cyanogenic content, however, makes Zygaena a slightly risky seasonal delicacy (Zagrobelny et al. 2009). The California Department of Health Services in November 2003 reported cases of lead poisoning after consumption of Chapulines. Severe allergic reactions are also reported which may be a possible hazard.

3.4.3 Allergic Reaction

It is reported that many people suffer from allergic reaction after consuming insects. However, it is reported in only few cases. According to some researchers current data suggest that arthropod pan-allergens such as tropomyosin or arginine kinase play an important role in food allergy (Ribeiro et al. 2018). It is advised that people with high protein allergies should avoid eating edible insects.

References

Ashiru MO (1988) The food value of the larva of Anaphe venata Buttler. Lepidoptera; Notodontidae. Ecol Food Nutr 22:313–320

Bodenheimer FS (1951) Insects as human food. The Hague: W. Junk. 352 pp.

Cerritos R (2009) Insects as food: an ecological, social and economical approach. Cab Rev 4(10)

Das K, Bardoloi S, Mazid S (2019) A study on the prevalence of entomophagy among the Koch-Rajbongshis of North Salmara subdivision of Bongaigaon district. Int J Basic Appl Res 9(3): 382–388

DeFoliart GR (1989) Insect as a source of protein. Bull Entomb Soc Am 35:22–35

DeFoliart GR (1992) Insects as human food: Gene Defoliart discusses some nutritional and economic aspects. Crop Prot 11(5):395–399

DeFoliart GR (1999) Insects as food: why the western attitude is important. Annu Rev Entomol 44:21–50

Ene JC (1963) Insects and man in West Africa. Ibadan University Press, Ibadan, pp 16–26

Food and Agriculture Organization of the United Nations (2013) Six-legged livestock: edible insect farming, collection and marketing in Thailand (PDF). Food and Agriculture Organization of the United Nations, Bangkok. ISBN 978-92-5-107578-4

Gahukar RT (2011) Entomophagy and human food security. Int J Tropic Insect Sci 31(3):129–144

Gordon DG (1998) The eat-a-bug cookbook. Ten Speed Press, Berkeley, CA, p xiv. ISBN 978-0-89815-977-6

Kariuki PW, White SR (1991) Malnutrition and gender relations in Western Kenya. Health Transit Rev 1:2

Meyer-Rochow VB (1975) Can insects help to ease the problem of world food shortage? ANZAAS J 6(7):261–262

Narzari S, Jatin S (2015) A study on the prevalence of entomophagy among the Bodos of Assam. J Entomol Zool Stud 3(2):315–320

Oonincx DGAB, Van Itterbeeck J, Heetkamp MJW, Van den Brand H, Van Loon JJA, Van Huis A (2010) An exploration on greenhouse gas and ammonia production by insect species suitable for animal or human consumption. PLoS One 5(12):e14445. https://doi.org/10.1371/journal.pone.0014445.

Paoletti MG (2005) Ecological implications of minilivestock: potential of insects, rodents, frogs, and snails. Science, Enfield, p 648. ISBN 978-1-57808-339-8. Retrieved 7 May 2010

Premalatha M, Abbasi T, Abbasi T, Abbasi SA (2011) Energy-efficient food production to reduce global warming and ecodegradation: the use of edible insects. Renew Sust Energ Rev 15(9):4357–4360. https://doi.org/10.1016/j.rser.2011.07.115

Ribeiro JC, Cunha LM, Sousa-Pinto B, Fonseca J (2018) Allergic risks of consuming edible insects: a systematic review. Mol Nutr Food Res 62(1). https://doi.org/10.1002/mnfr.201700030

Rumpold BA, Schlüter OK (2013) Nutritional composition and safety aspects of edible insects. Mol Nutr Food Res 57(5):802–823. https://doi.org/10.1002/mnfr.201200735

Thakur A, Thakur KS, Thakur NS (2017) Entomophagy (insects as human food): a step towards food security. https://doi.org/10.13140/RG.2.2.29644.72327

van Huis A, Itterbeeck JV, Klunder H, Mertens E, Halloran A, Muir G, Vantomme P (2013) Edible insects: future prospects for food and feed security. FAO FORESTRY PAPER; ISSN 0258-6150

Walsh B (2008) Eating bugs. Time. Archived from the original on 26 September 2012.

Zagrobelny M, Dreon AL, Gomiero T, Marcazzan GL, Glaring MA, Møller BL, Paoletti MG (2009) Toxic moths: source of a truly safe delicacy. J Ethnobiol 29:64–76. https://doi.org/10.2993/0278-0771-29.1.64.

Chapter 4
Microbiology of African Edible Insects

Nils Th. Grabowski

Abstract Each foodstuff hosts a specific micro- and mycobiome during its passage from primary production to the final product to be consumed, and edible insects are no exceptions to the rule. Being so, however, this microbiological profile varies with the species and the environment it is placed in. Taxonomically, the species' micro/mycobiome contains several species presumably present in all insects, some species which appear to be shared by most if not all insect species belonging to the same order, the same family, and the same genus. The specific array of bacteria and fungi is determined by this taxonomical setup, the species' instar, and the environment, the latter including the place of origin (wild range resp. farm), and the microbiological features affecting the animal resp. the product during processing and storage. Insect consumption and activities linked to it is backed by a solid tradition seeking to minimize consumer risks. However, changes in this habit, e.g. farming, packaging, transport, and "novel" storage conditions, may pose risks not contemplated by tradition. In this way, it is recommended to re-evaluate these traditions scientifically and adapt traditions to these novel situations.

Keywords Microbiome · Mycobiome · Entomophagy · Food tradition · Food safety Food-borne diseases · Food spoilage

4.1 Introduction

The history of consuming insects is as old as mankind itself. Embracing entomophagy—as the consumption of any foodstuff—has been a question of availability, food safety, and cultural framework. Nowadays, more than 2.8 billion people around the planet consume insects on a regular base, in most cases out of a given tradition.

N. T. Grabowski (✉)
Institute of Food Quality and Food Safety, Hannover University of Veterinary Medicine, Hannover, Germany
e-mail: Nils.Grabowski@tiho-hannover.de

© Springer Nature Switzerland AG 2020

A. Adam Mariod (ed.), *African Edible Insects As Alternative Source of Food, Oil, Protein and Bioactive Components*, https://doi.org/10.1007/978-3-030-32952-5_4

Over the millennia, man learned how to handle the resource "food insect" by trial and error. In this way, methods to ensure food safety (or, at least, keeping risks as low as possible) have been developed, and the fact that the scientific literature has presented few cases of food insect-related problems over the years is not only a sign of an undisputed previous arrogance of the scientific community towards "primitive foodstuffs", but does also appear to reflect that there are, in fact, only few cases occurring. This, in turn, means that applying traditional methods to edible insects is more or less effective in keeping risks at a low level. From this point of view, addressing food insect microbiology may seem superfluous.

However, the scenario is changing.

- Insects are becoming an increasingly popular foodstuff, and even if many people could imagine themselves actually eating them, the issue has raised their interest.
- This increase in demand is prone to mobilize certain areas of the food sector, including those that may not have worked with edible insects before, ignoring the potential risks associated with these feedstuffs.
- When entomophagy becomes appealing for more persons, traditional processing and marketing methods and strategies are likely to become adapted to this new situation, e.g. more and new products, selling in containers rather than as bulk ware, larger distances, and exportation. However, the question remains whether the traditional ways of ensuring food safety will still be effective.
- Increased use of edible insects leads to either increased trapping from the wild (with a definite risk of affecting natural feeding systems by exploiting insect populations) or starting insect farms (which ecologically seem the better solution). However, farming conditions may differ from natural habitats, changing the microbiological profile of the insects, and this may raise risks that simply did not exist when insect consumption was based on collecting and trapping.
- Farming insects also opens another dimension since farming systems have been developed for some typical insect species, e.g. crickets (like *Acheta domesticus* and *Gryllodes sigillatus*) and darkling beetles (like *Tenebrio molitor* and *Alphitobius diaperinus*) which, however, may not be native to a given area, and their introduction may affect the microbial community of the zone, and the local microbiome may in turn affect these newly introduced species. In fact, this situation is comparable to any other in which domestic animals from one locality are moved and reared in another one.

The conciseness of this chapter is affected by two major limitations: on one hand, by the large amount of edible insect species (expected to be approx. 2000–3000), and by the lack of information about most of them. This also includes the vast array of entomophagous traditions in Africa; however, while documentation of African insect consumption is plentiful, concise implications on food safety are not. Extrapolation is a frequent method to overcome this kind of shortages, but in this case, extrapolating must be done with extreme caution since each species seems to have a specific microbiome, and production conditions also vary strongly among species, production system, and locality.

Dealing with the microbiology of edible insects is, thus, a complex matter, caught between highly specific scientific data (e.g. the bacteriological findings of a traditional product made from a given insect species of a well-defined region) and possibly undue generalizations. An attempt will be made to provide a general frame with specific, merged data when appropriate.

As in any living species and in every food product, its microbiology is the result of a complex interaction of factors, starting from the primary production and modified by processing and logistics until the kitchen hygiene of anyone preparing a meal from the ingredients. In the following sections, these steps will be illustrated.

In this chapter, scientific names of insects will be used whenever possible. The appendix subsumes all relevant taxonomic information and provides, if possible, vernacular names.

Finally, few areas of food sciences are developing at the speed currently experienced for edible insects. After the FAO stated that insects are one of the solutions to fight global hunger and to mitigate the climate change, the rate of new publications in this sector has been increasing strongly. In this way, the present contribution cannot be more than a snapshot of the status quo at the moment of writing it.

4.2 Traditional Gathering and Primary Production Farming

Figure 4.1 sums up the most important sources for microorganisms to be found in and outside edible insects, i.e. intrinsic flora, environment, gathering, and farming. All contribute to the microbiome of the insects, which is encountered as such when submitting insects and products made of them to microbiological analyses. Determining the exact source of each of them is a task not undertaken yet to the knowledge of the author. However, some basic observations made in other foodstuffs may provide clues also for edible insects.

Traditionally, insects have been gathered from the wild in Africa (Raheem et al. 2018). This is relatively well documented, showing which insects are consumed in which countries and/or areas, e.g. Riggi et al. (2013), Tchibozo et al. (2016) and van Huis (2003). While a simple correlation between ethnicity and insect was the main focus by anthropologists, modern entomophagy research regarding traditions also includes many other aspects, e.g. ecology (e.g. Okeke et al. 2019), composition (e.g. Alamu et al. 2013), economy (e.g. Adeoye et al. 2014) and society (e.g. Stull et al. 2018).

The intrinsic flora comprises a highly diversified array of microorganisms, among them bacteria, fungi and stramenopiles. Reviews done on insect taxa containing edible species (Grabowski and Klein 2017a; Grabowski et al. 2017) show

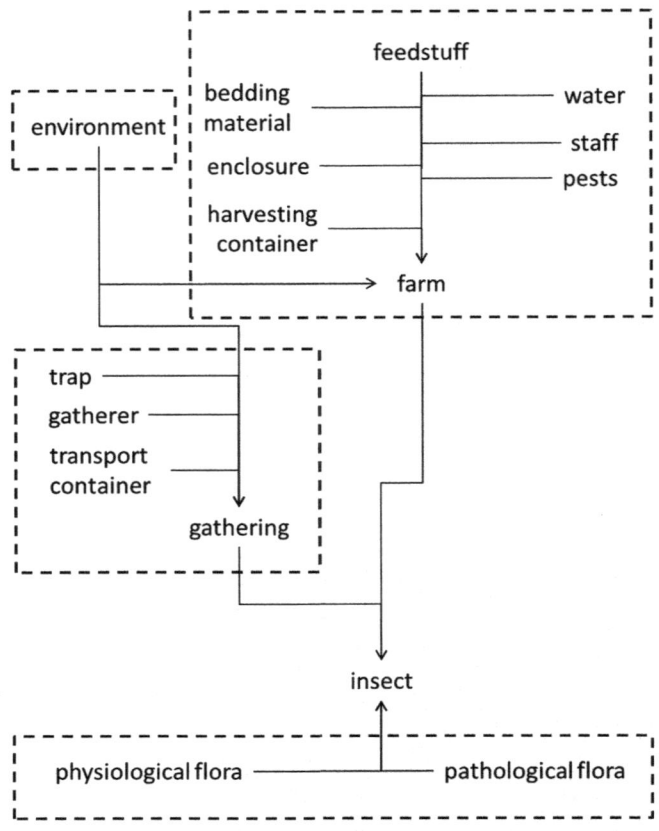

Fig. 4.1 Factors affecting the microbiome of insects during primary production

that the micro- and mycobiome are taxon-related. There seems to be a basic stock of bacteria from the families Bacillaceae (e.g. *Bacillus thuringiensis*), Enterobacteriaceae (e.g. *Proteus vulgaris, Serratia marcescens*), Enterococcaceae (e.g. *Enterococcus faecalis*), and Pseudomonadaceae (e.g. *Pseudomonas fluorescens*), and, in terms of fungi, Cordycipitaceae (e.g. *Beauveria bassiana*), Saccharomycetaceae (e.g. *Saccharomyces cerevisiae*), Trichocomaceae (e.g. *Aspergillus* spp., *Penicillium* spp.), and Saccharomycetales (e.g. *Candida albicans*), to cite the most prominent ones. These organisms occur in most if not all orders. Following taxonomy, there is a common stock for each family, genus, and species, respectively. For a more comprehensive list, see Grabowski and Klein (2017a) and Grabowski et al. (2017). On species level, microbiome and mycobiome also vary according to the locality, i.e. the environment. To give an example, the African palm weevil (*Rhynchophorus phoenicis*) is one of the better-studied African species used in traditional consumption. Table 4.1 shows the diversity of bacteria encountered in relation to that species. It contains data of both fresh and processed samples.

Table 4.1 Microbiology of the African palm weevil (*Rhynchophorus phoenicis*)

Instar	Product	Parameter	Value (cfu/g)	References
Imago	Skin	Total bacterial count	9.2×10^5	Amari and Kiin-Kabari (2016)
		Coliforms	5.3×10^5	
		Fungal count	7.3×10^5	
		Bacillus spp., *Klebsiella* spp., *Pseudomonas* spp., *Serratia* spp., *Staphylococcus* spp., *Saccharomyces* spp.		
	Gut	Total bacterial count	1.8×10^7	Amari and Kiin-Kabari (2016)
		Coliforms	4.2×10^6	
		Fungal count	3.6×10^6	
		Bacillus spp., *Enterobacter* spp., *Serratia* spp., *Staphylococcus* spp.		
Larva	Fresh	*Bacillus* spp., *Enterobacter* spp., *Serratia* spp., *Staphylococcus* spp.		Amari and Kiin-Kabari (2016)
		Bacillus cereus, *Escherichia coli*, *Pseudomonas* spp., *Proteus* spp., *Aspergillus* spp., *Fusarium* spp., *Penicillium* spp.		Mutunagi et al. (2017)
		Escherichia coli, *Klebsiella aerogenes*		Opara et al. (2012)
	Fried	*Bacillus* spp., *Staphylococcus* spp.		Amari and Kiin-Kabari (2016)
	Roasted	*Bacillus cereus*, *Enterococcus faecalis*, *Escherichia coli*, *Pseudomonas aeruginosa*, *Staphylococcus aureus*, *Aspergillus* spp., *Mucor* spp., *Rhizopus* spp.		Amari and Kiin-Kabari (2016)
	"Heat-treated"	*Staphylococcus* spp.		Opara et al. (2012)
	"Processed"	Total bacterial count	1.6×10^5	Braide and Nwaoguikpe (2011)
		Fungal count	9.2×10^2	
		Bacillus subtilis, *Lactobacillus plantarum*, *Proteus vulgaris*, *Pseudomonas aeruginosa*, *Staphylococcus aureus*, *Aspergillus flavus*, *Cladosporium* spp., *Fusarium poae*, *Penicillium verrucosum*, *Saccharomyces cerevisiae*		

Table 4.1 shows:

- that bacterial counts and microbiome vary according to the instar (imago vs. larva) and localisation (skin vs. gut which are the main sources of microbes),
- the extent of the study and the methods used,
- that processing may (or may not) reduce bacterial and fungal counts,
- that microbiome and mycobiome include saprophytic, physiologically necessary, and entomopathogenic and zoonotic microorganisms, and…,
- that (re)contamination may take place after harvesting.

The intrinsic flora of insects may also be physiological and pathological. On one hand, many bacteria and fungi fulfil important functions for the insect physiology, acting as probiotics. On the other hand, there are many other bacteria and, moreover, fungi that are known entomopathogens. In fact, some of them, e.g. certain strains of *Bacillus thuringiensis* have been used to fight pest insects (Tanada and Kaya 1993). A relatively new issue is the role of insect-associated microorganisms that contain antibiotic resistance genes (Osimani et al. 2018).

It should, however, be kept in mind that these results reflect the status quo of sampling after harvesting, meaning that the influence factors shown in Fig. 4.1 already affected the product, and some parts of the micro- and mycobiome are thought to have originated while trapping, transporting, and processing the insects. In their review, Mutunagi et al. (2017) mentioned a case of cross contamination of harvested stinkbugs with aflatoxins due to cross-contamination originating from the containers the insects were kept in, while Opara et al. (2012) attributed staphylococci found in processed *R. phoenicis* larvae to the hands of gatherers and retailers, and *Klebsiella* spp. and *E. coli* detected on fresh larvae on the skewers they were put on to.

In short and in terms of epidemiology, each microbiological analysis must be evaluated closely regarding the level the sample was taken, and results are expected to differ among batches.

Table 4.2 summarises microbiological findings of some common African and non-African edible insect species. It must be made clear that this is patchwork table comprising the results of many different research papers, and the list is, in no way, complete. However, it shows the marked variety of findings.

Tables 4.1 and 4.2 contain a series of well-known food-borne pathogens and spoilage flora. *Bacillus cereus*, *Staphylococcus aureus*, coliforms, salmonellae or *Aspergillus* spp. pose serious threats to the human health, particularly if the patients of the YOPI (young, old, pregnant, immunocompromised) are affected. However, other microorganisms were found that, like *Buttiauxella* sp., may be opportunistic pathogens.

By themselves, these findings are not truly spectacular as they occur in all kinds of foodstuffs, because they originate from the environment the foodstuff is generated, processed, transported or stored. Some of them can be found on the ground, some on plants, some in the air, some on livestock, some on equipment and materials, and some on human beings (Braide et al. 2011; Mutunagi et al. 2017).

Along with the actual findings, it is also noteworthy to consider what has *not* been found. To give an example, Amiri (2017) analysed raw and boiled samples *of Locusta migratoria, Tenebrio molitor*, and *Zophobas atratus*, and did neither encounter *Campylobacter* spp., Yersinia spp., *Vibrio* spp., nor *Shigella* spp. The reasons for that may be multiple, and among them, one seems to be that certain insects are capable of inactivating certain microorganisms and even detoxify certain mycotoxins. This shows that insects may be able to cope with contamination from outside.

Another factor of influence is the method applied for the bacteriological analysis. Typically, samples are treated using cultural methods. Applying molecular biological ones, a wider array of bacteria opens up. In this way, it was seen that *Acheta domesticus* samples also contained low DNA reads associated with *Bacillus* spp., *Listeria*

Table 4.2 Microbiology of selected African and non-African edible insect species; for *Rhynchophorus phoenicis*, see Table 4.1

Species	Instar	Product	Microbiological findings
Acheta domesticus	Imago	Powder	*Bacillus cereus, Clostridium thermopalmarium, Pediococcus lolii, Aspergillus* spp., *Eurotium* spp.
		Processed	*Bacillus cereus, Bacteroides* spp., *Clostridium perfringens, Citrobacter* spp., *Fusobacterium* spp., *Yersinia* spp.
	Nymph	Dried	*Acinetobacter* spp., *Bacillus* spp., *Citrobacter* sp., *Erwinia* sp., *Streptomyces glaucusporus, Tetrapisispora* spp., *Wallemia muriae*
Alphitobius diaperinus	Larva	Raw	Coliforms, *E. coli, Micrococcus* spp., *Salmonella* spp., *Staphylococcus* spp., *Streptococcus* spp.
Bematitistes alcinoe	Larva	Raw	*Bacillus* spp., *Staphylococcus* spp.
Bunaea alcinoe	Larva	Processed	*Acinetobacter* spp., *Bacillus* spp. (incl. *B. cereus*), *E. coli, Klebsiella* spp., *Proteus* spp., *Pseudomonas* spp., *Serratia* spp., *Staphylococcus* spp. (incl. *S. aureus*), *Aspergillus* spp., *Fusarium moniliforme, Penicillium caseicolum, Saccharomyces cerevisiae*
Gonimbrasia belina	Larva	Raw	*Aspergillus* spp., *Chaetomium* spp., *Cladosporium* spp., *Fusarium* spp., *Mucor racemosus, Penicillium* spp., Phycomycetes spp.
Gryllodes sigillatus	Imago	Raw	*Aspergillus* spp.,*Candida* spp., *Debaryomyces* spp., *Kodamaea* spp., *Lichtheimia* spp., *Tetrapisispora* spp., *Trichoderma* spp., *Trichosporon* spp.
Gryllotalpa africana	Imago	Raw	*Bacillus* spp., *Corynebacterium* spp., *Proteus* spp., *Staphylococcus* spp.
Hermetia illucens	Pupa	Powder	*Enterobacteriaceae* spp., *E. coli, Salmonella* spp., *Alternaria* spp., *Aspergillus* spp., *Penicillium* spp.
Locusta migratoria	Imago	Dried	*Enterococcus* sp., *Clostridium thermopalmarium, Leuconostoc mesenteroides, Weissella* spp., *Aspergillus* spp., *Wallemia muriae*
		Raw	*E. coli, Staphylococcus* spp., *Candida* spp.
	Nymph	Raw	*Enterobacteriaceae* sp., *Enterococcus* sp., *Haemophilus* sp., *Lactococcus* sp., *Pseudomonas* spp., *Weissella* sp., *Yersinia/Rahnella* sp.
Musca domestica	Pupa	Raw	*Bacillus cereus, E. coli, Klebsiella aerogenes, Pseudomonas aeruginosa, Staphylococcus aureus, Cladosporium* spp., *Fusarium* spp., *Moniliella* spp., *Mucor* spp., *Penicillium* spp.
Oryctes monoceros	Imago	Raw	*Acinetobacter* spp., *Bacillus* spp., *E. coli, Proteus* spp., *Staphylococcus* spp., *Aspergillus* spp., *Fusarium* spp., *Mucor* spp., *Penicillium* spp.
	Larva	Raw	*Bacillus* spp. (incl. *B. cereus*), *E. coli, Klebsiella aerogenes, Pseudomonas aeruginosa, Staphylococcus aureus*

(continued)

Table 4.2 (continued)

Species	Instar	Product	Microbiological findings
Ruspolia differens	Imago	Raw	*Acinetobacter* spp., *Bacillus* spp., *Buttiauxella* spp., *Campylobacter* spp., *Clostridium* spp., Enterobacteriaceae spp., lactic acid bacteria, *Staphylococcus* spp., *Pseudomonas* spp., *Neisseria* spp., spore-formers, yeasts and moulds
Tenebrio molitor	Larva	Dried	*Bacillus cereus, Clostridium thermopalmarium, Enterobacter* spp. (incl. *E. cloacae), Erwinia* sp., *Loktanella marincola, E. coli, Klebsiella* spp. (incl. *K. oxytoca), Pantoea agglomerans, Salmonella* sp., *Staphylococcus aureus, Vibrio* sp., *Candida zeylanoides, Debaryomyces hansenii, Pichia* sp.
		Processed	*Bacillus* spp., *Eikenella corrodens, Exiguobacterium* sp., *Neisseria shayeganii, Staphylococcus* spp.
		Raw	*Acidovorax* sp., *Clostridium* sp., Enterobacteriaceae sp., *E. coli, Haemophilus* sp., *Lactobacillus* sp., *Propionibacterium* sp., *Pseudomonas* spp., *Staphylococcus* spp., *Streptococcus* sp., *Varibaculum* sp., *Candida* spp.
Zophobas atratus	Larva	Raw	*Bacillus cereus, E. coli, Candida* spp.

Amadi and Kiin-Kabari (2016), Amiri (2017)), Banjo et al. (2006), Braide et al. (2011), Fernández Cassi et al. (2018), Milanović et al. (2018), Mutunagi et al. (2017), Ng'ang'a et al. (2019), Ogbalu and Williams (2015), Osimani et al. (2017), Ssepuya et al. (2019), Stoops et al. (2016), Wanjiku (2018)

spp., and *Staphylococcus* spp. that were not detected while culturing, suggesting that bacteria were already dead or not culturable by routine methods (Fernández Cassi et al. 2018). Likewise, Vandeweyer (2018) undertook an extensive study of the microbiome of *Acheta domesticus, Gryllodes sigillatus* and *Tenebrio molitor.*

Insect farming has started to become an option, particularly in (sub)tropical regions, because climate conditions do not have to be adapted to native (or other tropical) species, and the higher degree of consumer acceptance in comparison to farming insects in areas with no resp. lost entomophagous traditions, e.g. Europe (Grabowski 2017). Insect farms can be grouped into four basic types (Table 4.3). Some insect species can be reared on more than one system, e.g. *Hermetia illucens* which can be produced in both xiroculture and hygroculture. For some species, a combination of production types may be necessary, e.g. if the life cycle includes terrestrial and aquatic phases. Insect farms may serve different purposes, e.g. food-stuff, feedstuff, industrial use of a certain substance extracted from the insect (e.g. chitin, oils, proteins), waste reduction, or a combination of those, e.g. rearing cal-liphorid flies on slaughter wastes for feed production. Each of these production systems will create a specific environment and thus, specific microbiological conditions.

Farming implies an environment that may be different from the natural habitat of the insect species reared. Thus, the microbiological conditions may also change. Debrah (2017) studied different rearing media for *Rhynchophorus phoenicis* in Ghana

Table 4.3 Insect farming systems

Type	Brief description	Examples
Xiroculture	Dry environment needed for many terrestrial species, providing dry feedstuffs and sufficient water or water-containing feedstuffs	Orthopterans (crickets, locusts, grasshoppers), mealworms, lepidopterans (silk and wax moths), hymenopterans (bees, wasps, ants)
Hygroculture	Humid environment preferred by detritivores feeding on humid wastes and by-products and manure	Dipterans (flies)
Aquaculture	Freshwater farming, being either the only environment or part of a specific phase of the life cycle	Edible aquatic species (water bugs, water beetles, flies ovipositing in water)
Xyloculture	Rearing xylophages either in logs or in boxes containing wood shaves of their host tree	Coleopterans (weevils)

and demonstrated that palm heart was most suitable for oviposition while sugarcane slices were better for pupation. Each of these substrates comes with its own microbial community, and although no bacteriological analyses were made in this study, it is likely that the substrate will contribute to the grubs' and pupae's micro- and mycobiome. Vandeweyer (2018) studied the microbiology of three species (*Acheta domesticus*, *Gryllodes sigillatus*, and *Tenebrio molitor*) evaluating raw and processed samples from different companies, also found differences among bacterial counts and array of bacteria present in the samples. The mycobiota of *Hermetia illucens* reared for feed was analysed according to the feedstuff the larvae were provided, and the most prominent genus in larvae feed on vegetable waste was *Pichia*, while when feeding exclusively chicken feed, other genera (*Geotrichum, Rhodotorula,* and *Trichosporon*) became more frequent (VarottoBoccazzi et al. 2017). Another important factor is the localisation of the farm. In the vicinity of livestock farms, farmed insects may acquire salmonellae and *Campylobacter* spp., making the reared insects potential vectors for these pathogens (Belluco et al. 2015; Dobermann et al. 2017).

All this shows that the microbiology of harvested insects, be it from the wild, be it by means of farming, is the result of the reflecting the microbiome of their environment modified by their own physiological and pathological flora.

4.3 Processing

Foodstuff processing is done, among others, to grant food safety, extend shelf life, and to provide variety in flavours. The former two are particularly important since insects may contain zoonotic pathogens and dead insects spoil easily, especially under ambient conditions. This makes spoilage a special problem in tropical and subtropical areas with relatively high humidity, because the amount of available food is reduced (as is the income, if foodstuff was produced or obtained using good exchange) and the risk for food-borne infections increases. Regarding variety, many insects have a relatively neutral taste. This may seem stodgy at first glance, but opens up a huge variety of preparing methods, a feature that among other animal-based foodstuffs is only matched by dairy and egg products.

Like with other foodstuffs in general and foodstuffs of animal origin in particular, good hygienic practice is the key element to promote and establish food safety from the microbiological point of view. Traditional methods may have been efficient to avert major food safety problems—if not, they would have been discouraged already—but it should be kept in mind that the current practices have been developed for a specific goal which is, in many cases, the consumption shortly after processing.

If, however, the situation changes in that way that edible insects become farmed livestock, and products made from them are traded beyond local markets and are sent over larger distances (Kelemu et al. 2015), these traditional techniques may not suffice any longer. This is where research enters at latest, because together with shifts in the animals' flora due to farming, these changes in the marketing may also require adapting those techniques to these novel challenges. As this is already done on a worldwide base, scientific exchange will facilitate these adaptions.

There are relatively few cases worldwide in which insects are consumed directly and without any processing. In most cases, processing includes (a) dressing, (b) heat treatment, (c) cleaning, (d) heat treatment, (e) consumption, (f) milling and (g) storage. Killing is not always performed as a separate working step, but part of other treatments like dressing, heat treatment or milling. The steps do not necessarily have to occur in this order; sometimes, cleaning is done before dressing, sometimes after. Table 4.4 provides an overview of the best-documented processing steps according to Mutunagi et al. (2017).

In Table 4.4, the specific data provided by Mutunagi et al. (2017) was merged, i.e. different species with different processing techniques from different countries. This means that not all steps are performed in all species of that given taxon. However, this overview shows the wide array of processing methods applied traditionally. Tables 4.1 and 4.2 also contain data obtained from processed insects, showing that, in some cases, neither lower bacterial counts nor the inactivation of zoonotic pathogens were achieved. This may be due to the processing technique itself or the lack of hygienic conditions.

Dressing refers to taking off a certain portion of the insect. From the hygienic point of view, this step means the rupture of the exoskeleton which protects the inner structures from environmental contamination, even more, when done with bare hands and with large amounts of insects. By means of illustration, Fig. 4.2 presents the development of bacterial counts of a Tanzanian tettigoniid grasshopper.

It shows that dressing ("plucking") increases bacterial counts, rinsing does not eliminate all those bacteria, and heating techniques affect bacterial growth to a varying degree.

Heat treatment is a relatively effective way to reduce bacterial counts or kill pathogens. Amiri (2017) reported that boiling killed *E. coli*, staphylococci, and *Candida* spp. in *Tenebrio molitor, Zophobas atratus,* and *Locusta migratoria.* However, finding the optimum combination of temperature and residence time may

Table 4.4 Traditional insect processing in Africa (based on Mutunagi et al. 2017)

Basic	Specific	Termites	Bugs	Cockroaches	Grasshoppers	Crickets	Beetles		Moths	Ants		Bees		Lake flies
		Adult	Adult	Adult	Adult	Adult	Larva	Adult	Larva	Eggs	Adult	Larva	Pupa	Adult
Killing		+	+	.	.	.	.	.	+	.	.	.	.	.
Dressing	Antennae	.	.	.	+	.	.	.	.	.	.	.	.	.
	Gaster	.	.	.	.	.	.	.	.	.	+	.	.	.
	Guts	.	.	.	+	+	.	.	+	.	.	.	.	.
	Head	.	+	.	+	.	.	+	.	.	.	.	.	.
	Wings	+	.	+	+	.	.	+	.	.	.	.	.	.
Cleaning	Soak	.	+	.	.	.	+	.	+	.	.	.	.	.
	Wash	+	+	.	.	+	+	.	+	.	.	.	.	.
Heat treatment	Boil	+	+	.	+	+	+	.	+	.	.	.	.	.
	Dry	+	+	.	+	.	+	.	.	.	.	.	.	.
	Dry-fry	.	.	.	.	.	.	.	.	.	.	.	.	.
	Extract oil	.	+	.	.	.	.	.	+	.	.	.	.	.
	Fry	+	+	+	+	.	+	.	+	+	+	+	.	.
	Grill	.	.	.	.	.	+	.	.	.	.	.	+	.
	None	+	+	+	+	.	+	.	.	+	.	+	.	.
	Roast	+	+	.	+	.	+	+	+	.	.	+	.	.
	Salt	+	.	.	.	.	+	.	+	.	.	.	.	.
	Smoke	.	.	.	.	.	.	.	.	.	+	.	.	.
	Smoke-dry	+	.	.	.	.	.	.	.	.	.	.	.	.
	Steam	+	.	.	.	.	.	.	.	.	+	.	.	.
	Stew	.	.	+	+	.	+	.	+	.	.	.	.	.
	Sun-dry	+	+	.	+	.	.	.	+	.	.	.	.	+
	Toast	+	+	.	+	.	.	.	.	.	.	.	.	.
Milling	Grind	+	+	.	+	.	+	.	+	.	.	.	.	.
	Pound	+	+	.	.	.	+	+	.	.	+	.	.	+

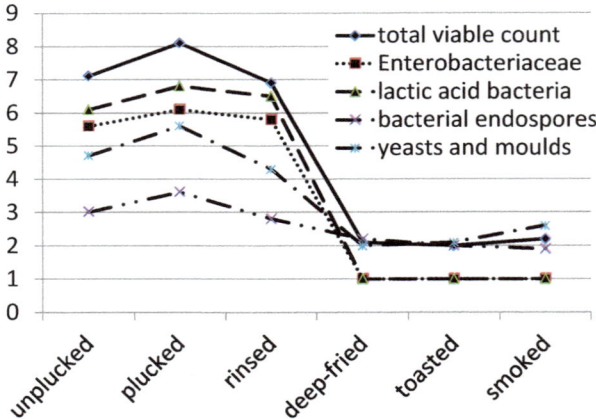

Fig. 4.2 Development of bacterial counts [log cfu/g] during processing of nsenene (*Ruspolia differens*) from a single Tanzanian locality (adapted from Ng'ang'a et al. 2019); standard deviations (<0.3 log cfu/g) are omitted, and values = 1 refer to "<1" in the original publication

a delicate task, moreover if in the same moment, the original nutritional composition is to be kept, and the flavour not to be impaired beyond the tolerable. Heat treatment is known to affect the nutritional composition of insects, because nutrients are:

- concentrated by the water loss when frying, roasting or toasting,
- lost when passing over to water while boiling (this provides the opportunity to use the broth as an additional product),
- denatured because of the heat and reduced in bioavailability (Mutungi et al. 2017).

This by itself is nothing new and has been practised over centuries, e.g. in milk and other dairy products. However, the perfect balance, i.e. producing a microbiologically safe product with the composition and the taste of the raw product, is impossible; compromises must be found. It was seen that different heating techniques yielded highly variable results for the different insect species and the different microorganisms, be it among techniques, be it within techniques, with different temperature and exposure times. In this way, a technique efficient to reduce bacterial counts in one insect species may fail to be effective in another insect species, and a technique that may reduce the total bacterial count in one species is not that efficient in reducing Enterobacteriaceae of the same insect species (Mutunagi et al. 2017). To illustrate this, Fernández Cassi et al. (2018) evaluated *Acheta domesticus* and reviewed bacterial counts for different products derived from this species (Table 4.5) where the high degree of variance among treatments becomes apparent. So, individual solutions have to be found.

Table 4.5 Total bacterial counts for products derived from the house cricket (*Acheta domesticus*), based on the review of Fernández Cassi et al. (2018, p. 11)

Product	Total aerobic count [cfu/g][a]							
	10^1	10^2	10^3	10^4	10^5	10^6	10^7	10^8
Blanching (4 min)	.	.	.	2.5	.	.	.	.
Boiled	.	4.0	.	.	.	.	.	.
Boiled (5 min)	5.0	.	.	.	.	.	.	.
Freeze-dried	.	.	.	1.1	.	.	.	.
Frozen	.	2.5	.	.	.	.	.	.
Oven-dried	.	.	.	2.0	.	.	.	.
Powder	.	.	.	3.6, 8.2	1.0	.	.	.
Smoked and dried	.	.	.	.	.	.	7.9	.
Sterilized (16 min, 120 °C)	.	.	5.5	.	.	.	.	.
Stir-fried (5 min)	.	5.0	.	.	.	.	.	.
Whole	.	.	.	1.4, 1.6, 2.1	.	.	1.6, 3.2, 8.9	2.1, 3.2
Whole, dead	.	.	.	.	5.0	.	.	.

[a]Read as e.g. "5.0×10^1" in the case of "Boiled (5 min)"

Milling insects to a paste or a powder (insect meal) is a typical step to make the animals' bodies more manageable. For many consumers not accustomed to consume insects, this step is of fundamental importance as it dissolves the shape of insects which is frequently associated with disgust. However, this step also creates and additional risk for microbial growth. It seems that if animals are kept in their original state, the flora can only grow inside the gut and on the exoskeleton which act as natural barriers. By crushing the animals, these barriers are lost, and bacteria can start to grow. In a survey conducted with several insect-based products it was seen that cooked and deep-fried insects displayed markedly lower bacterial counts than products that were dried and milled. This could add to a more appropriate evaluation of different kinds of products, granting high-count products more tolerance and demanding more hygiene in low-count products (Grabowski and Klein 2017b, c). So if milling is part of the processing, special caution must be placed on ensuring as low bacterial counts as possible.

Fermentation is a typical Asian technique to preserve small animals, e.g. as nước mắm (fish sauce), terasi (prawn paste) or prahok (fish paste). To the author's knowledge, these techniques have not been used for insects in Africa so far. However, Klunder et al. (2012) demonstrated that lactic acid fermentation also contributes to reduce bacterial growth. Besides, modern food preservation techniques using plasma and high hydrostatic pressure could also be used (Rumpold et al. 2014).

Even if the processing by itself was successful in controlling microbial risks, special caution must be exercised that no recontamination may occur. That is where good hygienic practice is very important.

4.4   Storage and Logistics

The need to preserve insect foodstuff has been depending on availability (needful for insects with seasonal life cycles) and the amount harvested at a given moment (if there was enough foodstuff worth preserving beyond one or two meals).

As mentioned previously, dead, unprocessed insects spoil easily (max. 1 or 2 days after killing) at ambient temperatures. Heating increases shelf life, but since the microbiome of each insect population is special, heating regimes must be adapted to these particularities in order to ensure food safety (see above). Klunder et al. (2012) studied the bacterial growth in *Acheta domesticus* during storage over 16 days. While boiling the animals for 1 min and keeping them in refrigeration (5–7 °C) was sufficient to ensure low total bacterial, Enterobacteriaceae, and spore-formers' counts, storing them at ambient temperatures lead to spoilage after 1 day. Boiling them for 5 min, drying and grinding them to keep them at ambient temperature also proved to be effective, though not as effective as using refrigeration. Recently, research has started to focus on the storage of edible insects and their products, particularly dried insects and insect meals. Examples of research made in Africa are:

- Dry-frying *Ruspolia differens* stored in opaque vacuum packages or transparent plastic containers at ambient temperature lead to 12–22 weeks shelf life with retained acceptability and low bacterial counts (Ssepuuya et al. 2016).
- Storage of *Musca domestica* meal dried, ground, and stored in nylon bags for 9 months at ambient temperature resulted in threefold increase of bacterial counts, 18-fold increase of fungal counts and growth of pathogenic and toxinogenic microorganisms (Awoniyi et al. 2004).
- *Acheta domesticus* and *Hermetia illucens* were boiled, sun-dried, ground and stored in different plastic containers (of which polypropylene was less efficient to maintain low bacterial counts than polyethylene) and over a 6-month period with regular testing each 45 days it was seen that *Hermetia* illucens degradation occurred in in a sigmoidal pattern (days 45–90 and 135–180), while drastic changes in *Acheta domesticus* were detected between days 0 and 45 (Wankiju 2018).

The few examples show clearly that an adequate storage of edible insects is a complex matter. Just as with processing and even within one and the same species, a specific temperature/residence time combination may decrease one microbiological parameter while it favours, at the same time, another one. Bacterial endospores frequently survive preservation and can re-emerge, even if sampling of previous processing stages had revealed low numbers. This may be due to re-contamination or the outgrowth of the spores (Vandeweyer 2018). Freezing and drying may be more effective in some species–product combinations.

For the future, tailored solutions will be necessary, balancing food safety with nutritious value and sensorial appreciation. This is particularly important if traditional marketing strategies are modified, e.g. by extending the scope from offering bulk goods at local markets to packaging the product and sending them to other markets, supermarkets, or exporting them. It was seen that, like other foodstuffs, packaging material vary in their efficiency to maintain quality during storage of insect-based products.

Finally, also the container used for storage can also affect the (microbiological) quality of the insect product, be it by contamination from previous uses, or by creating conditions that favour certain microorganisms like anaerobic bacteria or fungi (Mutunagi et al. 2017). Wanjiko (2018) concluded for *Acheta domesticus* and *Hermetia illucens* that shelf life may be extended by using "packages with low water vapour and gas permeability and at lower temperatures" (p. xx).

4.5 Kitchen Hygiene

As was seen in Table 4.1 and 4.2, edible insects may contain a series of pathogens. It is admittedly hard to correct flaws that occurred at earlier stages of the food production chain at the consumer level, but the kitchen is the final opportunity to do so. Regardless of whether these pathogens are "native" to a given species or whether they entered the food production chain at another stage, the main objective at kitchen level is to provide a meal which is, in the first place, safe, then nutritious, and, if possible, tasty. Although this is clearly wishful thinking, this ranking is necessary in order to ensure and promote public health. In this way, this includes private kitchens and food stalls, luncheonettes and restaurants.

As the range of insects, products made of them, and transport and storing conditions is so wide, so should the extent of precautions taken on kitchen level. Although eating some insects raw is a millennial tradition in some areas, it must be kept in mind that environmental conditions have been changing, and raw consumption may not be as safe as it had been at the days of the consumers' ancestors. This is not specific for insects in Africa; it also happens in other continents with other foodstuffs, e.g. the debate on raw milk consumption in Europe. Thus, as long as the microbiological safety of insects (farmed or taken from the wild) cannot be guaranteed, raw consumption should be discouraged in favour of an appropriate preservation treatment.

Traditional preserving methods may still function but should be put to the test and evaluated accordingly. In Western Africa, dried and degutted *Cirina forda* caterpillars are used in soups, and while bacterial counts can be high in these caterpillars, the need to rehydrate them in the boiling soup is prone to inactivate even spore-forming bacteria (Grabowski and Klein 2010).

Finally, kitchen hygiene is known to be more than mere heating sensitive foodstuffs. It comprises basic hygienic principles including avoiding cross-contamination. This, however, applies to all foodstuffs and not exclusively edible insects.

4.6 Public Health Monitoring

As in other parts of the world, the extent of legislation in Africa varies strongly, starting with the accessibility for citizens to consult it via internet to the combination of national and international rules to be observed (Bagumire et al. 2009; Kussaga et al. 2014; UEMOA 2007). To the knowledge of the author, in all African countries that have a food law that can be consulted via internet (e.g. EFDA 2010; Government of Sierra Leone 1960), no concise regulation on the microbiology has been published so far. Madagascar claims to consult Codex Alimentarius guidelines for foodstuffs, although the Codex does not mention insects specifically.[1] However, traditional regulation systems are developed when particular insects are relatively scarce so that usage of the resources must be regulated by the community, as described by Mbata et al. (2002).

Still, the lack of microbiological criteria for edible insects is currently the rule rather than the exception. Traditional entomophagy was either safe (having developed over the ages via trial and error) and/or practised on a small level so that impairments would not have affected a larger portion of a nation's population. The need for these parameters has emerged only recently along with the concept of insect farmed and anticipated public health issues that may be associated with it in the future.

One of the areas in which insect farming is developing after the first successes assessed in South Asia, is Europe, a continent with extensive food legislation. For EU member states, food law is twofold, consisting of a series of EU regulations that apply to all member countries and a series of national regulations that address issues not covered by EU law, so that rather than regulating a specific item twice, the combination of EU and national food law has a mosaic display. Non-member states either adapt, to some extent, EU regulations or have a basically similar, national framework.

Regarding insects, microbiological criteria to evaluate insect-based products have been issued so far only by means of national recommendations (Table 4.6). These parameters are, however, based on better-known products like minced meat, meat preparations, and seafood or have to be determined in every foodstuff anyway (in compliance to national and/or EU regulations). This raises several concerns:

- *Comparability of matrices.* While meat is pure muscle and connective tissue, insects are whole organisms, and while seafood originates from fresh or saltwater, most edible insects are terrestrial.
- *"One size fits all".* So far, microbiological criteria refer to "the" insect, regardless of the species and the product. As seen before, there are marked differences due to these factors of influence. In other foodstuffs, there is also a product-related differentiation, cf. fluid milk with cheeses or milk powders.
- *Criteria range.* Although classical food-borne pathogens have been found in insect products (intrinsic or by contamination), insects may also contain other pathogens which have not been considered as a public health concern so far. Future will tell if the present evaluation schemes will have to be adapted.

For a more detailed discussion of the current legal frame of edible insects in Europe, see Grabowski et al. (2019).

Like any other subject related to edible insects, any written contribution may become outdated soon, and regulative aspects are particularly subjected to this accelerated development. The reader is therefore strongly advised to consult the corresponding authorities on the matter at the moment of need.

---

[1] Dr. Séverin Tchibozo, Centre de Recherche pour la Gestion de la Biodiversité, *Cotonou, Bénin, pers. comm. 2019.*

Table 4.6 Microbiological safety parameters recommended in European countries (based on Grabowski et al., 2019, updated)

Parameter	Evaluation scheme	Austria	Belgium	Denmark	Finland	Germany	Netherlands	Switzerland
Aerobic colony count	None specified	+	.	+	.	.	.	.
	$n = 5$, $c = 2$, $m = 5 \times 10^5$ cfu/g, $M = 5 \times 10^6$ cfu/g*	.	+	.	.	.	+	.
	$n = 5$, $m = 1 \times 10^5$ cfu/g, $M = 1 \times 10^6$ cfu/g**	.	.	.	+	.	.	.
	10^4 cfu/g*	.	.	.	.	+[3]*	+	.
	10^6 cfu/g*	.	.	.	.	+[4]*	+	+
Enterobacteri-aceae	None specified	+	.	.	.	.	.	.
	10^2 cfu/g	.	.	.	.	.	.	+
	$n = 5$, $m = 1 \times 10^2$ cfu/g, $M = 1 \times 10^3$ cfu/g**	.	.	.	+	.	+	.
	10^3 cfu/g	.	.	.	.	+	.	.
Escherichia coli	$n = 5$, $c = 2$, $m = 1$ MPN/g, $M = 10$ MPN/g*	+	.	.	.	.	+	.
	230 MPN/100 g**	.	+	.	.	.	.	.
	$n = 5$, $c = 2$, $m = 50$ cfu/g, $M = 500$ cfu/g*	.	+	.	.	.	+	.
	$n = 5$, $c = 2$, $m = 500$ cfu/g, $M = 5000$ cfu/g*	.	.	.	+	.	+	.
	10^1 cfu/g	.	.	.	.	+	.	.
Staphylococcus aureus	$n = 5$, $c = 2$, $m = 10^2$ cfu/g, $M = 10^3$ cfu/g*	+	.	.	+	.	+	.
	10^2 cfu/g*	.	.	.	.	.	.	+
	10^5 cfu/g**	.	.	.	.	.	+	.
	10^3 cfu/g	.	.	.	.	+	.	.
Bacillus cereus	None specified	+	.	.	.	.	.	.
	10^3 cfu/g*	.	.	.	.	.	.	+
	10^5 cfu/g**	.	.	.	.	.	+	.
Bacillus cereus group	10^4 cfu/g	.	.	.	+	.	.	.
	$n = 5$, $m = 1 \times 10^2$ cfu/g, $M = 1 \times 10^3$ cfu/g**	.	.	.	+	+	.	.
	10^3 cfu/g	.	.	.	.	.	.	.

(continued)

Table 4.6 (continued)

Parameter	Evaluation scheme	Austria	Belgium	Denmark	Finland	Germany	Netherlands	Switzerland
Clostridium perfringens	None specified	+	.	.	.	.	.	.
	10^5 cfu/g**	.	.	.	.	.	+	.
Sulphite-reducing clostridia	n = 5, m = 10 cfu/g, M = 1 x 10^2 cfu/g**	.	.	.	+	.	.	.
Salmonella spp.	n = 5, c = o, absent, 25 g**	+	.	.	+	.	.	.
	Absent in 10 or 25 g**	.	+	.	.	.	+	.
	Absent in 25 g**	.	.	.	.	+	.	+
Campylobacter spp.	None specified; 25 g	+	+	.	.	.	.	.
	Absent in 25 g**	.	.	.	.	.	+	.
Listeria monocytogenes	None specified; 25 g	+	.	.	.	+	.	.
	Absent in 25 g	.	.	.	+	.	.	.
	10^2 cfu/g**	.	+	.	.	.	+	+
Yeasts	n = 5, m = 1 x 10^3 cfu/g, M = 1 x 10^4 cfu/g**	.	.	.	+	.	.	.
Moulds	n = 5, m = 1 x 10^3 cfu/g, M = 1 x 10^4 cfu/g**	.	.	.	+	.	.	.

* = process hygiene criterion according to Regulation (EU) 2073/2005 in the current, consolidated version, and national guidelines derived from it

** = food safety criterion according to Regulation (EU) 2073/2005 in the current, consolidated version, and national guidelines derived from it

3* = cooked and/or deep-fried, whole insects

4* = all other insect products

4.7 Conclusion

Technically, the microbiology of edible is subjected to the same factors as the one of other foodstuffs, i.e. the entry of certain microorganisms into a given foodstuff, and the multitude of factors that either favours or inhibits the growth of these microorganisms. Evaluating this system is, however, far more complex on the level of the umbrella term "edible insects":

- The category "edible insect" comprises 2000–3000 different species, and each species is truly different from the others. So talking about edible insects is like talking about eating "mammals", "birds" or "fish", only that there are far less species in those categories.
- Since all the species are different, so are their micro- and mycobiome which are in fact reflections of this species' surroundings. It must be stressed that these microbial communities, in terms of food safety, also react differently to processing and preserving techniques and present a minor or major risk for the consumer.
- Entomophagous traditions have developed over the millennia, but recent environmental changes may have started to impair their efficiency. This makes it necessary to (re)evaluate them, especially if the original scope of this tradition is modified, e.g. if a product is processed/packaged in a different way or subjected to longer distances for marketing. Raw consumption should be discouraged if research uncovers public health concerns.
- Farming and novel food development may change the microbiological communities of the farmed insects. Also here, a through monitoring of the microbiology in terms of spoilage and food pathogens is advised.
- If edible insects are to gain more importance in the nutrition of a nation's population, then a stronger commitment of the public health authorities is needed. In terms of food microbiology, this translates into an appropriate selection of safety and processing criteria, adapted to the species and the product, respecting the diversity of products as done also with other foodstuffs.

Appendix: List of Scientific Names of Edible Insects

Species	Order	Family	English name	Local names
Acheta domesticus	Orthoptera	Gryllidae	House cricket	صرصور الحقل الأليف (Arabic), grillon domestique (French), chan (Kasem), grilo-doméstico (Portuguese), grillo doméstico (Spanish)
Alphitobius diaperinus	Coleoptera	Tenebrionidae	Buffalo worm	خنفساء القمامة (Arabic)

(continued)

Species	Order	Family	English name	Local names
Bematitistes alcinoe	Lepidoptera	Nymphalidae		.
Bunaea alcinoe	Lepidoptera	Saturniidae	Cabbage tree emperor moth	Kiepersol-pouoogmot (Afrikaans), egu (Igbo)
Gonimbrasia belina	Lepidoptera	Saturniidae	Mopane moth	Mopanimot (Afrikaans), mingolo (Kongo), omagungu (Ovambo), phane (Tswana), mashonzha (Venda)
Gryllodes sigillatus	Orthoptera	Gryllidae	Tropical/ Indian house cricket, banded cricket	.
Gryllotalpa africana	Orthoptera	Gryllotalpidae	African mole cricket	.
Hermetia illucens	Diptera	Stratiomyidae	Black soldier fly, BSF	Mouche soldat (French), mosca soldado negra (Spanish)
Locusta migratoria	Orthoptera	Acrididae	Migratory locust	አምበጣ (Amharic), gbamedor (Ewe), criquet migrateur (French), gafanhoto-migratório (Portuguese), langosta migratoria (Spanish), panzi (Swahili)
Musca domestica	Diptera	Muscidae	Housefly	Huisflieg (Afrikaans), የቤትዝንብ (Amharic), منزلية ذبابة (Arabic), nhunzi (ChiShona), mouche domestique (French), kwazi (Hausa), mosca-de-casa (Portuguese), mosca común (Spanish)
Oryctes monoceros	Coleoptera	Scarabeidae	African rhinoceros beetle	عريقطة (Arabic), scarabée rhinoceros africain (French), akpaagbugbo (Igbo), escarabajo rinoceronte africano (Spanish)
Rhychophorus phoenicis	Coleoptera	Curculionidae	African palm weevil	Elulungwo (Igbo), akonono (Twi)
Ruspolia differens	Orthoptera	Tettigoniidae	Nsenene	Nsenene (Luganda)
Tenebrio molitor	Coleoptera	Tenebrionidae	(common) mealworm	Ténébrion meunier (French), larva-da-farinha (Portuguese), gusano de la harina (Spanish)
Zophobas atratus[a]	Coleoptera	Tenebrionidae	Superworm, morio worm	.

[a]Formerly known as *Z. morio*

References

Adeoye OT, Alebiosu BI, Akinyemi OD, Adeniran OA (2014) Socio economic analysis of forest edible insects species consumed and its role in the livelihood of people in Lagos State. J Food Stud 3:103–120

Alamu OT, Amao AO, Nwokedi CI, Oke OA, Lawa IO (2013) Diversity and nutritional status of edible insects in Nigeria: a review. Int J Biodiv Conserv 5:215–222

Amadi EN, Kiin-Kabari DB (2016) Nutritional composition and microbiology of some edible insects commonly eaten in Africa, hurdles and future prospects: a critical review. Food Microbiol Saf Hyg 1:1, 1000107

Amiri A (2017) An investigation of some edible insects as source of human food and animal feed. Graduate School of Applied Sciences of Near East University, Nicosia/CY, thesis, 60 pp.

Awoniyi TAM, Adetuyi FC, Akinyosoye FA (2004) Microbiological investigation of maggot meal, stored for use as livestock feed component. J Food Agricult Environ 2:104–106

Bagumire A, Todd ECD, Muyanja C, Nasinyama GW (2009) National food safety control systems in Sub-Saharan Africa: does Uganda's aquaculture control system meet international requirements. Food Policy 34:458–467

Banjo AD, Lawal OA, Adeyemi AI (2006) The microbial fauna associated with the larvae of *Oryctes monocerus*[sic!]. J Appl Sci Res 2:837–843

Belluco S, Losasso C, Maggioletti M, Alonzi C, Ricci A, Paoletti MG (2015) Edible insects: a food security solution or a food safety concern? Anim Front 5:25–29

Braide W, Nwaoguikpe RN (2011) Assessment of microbiological quality and nutritional values of a processed edible weevil caterpillar (*Rhynchophorus phoenicis*) in Port Harcourt, southern Nigeria. Int J Biol Chem Sci 5:410–418

Braide W, Oranusi S, Udegbunam LI, Oguoma O, Akobondu C, Nwaoguikpee RN (2011) Microbial quality of an edible caterpillar of an emperor moth, *Bunaea alcinoe*. J Ecol Nat Environ 3:178–180

Debrah SK (2017) Edible insect as a traditional food source among the Akans in southern Ghana. University of Ghana, Accra/GH, thesis, 94 pp.

Dobermann D, Swift JA, Field LM (2017) Opportunities and hurdles of edible insects for food and feed. Nutr Bull 42:293–308

EFDA, Ethiopian Food and Drug Administration (2010) Proclamation no. 1: Proclamation on the Food and Drug Administration. Proclamation no. 661/2009, a proclamation to provide for food, medicine and health care administration and control. Federal Negarit Gazeta of the Federal Democratic Republic of Ethiopia 16:5157–5191

Fernández Cassi X, Supeanu A, Jansson A, Boqvist S, Vagsholm I (2018) Novel foods: a risk profile for the house cricket (*Acheta domesticus*). EFSA J 16(S1):e16082. 13 pp

Government of Sierra Leone (1960) Public health ordinance. www.sierra-leone.org/Laws/1960-23.pdf

Grabowski NT (2017) Speiseinsekten. Behr's Verlag, Hamburg, 77 pp

Grabowski NT, Ahlfeld B, Lis KA, Jansen W, Kehrenberg C (2019) The current legal status of edible insects in Europe. Berliner und Münchener Tierärztliche Wochenschrift 132:295–311

Grabowski NT, Klein G (2010) Mikrobiologische und chemische Analyse exotischer Lebensmittel tierischer Herkunft in Europa. Arch Leb 61:63–74

Grabowski NT, Klein G (2017a) Bacteria encountered in raw insect, spider, scorpion, and centipede taxa including edible species, and their significance from the food hygiene point of view. Trends Food Sci Technol 63:80–90

Grabowski NT, Klein G (2017b) Microbiology of processed edible insect products—results of a preliminary survey. Int J Food Microbiol 243:103–107

Grabowski NT, Klein G (2017c) Microbiology of cooked and dried edible Mediterranean field crickets (*Gryllus bimaculatus*) and superworms (*Zophobas atratus*) submitted to four different heating treatments. Food Sci Technol Int 23:17–23

Grabowski NT, Pingen S, Grootaert P, Klein G (2017) Fungi in raw insect and arachnid taxa containing species used in human entomophagy: a review. Arch Leb 68:39–47

Kelemu S, NIassy S, Torto B, Fiaboe K, Affognon H, Tonnang H, Maniania NK, Ekesi S (2015) African edible insects for food and feed: inventory, diversity, commonalities and contribution to food security. JIFF 1:103–119

Klunder H, Wolkers-Rooijackers J, Korpela J, Nout M (2012) Microbiological aspects of processing and storage of edible insects. Food Control 26:628–631

Kussaga JB, Jacxsens L, Tiisekwa BPM, Luning PA (2014) Food safety management systems performance in African food processing companies: a review of deficiencies and possible improvement strategies. J Sci Food Agric 94:2154–2169

Mbata KJ, Chidumayo EN, Lwatula CM (2002) Traditional regulation of edible caterpillars exploitation in the Kopa area of Mpika district in northern Zambia. J Insect Conserv 6:115–130

Milanović V, Osimani A, Roncolini A, Garofalo C, Aquilanti L, Pasquini M, Tavoletti S, Vignaroli C, Canonico L, Ciani M, Clementi F (2018) Investigation of the dominant microbiota in ready-to-eat grasshoppers and mealworms and quantification of carbapenem resistance genes by qPCR. Frontier Microbiol 9:article 3036. 11 pp

Mutungi C, Irungu FG, Nduko J, Mutua F, Affognon H, Nakimbugwe D, Ekesi S, Fiaboe KKM (2017) Postharvest processes of edible insects in Africa: a review of processing methods, and the implications for nutrition, safety and new products development. Crit Rev Food Sci Nutr. https://doi.org/10.1080/10408398.2017.1365330. 23 pp

Ng'ang'a J, Imathiu S, Fombing F, Ayieko M, Broeck JV, Kinyuru J (2019) Microbial quality of edible grasshoppers *Ruspolia differens* (Orthoptera: Tettigoniidae): from wild harvesting to fork in the Kagera region. Tanzania. J Food Safety 39:e12549

Ogbalu OK, Williams JO (2015) The edibility, distribution and damage indices of *Oryctes monoceros* oliv. [Coleoptera: Scarabeidae] an edible larva of the oil palms [Elaise Guineensis] and associated microorganisms [sic!]. J Pharma Biol Sci 10:118–125

Okeke TE, Ewuim SC, Akanne CE, Onnoye BU (2019) Survey of edible insects in relation to their habitat and abundance in awka and environ [sic!]. Int J Entomol Res 4:17–21

Opara MN, Sanyigha FT, Ogbuewu IP, Okoli IC (2012) Studies on the production trend and quality characteristics of palm grubs in the tropical rainforest zone of Nigeria. J Agricult Technol 8:851–860

Osimani A, Garofalo C, Milanović V, Taccari M, Cardinali F, Aquilanti L, Pasquini M, Mozzon M, Raffaelli N, Ruschioni S, Riolo P, Isidoro N, Clementi F (2017) Insight into the proximate composition and microbial diversity of edible insects marketed in the European Union. Eur Food Res Technol 243:1157–1171

Osimani A, Milanović V, Cardinali F, Garofalo C, Clementi F, Ruschioni S, Riolo P, Isidoro N, Loreto N, Galarini R, Moretti S, Petruzzelli A, Micci A, Tonucci F, Aquilanti L (2018) Distribution of transferable antibiotic resistance genes in laboratory-reared edible mealworm (*Tenebrio molitor* L.). Front Microbiol 9:article 2702. 11 pp

Raheem D, Carrascosa C, Oluwole OB, Nieuwland M, Saraiva A, Millán R, Raposo A (2018) Traditional consumption of and rearing edible insects in Africa. Asia Eur Critic Rev Food Sci Nutr. https://doi.org/10.1080/10408398.2018.1440191. 20 pp

Riggi L, Versomesi M, Verspoor R, MacFarlane C (2013) Exploring entomophagy in Northern Benin—practices, perceptions and possibilities. Bugs Life:47

Rumpold BA, Fröhling A, Reinecke K, Knorr D, Boguslawski S, Ehlbeck J, Schlüter O (2014) Comparison of volumetric and surface decontamination techniques for innovative processing of mealworm larvae (*Tenebrio molitor*). Innov Sci Emerg Technol 26:232–241

Ssepuuya G, Aringo R, Mukisa I, Nakimbugwe D (2016) Effect of processing, packaging and storage temperature based hurdles in the shelf-stability of sautéed ready-to-eat *Ruspolia differens*. JIFF 2:245–253

Ssepuya G, Wynants E, Verreth C, Crauwels S, Lievens B, Claes J, Nakimbugwe D, Van Campenhout L (2019) Microbial characterisation of the edible grasshopper *Ruspolia differens* in raw condition after wild-harvesting in Uganda. Food Microbiol 77:106–117

Stoops J, Crauwels S, Waud M, Claes J, Lievens B, Van Campenhout L (2016) Microbial community assessment of mealworm larvae (*Tenebrio molitor*) and grasshoppers (*Locusta migratoria migratorioïdes*) sold for human consumption. Food Microbiol 53:122–127

Stull VJ, Wamulume M, Mwalukanga MI, Banda A, Bergmans RS (2018) "We like insects here": entomophagy and society in a Zambian village. Agric Hum Values 35:867–883

Tanada Y, Kaya HK (1993) Insect pathology. Academic, San Diego

Tchibozo S, Malaisse F, Mergen P (2016) Insectes consommés par l'homme en Afrique occidentale francophone. Geo-Eco-Trop 40:105–114

UEMOA, Union Économique et Monétaire Ouest Africaine (2007) Règlement N° 007/2007/CM/UEMOA relatif à la sécurité sanitaire des végétaux, des animaux et des aliments dans l'UEMOA. www.droit-afrique.com. 42 pp.

Van Huis A (2003) Insects as food in Sub-Saharan Africa. Insect Sci Appl 23:163–185

Vandeweyer D (2018) Microbiological quality of raw edible insects and impact of processing and preservation. KU Leuven/B, thesis, 177 pp.

VarottoBoccazzi I, Ottoboni M, Martin E, Comandatore F, Vallone L, Spranghers T, Eeckhout M, Mereghetti V, Pinotti L, Epis S (2017) A survey of the mycobiota associated with larvae of the black soldier fly (*Hermetia illucens*) reared for feed production. PLoS One 12:e0182533. 15 pp

Wanjiko EK (2018) Physicochemical and microbiological stability of semi-processed edible crickets (*Acheta domesticus*) and black soldier fly larvae (*Hermetia illucens*) during storage. Jomo Kenyatta University of Agriculture and Technology, Nairobi/KE. 116 pp

Chapter 5
Food Safety of Edible Insects

Miklós Mézes and Márta Erdélyi

Abstract Insects are alternative dietary protein sources for humans, but currently there are no specific regulations in most of the countries on the breeding and marketing of insects destined for human consumption. The insects used for human consumption has some potential risks, even when reared in controlled environment and substrate. There are about 2000 edible insect species known and, in certain regions, insects are already eaten by humans for centuries. Nevertheless, there is only little scientific literature available on the food safety of insects. The present chapter summarizes the potential microbial, chemical, and toxicological hazards specifically related to the consumption of insects. Insect proteins may induce allergic reactions, but only few studies are available. These hazards depend on the insect species, the cultivation conditions (substrate and environment), and the subsequent processing technology.

Keywords Food safety regulations · Microbial safety · Molds and mycotoxins · Parasites · Allergic reactions · Toxic substances · Metals and organic pollutants

5.1 Insects as Alternative Protein Sources

There is discussion about the insects as a potent sustainable protein source for the rapidly growing world population. The future demand for food and animal-derived protein will require environmentally friendly, novel food sources. Insects might be good alternatives, both for animals and humans. The United Nations have been promoting insect-eating as a promising, protein-packed, and nutritious way to feed billions of people predicted to live on earth in 2050 (Halloran and Vantomme 2013). The beneficial nutritive effects of insect proteins are well known, and it has a long tradition in human history as a food, which dates back to prehistoric times

M. Mézes (✉) · M. Erdélyi
Department of Nutrition, Szent István University, Gödöllő, Hungary
e-mail: Mezes.Miklos@mkk.szie.hu; Erdelyi.Marta@mkk.szie.hu

© Springer Nature Switzerland AG 2020
A. Adam Mariod (ed.), *African Edible Insects As Alternative Source of Food, Oil, Protein and Bioactive Components*, https://doi.org/10.1007/978-3-030-32952-5_5

(Tommaseo-Ponzetta 2005). Total number of edible insect species around the world is not exactly known, the estimations have increased in time from 1000 (DeFoliart 1997), through 1681 (Ramos-Elorduy 2005) up to 1900 (Jongema 2012). According to a global assessment (Van Huis et al. 2013), the insects most commonly consumed by humans are beetles (31%), caterpillars (18%), bees, wasps and ants (14%). Consumption of grasshoppers, locusts, and crickets comprises 13%, followed by cicadas, leafhoppers, plant hoppers, scale insects, and true bugs (10%), termites (3%), dragonflies (3%), flies (2%), and others (5%).

According to certain research data potential food safety hazards might be associated with the use of insects or insect proteins. These include different contaminants, such as heavy metals, mycotoxins, pesticide residues, antinutritive factors, and pathogens (Van Huis et al. 2013). On the other hand, in regions where insect consumption has a long-lasting tradition and has a "history of safe use," insects are considered to be nutritive substances having no health risk (Anankware et al. 2015). However, this has not been scientifically proven.

Although the technologies used to isolate and prepare insects and insect proteins in food processing have a potential to remove the undesirable insect-specific components and contaminants in the different fractions, further detailed hazard analysis should be performed in the entire processing protocol to identify all the possible (physical, microbial, allergenic, and chemical) risks. According to a recent review, there is a lack of scientifically based knowledge about food safety issues in insect processing (Schlüter et al. 2017). Only few data are available on the health hazards of the extraction of protein and lipid fractions or enzymes from insects, which removes the microbial contamination of the raw material. However, these cannot be applied to other risks automatically. Therefore, it is essential to identify the critical control points for the whole production and processing chain of insects (Stoops et al. 2016).

European Food Safety Authority has released an initial risk assessment of using insects as human food (EFSA 2015), but depending on the insect species intended to be used species-specific safety aspects must be considered. Different stages of development, such as complete or incomplete metamorphosis also should be considered as a potential safety aspect of use.

5.2   Food Safety Regulations of Insects

Studies on food safety of insects are limited, and some insects considered to be safe can turn out to be unhealthy as they might contain allergens or as they are possibly fed with plants originated from a polluted area. There is a long-term tradition of using insects as food outside the EU. In those countries the use of insects is operated rather according to customs than to a controlled protocol. Anyway, it is generally accepted that the existing evidences are adequate to consider insect food as safe for human consumption (DeFoliart 1992), and at least two billion people worldwide is estimated to eat insect regularly. Although no significant health problems have arisen so far from the consumption of edible insects (Banjo et al. 2006), consumers' confidence is strongly correlated with the perceived safety of a given product. However, many low-income countries do not have policies controlling the use of agrochemicals in areas where people collect edible insects, and the consumers have little or no knowledge about the potential consequences of eating chemically treated insects (Kinyuru et al. 2012).

There is a regulation in the European Union (EU 2015) about the insects as novel food, which is partially based on a recent EFSA Opinion on the safety of insect as food (EFSA 2015). The regulation specifies those insects, which can be used as novel food, but other species might also be used for human consumption in the EU countries (Schlüter et al. 2017). It is declared in the regulation that insects as human food "should not be pathogenic or have other adverse effects on plant, animal, or human health; they should not be recognized as vectors of human, animal or plant pathogens and they should not be protected or defined as invasive alien species." There is a European Union regulation on how animal origin materials might be introduced into the food chain, which should be applied for insects and it should comply with a previous regulation, as well (EU 2004a). Furthermore, the whole process of insect protein production should meet the requirements outlined in the Food Hygiene Regulations of the European Union (EU 2004a, b). To date, European law is not conclusive on several issues regarding the use of novel protein sources, such as insects, in food products (Belluco et al. 2013). Additionally, special rules should be introduced for processing, storage, and transport of insects in order to meet the food safety criteria (Schabel 2008). Recently, a new European Union regulation was announced, which contain the provisions for general criteria of insect and insect protein production, with particular attention to the materials used for feeding the insects and their larvae (EU 2017).

Insects are not listed in the Codex Alimentarius as foodstuffs, they are mentioned only as impurities that contaminate food (Van Huis et al. 2013). This is also a problem in the USA, where insects are described as a "defect" in food, but they are not explicitly defined as food (Food and Drug Administration 2010). However, insects are currently sold in the USA and in other countries as novelty foods for human consumption, while in other regions they are considered as traditional food.

5.3 Microbial Safety

Insect-specific pathogenic microorganisms are harmless to humans because they are taxonomically different (Van Huis et al. 2013) and they have tissue specificity thus they colonize most probably only in the tissues of insects (Eilenberg et al. 2015). Insects are taxonomically distant from humans, so the risk of zoonotic infections is assessed to be low. The risk of zoonotic infections could rise with the careless use of waste products, unhygienic handling, and direct contact between farmed insects and insects outside the farm due to weak biosecurity (EFSA 2012). Transfer of pathogenic bacteria from substrate to insects is not thoroughly studied, but a recent report suggests that when present at a low level in the substrate, *Salmonella* sp. is not retained by the larvae, possibly due to competitive exclusion by the endogenous larval microbiota and/or because of antibacterial activity of the larvae (Wynants et al. 2019).

However, insects might act as vectors of microorganisms either as a result of mechanical contact with the body surface (Graczyk et al. 2005) or due to bacterial growth inside the insect (Wasala et al. 2013), even without any sign of disease. It should be mentioned that it is not possible to remove the gut with its microbiota in the processing, therefore all the microbes living in the insect get into the final product. However, the presence of pathogenic bacteria depends on the hygienic conditions of the substrate and the rearing conditions. Different feed substrates also can alter the microbial composition of the insects' gut microbiota. For this reason EFSA has proposed a classification of substrates for insect breeding according to the level of hazard (EFSA 2015). The composition of the insect microbiota also depends on the species and on the developmental stage of the insect (Yun et al. 2014). Both insects collected in the nature or raised on farms may be infected by pathogenic microorganisms, and insect species have autochthonous microbiota (Schlüter et al. 2017), including bacteria (*Staphylococcus*, *Bacillus*, *Campylobacter*, *Pseudomonas*, *Micrococcus*, *Acinetobacter*, *Proteus*, *Escherichia*, *Enterobacteriaceae*, and other spore-forming bacteria), viruses, fungi, and protozoa (Klunder et al. 2012; Vega and Kaya 2012). However, in case of most pathogens no active growth occurs in the intestinal tract of insects (Hazeleger et al. 2008). Otherwise, herbivore insects, such as larvae of black soldier fly, may contain *Enterobacteriaceae* which can cause problem when the insect is ingested in raw form (EFSA 2015). Also *Clostridium botulinum* was reported to cause fatalities in poorly stored termites in Kenya (Brickey and Gorham 1989). Insects are definitely not biological vectors of prions (EFSA 2015), but in an experiment with the corpse fly (*Sarcophaga carnaria*) grown on substrate containing brains from scrapie-infected hamsters, infection has occurred in hamsters eating the larvae of these flies (Post et al. 1999). Studies have also revealed that insects fed with the nervous system of ruminants infected with the scrapie agent were themselves a source of contamination, and ectoparasites, such as Hypoderma bovis and Oestrus ovis also can be transmission factors of prions (Lupi 2006).

The risk of transmission of bacteria could be mitigated through strict hygienic processing. A study on four commercial insect species (*Zophobas morio*, Tenebrio molitor, *Galleria mellonella*, and Acheta *domesticus*) showed high total microbial content (10^5 to 10^6 CFU g^{-1}) on samples originating from a closed-cycle farm. The microbial population was mainly composed of gram-positive bacteria, and coliforms. Within the gram-positive population *Micrococcus* and *Lactobacillus* spp. (10^5 CFU g^{-1}), and *Staphylococcus* spp. (approximately 10^3 CFU g^{-1}) were the dominant, while *Salmonella* spp. and Listeria monocytogenes were not isolated from the tested samples (Giaccone 2005). Spores of various microorganisms may be present on insect cuticles, including those microorganisms that grow saprotrophically on edible insect products.

Strict control of insect farming and processing are the key issue to ensure safety of insect products. Hygienic farming practices, safe feed sources for insects and proper processing techniques (preparation, storage) can mitigate potential microbiological hazards. The importance of hygienic handling and correct storage was investigated in laboratory experiments (Klunder et al. 2012). Spore-forming bacteria might occur in the gut and on the skin of insects and have the potential to cause health problems in human. The most common preparation method for insects, the roasting alone does not kill all *Enterobacteriaceae*. The microbe content of yellow mealworm larvae (Tenebrio molitor) and house crickets (Acheta domesticus) can be partially eliminated by boiling in water for a few minutes, so boiling is suggested before roasting. However, spores may survive this process, and can germinate, then bacteria can grow under favorable conditions. Alternative preservation techniques are drying and acidifying. Lactic acid fermentation resulted in successful acidification, and it was effective to prolong shelf life and maintain safety by controlling the growth of *Enterobacteriaceae* and survival of bacterial spores (Klunder et al. 2012).

5.4 Molds and Mycotoxins

Pathogenic molds, like *Aspergillus*, *Penicillium*, *Mucor*, *Rhizopus*, *Alternaria*, and *Candida* can infect host insects via feed, and in this case as part of the microbiota on the surface and gut of insects these can be directly contagious to humans or can be harmful due to their secondary substances having toxic or allergenic effects (Banjo et al. 2005; Suh et al. 2008). Approximately 1000 mold species are known to infect insects (Vega et al. 2012), but only few of them produce mycotoxins. In humid areas, dried insects are susceptible to absorb moisture, which enables mold growth, possibly resulting in safety risk (Amadi et al. 2005).

Aspergillus mold infection and aflatoxin contamination were checked in edible grasshopper (*Imbrasia belina*), and the level of total aflatoxins varied from 0 to 50 μg kg^{-1} product (Mpuchane et al. 1996). The maximum safe level of aflatoxins set by EU is 10–15 μg kg^{-1} for foodstuffs of plant origin, but there is no regulation for foodstuffs of animal origin, including insects. However, frequent consumption of contaminated foods over long periods is likely to pose health risks. Larvae of two other edible insect species, black soldier fly (*Hermetia* illucens) and yellow mealworm (Tenebrio molitor), were grown on poultry feed loaded with aflatoxin B1, and high tolerance was found to it in both species. They accumulate aflatoxin B1 only at approximately 10% level of its aforementioned European Union's legal limit for food materials (Bosch et al. 2017). Black soldier fly larvae also did not accumulate other mycotoxins (Purschke et al. 2017). Mealworms have been shown to be able to metabolize zearalenone partially to its more toxic metabolite, to alpha-zearalenol. However, this is not hazardous for animals, due to the low level of its accumulation in mealworm larvae (Hornung 1991).

5.5   Parasites

Parasites also represent a potential hazard in relation to insect consumption. In Southeast Asia, insects infected with metacercariae were found to be responsible for fluke in human (Chai et al. 2009). *Plagiorchid* parasites, a new parasitic species in humans, probably originate from insect larvae consumption (Chai et al. 2009). Infections of a nematode parasite, *Gongylonema pulchrum*, have been reported in humans in many places around the world (Molavi et al. 2006), because the interme- diate host of this parasite is the cockroach. Myiasis is the infestation of humans with dipterous larvae, which occur when fly eggs reach the gastrointestinal tract, but common housefly is rarely reported as the cause of myiasis. Among edible insects, Hermetia illucens can be a potential vector (Sehgal et al. 2002). Insects can also be potential biological vectors of trypanosomiasis, which affects about ten million people worldwide (WHO 2010). Some cases have been reported linking the infec- tion with the ingestion of insects (Pereira et al. 2010). Among other potential food- borne pathogens, Entamoeba histolytica and Giardia lamblia have been isolated in cockroaches. Cryptosporidium parvum has also been found in insects, which is lethal for immunocompromised individuals (Graczyk et al. 2005). These parasites could be present also in edible insects, and should be considered in the case of insect consumption as food.

5.6   Allergic Reactions to Edible Insects

Like most protein-containing foods, arthropods can induce IgE-mediated allergic reactions insensitive humans (Mitsuhashi 2008). The main allergenic structures in insects are certain glycoproteins. Besides them, chitin and its derivatives are also presumed to be allergens, which enhances the formation of allergen specific IgE antibodies (Muzzarelli et al. 2001). Chitin also enhances the allergic reaction pro- voked by *Aspergillus fumigatus* (Dubey et al. 2015).

These allergens may cause eczema, dermatitis, rhinitis, conjunctivitis, conges- tion, angioedema, and bronchial asthma. While some people have a history of atopy (allergic hypersensitivity), development of allergic sensitivity due to long-term exposure is also possible. The majority of cases are caused by inhaling or contacting allergens in nature (Phillips and Burkholder 1995; Barletta and Pini 2003). Tropomyosins from cockroaches, mites, and shrimps have been reported to be aller- genic (Reese et al. 1999). Literature data suggest that people with seafood allergy also have allergic reactions when they consume edible insects (Reese et al. 1999).

However, few studies have been published on allergic reactions due to insect ingestion, and they did not prove the direct hypersensitivity to insect proteins. Therefore, the risk of food allergy associated with insect consumption needs further investigation, and greater attention is required in distinguishing toxic and allergic symptoms (Pitetti et al. 1999). For those people, who have no history of arthropod or insect allergen sensitivity, usually have no acquired sensitivity for allergic reactions, even if they eat and/or are exposed to insects through long-term exposure in sufficient quantities.

5.7 Toxicity

Some insect species are considered to be toxic, due to their toxin and antinutrient content. Among antinutrients, oxalate, tannins, phytate (Ekop et al. 2010) and thiaminases (Nishimune et al. 2000) may have importance, but they should be distinguished according to the fact that they are synthesized by the insect or they are acquired from the feed. However, only a few data are available on their adverse effects caused by the consumption of insects. For instance, cases of ataxia syndrome were reported after consumption of the seasonal silkworm (*Anaphe venata*) in Nigeria, possibly due to its thiaminase activity and the consequent thiamine deficiency (Adamolekun et al. 1997). Sexual steroids, such as testosterone and dihydrotestosterone, were found in beetles (*Dytiscidae* family), which can cause growth retardation, hypofertility, masculinization in females, edema, jaundice, and liver cancer in human. Cyanogenic glycosides are present in some insects (*Coleoptera* and *Lepidoptera*), causing inhibition of succinate dehydrogenase and carbonic anhydrase, and they prevent oxidative phosphorylation (Nahrstedt 1988). Spanish fly (Lytta vesicatoria) synthesizes cantharidin, atoxic monoterpene (2,6-dimethyl-4,10-dioxatricyclo-[5.2.1.02,6]decane-3,5-dione), which can cause difficulty in swallowing, nausea, and vomiting of blood in human (Till and Majmudar 1981). Darkling beetles (*Tenebrionidae*) contain quinones and alkanes (Brown et al. 1992), but possibly due to accumulation from plant substrates (Crespo et al., 2011). However, no data are available about the toxins in edible insects (Anankware et al. 2015).

5.8 Contamination with Metals and Organic Pollutants in the Environment

Insect derived food products may contain hazardous chemical residues (e.g., heavy metals, dioxins, pesticides, and veterinary drugs). However, published data on these substances in farmed insects as well as data on accumulation/excretion of chemical contaminants from the substrates are very limited. Heavy metals are not a negligible problem as they can be bio-accumulated in insect bodies. Harmful metals from the environment have been found in the different tissues and organs of insects, such as fat, exoskeleton, reproductive organs, and digestive tracts, where they accumulate. A study on the yellow mealworm (Tenebrio molitor) and black soldier fly (Hermetia illucens) larvae showed that these insects accumulate cadmium, lead, and arsenic when they are fed on contaminated substrates, such as organic matter in soils (Vijver et al. 2003), or from a herbicides (Van der Fels-Klerx et al. 2016), but do not accumulate chromium, arsenic, nickel, and mercury (Purschke et al. 2017), while zinc is incorporated at a decreasing rate as its concentration in the substrate has been increased (Diener et al. 2015).

An outbreak of lead poisoning reported in California caused extremely high lead content in dried chapuline grasshopper (*Sphenarium*). These insects acquired lead from tailings of silver mines in Mexico, and when they were dehydrated to preserve them lead concentration has increased further in the end product (Handley et al. 2007). It is also revealed that black soldier fly accumulates lead efficiently even up to higher concentration than it is present in the substrate (Gao et al., 2017). Similarly, extreme accumulation of selenium was found in Tenebrio molitor larvae (Hogan and Razniak 1991).

Pesticides used against agricultural pests are potentially dangerous for consumers, particularly when the insects and insect products originate from natural habitat rather than from controlled farming after spraying with insecticides, and in particular in those areas where no strict control for chemical application (van Huis et al. 2013). It is a real problem in some developing countries, where edible, even dead insects, mainly locusts and grasshoppers, are collected and consumed after insecticide treatment. For instance, in Mali the increasing use of pesticides in cotton farming has resulted in rapid decline in grasshopper consumption in rural areas (van Huis et al. 2013). According to a study in Kuwait, the captured locusts contained high amounts of organophosphorus pesticides, possibly due to the uncontrolled chemical use in that area (Saeed et al. 1993). Most of the pesticides will occur not only in the cultivated plants but in the beneficial biota, soil, and water, and consequently they can accumulate in the edible insects via the substrate used for feeding (Pimentel 1995). In contrast, black soldier fly larvae did not accumulate pesticides from the substrate (Lalander et al. 2016).

5.9   Physical Hazards

Locusts, crickets, and beetles often have powerful lower jaws, firm legs, wings and other appendages which can damage, even perforate the intestines or cause constipation (Schabel 2008). Therefore legs and wings of the insect have to be removed before consumption. Indigestible chitin also accumulates in different places in the colon and may result in constipation.

5.10   Conclusions

Insect consumption by humans has always been a worldwide practice, especially in those countries, where food is in short supply and food safety is not a concern. Insects, like other foodstuffs, may have some hazards like causing allergic symptoms.

Also they might harbor different kinds of pathogenic bacteria, parasites, and prions, but a properly managed insect farm could remain free from pathogens. Chemical risk is also a real issue, but toxicity can be reduced by avoiding consumption of insect synthetizing poisons and by reducing the bioaccumulation of chemicals, such as heavy metals and insecticides with minimizing the contamination of substrate as part of a controlled feeding program.

In conclusion, edible insect species have no more hazards than other animal products, and insects can be regarded as safe, when they are properly managed and consumed.

References

Adamolekun B, McCandless DW, Butterworth RF (1997) Epidemic of seasonal ataxia in Nigeria following ingestion of the African silkworm *Anaphe venata*: role of thiamine deficiency? Metab Brain Dis 12:251–258. https://doi.org/10.1007/BF02674669

Amadi EN, Ogbalu OK, Barimalaa IS, Pius M (2005) Microbiology and nutritional composition of an edible insect (*Bunaea alcinoe* Stoll) of the Niger Delta. J Food Safety 25:193–197 https://doi.org/10.1111/j.1745-4565.2005.00577.x

Anankware P, Fening K, Osekre E, Obeng-Ofori D (2015) Insects as food and feed: a review. Int J Agric Res Rev 3(1):143–151

Banjo AD, Lawal OA, Adeduji OO (2005) Bacteria and fungi isolated from housefly (Musca domestica L.) larvae. Afr J Biotechnol 4:780–784

Banjo AD, Lawal OA, Songonuga EA (2006) The nutritional value of fourteen species of edible insects in southwestern Nigeria. Afr J Biotechnol 5:298–301

Barletta B, Pini C (2003) Does occupational exposure to insects lead to species-specific sensitization? Allergy 58:868–870. https://doi.org/10.1034/j.1398-9995.2003.00278.x

Belluco S, Losasso C, Maggioletti M, Alonzi CC, Paoletti MG, Ricci A (2013) Edible insects in a food safety and nutritional perspective: a critical review. Compr Rev Food Sci Food Saf 12:296–313. https://doi.org/10.1111/1541-4337.12014

Bosch G, van der Fels-Klerx HJ, de Rijk TC, Oonincx DGAB (2017) Aflatoxin B1 tolerance and accumulation in black soldier fly larvae (*Hermetia illucens*) and yellow mealworms (Tenebrio molitor). Toxins 9:185. https://doi.org/10.3390/toxins9060185

Brickey PM, Gorham JR (1989) Preliminary comments on federal regulations pertaining to insects as food. Food Insects Newsl 2:1–7

Brown WT, Doyen JT, Moore BP, Lawrence JF (1992) Chemical composition and taxonomic significance of defensive secretions of some Australian Tenebroidae (Coleoptera). Austr J Entomol 31:79–89. https://doi.org/10.1111/j.1440-6055.1992.tb00461.x

Chai JY, Shin EH, Lee SH, Rim HJ (2009) Foodborne intestinal flukes in Southeast Asia. Korean J Parasitol 47:S69–S102. https://doi.org/10.3347/kjp.2009.47.S.S69

Crespo R, Villaverde ML, Girotti JR, Gureci A, Juarez MP, De Bravo MG (2011) Cytotoxic and genotoxic effects of defense secretion of *Ulomoides dermestoides* on A549 cells. J Ethnopharmacol 136:204–209. https://doi.org/10.1016/j.jep.2011.04.056

DeFoliart GR (1992) Insects as human food. Gene DeFoliart discusses some nutritional and economic aspects. Crop Prot 11:395–399. https://doi.org/10.1016/0261-2194(92)90020-6

DeFoliart GR (1997) An overview of the role of edible insects in preserving biodiversity. Ecol Food Nutr 36:109–132. https://doi.org/10.1080/03670244.1997.9991510

Diener S, Zurbrügg C, Tockner K (2015) Bioaccumulation of heavy metals in the black soldier fly, Hermetia illucens and effects on its life cycle. J Insects Food Feed 1:261–270. https://doi.org/10.3920/JIFF2015.0030

Dubey LK, Moeller JB, Schlosser A, Sorensen GL, Holmskov U (2015) Chitin enhances serum IgE in *Aspergillus fumigatus* induced allergy in mice. Immunobiology 220:714–721. https://doi.org/10.1016/j.imbio.2015.01.002

EFSA (2012) The European Union summary report on trends and sources of zoonoses, zoonotic agents and foodborne outbreaks in 2010. EFSA J 10(3):2597. https://doi.org/10.2903/j.efsa.2012.2597

EFSA (2015) Scientific opinion of a risk profile related to production and consumption of insects as food and feed. EFSA J 13(10):4257. https://doi.org/10.2903/j.efsa.2015.4257

Eilenberg J, Vlak J, Nielsen-LeRoux C, Cappellozza S, Jensen AB (2015) Diseases in insects produced for food and feed. J Insects Food Feed 1:87–102. https://doi.org/10.3920/JIFF2014.0022

Ekop EA, Udoh AI, Akpan PE (2010) Proximate and antinutrient composition of four edible insects in Akwa Ibom State, Nigeria. World J Appl Sci Technol 2:224–231

EU (2004a) Regulation (EC) No 852/2004 of the European Parliament and of the Council of 29 April 2004 on the hygiene of foodstuffs. Off J Eur Union L139/1

EU (2004b) Regulation (EC) No 853/2004 of the European Parliament and of the Council of 29 April 2004 laying down specific hygiene rules for on the hygiene of foodstuffs. Off J Eur Union L139/55

EU (2015) Regulation (EU) 2015/2283 of the European Parliament and of the Council of EU 25 November 2015 on novel foods, amending Regulation (EU) No 1169/2011 of the European Parliament and of the Council and repealing Regulation (EC) No 258/97 of the European Parliament. Off J Eur Union L327/1

EU (2017) Commission Regulation (EU) 2017/893 of 24 May 2017 amending Annexes I and IV to Regulation (EC) No 999/2001 of the European Parliament and of the Council and Annexes X, XIV and XV to Commission Regulation (EU) No 142/2011 as regards the provisions on processed animal protein. Off J Eur Union L138/92

Food and Drug Administration (2010) Defect levels handbook. The food defect action levels: levels of natural or unavoidable defects in foods that present no health hazards for humans. Center for Food Safety and Applied Nutrition, US Food and Drug Administration, Washington, DC, USA. https://www.scribd.com/document/198781036/FDA-Handbook-on-Defect-Action

Gao Q, Wang XY, Wang WQ, Lei CL, Zhu F (2017) Influences of chromium and cadmium on the development of black soldier fly larvae. Environ Sci Pollut Res 24:8637–8644. https://doi.org/10.1007/s11356-017-8550-3

Giaccone V (2005) Hygiene and health features of minilivestock. In: Paoletti MG (ed) Ecological implications of minilivestock: potential of insects, rodents, frogs and snails. Science Publishers, Enfield, pp 579–598

Graczyk TK, Knight R, Tamang L (2005) Mechanical transmission of human protozoan parasites by insects. Clin Microbiol Rev 18:128–132. https://doi.org/10.1128/CMR.18.1.128-132.2005

Halloran A, Vantomme P (2013) The contribution of insects to food security, livelihoods and the environment. Food and Agriculture Organization of the United Nations, Rome, Italy. http://www.fao.org/docrep/018/i3264e/i3264e00.pdf

Handley MA, Hall C, Sanford E, Diaz E, Gonzalez-Mendez E, Drace K, Wilson R, Villalobos M, Croughan M (2007) Globalization, binational communities, and imported food risks: results of an outbreak investigation of lead poisoning in Monterey County, California. Am J Public Health 97:900–906. https://doi.org/10.2105/AJPH.2005.074138

Hazeleger WC, Bolder NM, Beumer RR, Jacobs-Reitsma WF (2008) Darkling beetles (Alphitobiusdiaperinus) and their larvae as potential vectors for the transfer of Campylobacter jejuni and Salmonella enterica serovar paratyphi B variant Java between successive broiler flocks. Appl Environ Microbiol 74:6887–6891. https://doi.org/10.1128/AEM.00451-08

Hogan GR, Razniak HG (1991) Selenium-induced mortality and tissue distribution studies in Tenebrio molitor (Coleoptera: Tenebrionidae). Environ Entomol 20:790–794. https://doi.org/10.1093/ee/20.3.790

Hornung B (1991) Die Bedeutung der Larven des Mehlkafers (Tenebriomolitor, L. 1758) als Ubertrager von Zearalenon in der Futterung von insektivorenVogeln und anderenHeimtieren, 81 pp

Jongema Y (2012) List of edible insect species of the world. Laboratory of Entomology, Wageningen University, Wageningen. www.ent.wur.nl/UK/Edible+insects/Worldwide+species+list/

Kinyuru JN, Konyole SO, Kenji GM, Onyango CA, Owino VO, Owuor BO, Estambale BB, Friis H, Roos N (2012) Identification of traditional foods with public health potential for complementary feeding in Western Kenya. J Food Res 1:148–158. https://doi.org/10.5539/jfr.v1n2p148

Klunder HC, Wolkers-Rooijackers J, Korpela JM, Nout MJR (2012) Microbiological aspects of processing and storage of edible insects. Food Control 26:628–631. https://doi.org/10.1016/j.foodcont.2012.02.013

Lalander C, Senecal J, Calvo MG, Ahrens L, Josefsson S, Wiberg K, Vinnerås B (2016) Fate of pharmaceuticals and pesticides in fly larvae composting. Sci Total Environ 565:279–286. https://doi.org/10.1016/j.scitotenv.2016.04.147

Lupi O (2006) Myiasis as a risk factor for prion diseases in humans. J Eur Acad Dermatol 20:1037–1045. https://doi.org/10.1111/j.1468-3083.2006.01595.x

Mitsuhashi J (2008) The future use of insects as human food. In: Durst PB, Johnson DV, Leslie RN, Shono K (eds) Forest insects as food: humans bite back. RAP Publication 2010/02. FAO, Chiang Mai, pp 115–122. http://www.fao.org/docrep/012/i1380e/i1380e00.pdf

Molavi GH, Massoud J, Gutierrez Y (2006) Human gongylonema infection in Iran. J Helminthol 80:425–428. https://doi.org/10.1017/JOH2006355

Mpuchane S, Taligoola HK, Gashe BA (1996) Fungi associates with Imbrasiabelina, an edible grasshopper. Botswana Notes Records 28:193–197

Muzzarelli RAA, Terbojevich M, Muzzarelli C, Miliani M, Francesangeli O (2001) Partial depolymerization of chitosan with the aid of papain. In: Muzzarelli RAA (ed) Chitin enzymology. European Chitin Society, Grottammare, pp 405–414

Nahrstedt A (1988) Cyanogenesis and the role of cyanogenic compounds in insects. Ciba Found Symp 140:131–150. https://doi.org/10.1002/9780470513712

Nishimune T, Watanabe Y, Okazaki H, Akai H (2000) Thiamin is decomposed due to Anaphe spp. entomophagy in seasonal ataxia patients in Nigeria. J Nutr 130:1625–1628. https://doi.org/10.1093/jn/130.6.1625

Pereira KS, Schmidt FL, Barbosa RL, Guaraldo AM, Franco RM, Dias VL, Passos LA (2010) Transmission of chagas disease (American trypanosomiasis) by food. Adv Food Nutr Res 59:63–85. https://doi.org/10.1016/S1043-4526(10)59003-X

Phillips JK, Burkholder WE (1995) Allergies related to food insect production and consumption. Food Insects Newsl 8(2):2–4

Pimentel D (1995) Amounts of pesticides reaching target pests: environmental impacts and ethics. J Agric Environ Ethic 8:17–29. https://doi.org/10.1007/BF02286399

Pitetti RD, Kuspis D, Krenzelok EP (1999) Caterpillars: an unusual source of ingestion. Pediatr Emerg Care 15(1):33–36

Post K, Riesner D, Walldorf V, Mehlhorn H (1999) Flylarvae and pupae as vectors for scrapie. Lancet 354:1969–1970. https://doi.org/10.1016/S0140-6736(99)00469-9

Purschke B, Scheibelberger R, Axmann S, Adler A, Jager H (2017) Impact of substrate contamination with mycotoxins, heavymetals and pesticides on growth performance and composition of black soldier fly larvae (Hermetiaillucens) for use in the feed and food value chain. Food Addit Contam Part A Chem Anal Control Expo Risk Assess 34:1410–1420. https://doi.org/10.1080/19440049.2017.1299946

Ramos-Elorduy J (2005) Insects: a hopeful food source. In: Paoletti MG (ed) Ecological implications of minilivestock; role of rodents, frogs, snails, and insects for sustainable development. Science Publishers, Enfield, pp 263–291

Reese G, Ayuso R, Lehrer SB (1999) Tropomyosin: an invertebrate pan-allergen. Int Arch Allergy Immun 119:247–258. https://doi.org/10.1159/000024201

Saeed T, Dagga FA, Saraf M (1993) Analysis of residual pesticides present in edible locusts captured in Kuwait. Arab Gulf J Sci Res 11:1–5

Schabel HG (2008) Forest insects as food: a global review. In: Durst PB, Johnson DV, Leslie RN, Shono K (eds) Forest insects as food: humans bite back. RAP Publication 2010/02. FAO, Chiang Mai, pp 19–22. http://www.fao.org/docrep/012/i1380e/i1380e00.pdf

Schlüter O, Rumpold B, Holzhauser T, Roth A, Vogel RF, Quasigroch W, Vogel S, Heinz V, Jäger H, Bandick N, Kulling S, Knorr D, Steinberg P, Engel K-H (2017) Safety aspects of the production of foods and food ingredients from insects. Mol Nutr Food Res 61(6):1600520. https://doi.org/10.1002/mnfr.201600520

Sehgal R, Bhatti HP, Bhasin DK, Sood AK, Nada R, Malla N, Singh K (2002) Intestinal myiasis due to Musca domestica: a report of two cases. Jpn J Infect Dis 55(6):191–193

Stoops J, Crauwels S, Waud M, Claes J, Lievens B, Van Campenhout L (2016) Microbial community assessment of mealworm larvae (Tenebrio molitor) and grasshoppers (Locustamigratoriamigratorioides) sold for human consumption. Food Microbiol 53:122–127. https://doi.org/10.1016/j.fm.2015.09.010

Suh SO, Nguyen NH, Blackwell M (2008) Yeasts isolated from plant-associated beetles and other insects: seven novel Candida species near Candida albicans. FEMS Yeast Res 8(1):88–102. https://doi.org/10.1111/j.1567-1364.2007.00320.x

Till JS, Majmudar BN (1981) Cantharidinpoisoning. South Med J 74:444–447

Tommaseo-Ponzetta M (2005) Insects: food for human evolution. In: Paoletti MG (ed) Ecological implications of mini livestock: potential of insects, rodents, frogs and snails. Science Publishers, Enfield, pp 141–161

Van der Fels-Klerx HJ, Camenzuli L, Van der Lee MK, Ooonincx DGAB (2016) Uptake of cadmium, lead and arsenic by Tenebrio molitorand Hermetiaillucensfrom contaminated substrates. PLoS One 11(11):e0166186. https://doi.org/10.1371/journal.pone.0166186

Van Huis A, Van Itterbeck J, Klunder H, Mertens E, Halloran A, Muir G, Van Tomme P (2013) Edible insects: future prospects for food and feed security. FAO Forestry, Paper 171. FAO, Rome. http://www.fao.org/docrep/018/i3253e/i3253e00.htm

Vega FE, Kaya HK (2012) Insect pathology. Academic, London, p 490

Vega FE, Meyling NV, Luangsa-Ard JJ, Blackwell M (2012) Fungal entomopathogens. In: Vega FF, Kaya HK (eds) Insect pathology, 2nd edn. Academic, San Diego, pp 171–220

Vijver M, Jager T, Posthuma L, Pijnenburg W (2003) Metal uptake from soils and soil–sediment mixtures by larvae of Tenebrio molitor L. (Coleoptera). Ecotox Environ Safety 54:277–289. https://doi.org/10.1016/S0147-6513(02)00027-1

Wasala L, Talley JL, Desilva U, Fletcher J, Wayadande A (2013) Transfer of Escherichia coli O157:H7 to spinach by house flies, Musca domestica (Diptera: Muscidae). Phytopathology 103:373–380. https://doi.org/10.1094/PHYTO-09-12-0217-FI

WHO (2010) Chagas disease (American trypanosomiasis). Fact sheet nr. 340. http://www.who.int/mediacentre/factsheets/fs340/en/

Wynants E, Frooninckx L, Van Miert S, Geeraerd A, Claes J, Van Campenhout L (2019) Risksrelatedto the presence of Salmonella sp. during rearing of mealworms (Tenebriomolitor) for foodorfeed: survival in the feeding substrate and transmission to the larvae. Food Control 100:227–234. https://doi.org/10.1016/j.foodcont.2019.01.026

Yun JH, Roh SW, Whon TW, Jung MJ, Kim MS, Park DS, Yoon C, Nam YD, Kim YJ, Choi JH, Kim JY, Shin NR, Kim SH, Lee WJ, Jin-Woo Bae JW (2014) Insect gut bacterial diversity determined by environmental habitat, diet, developmental stage, and phylogeny of host. Appl Environ Microbiol 80:5254–5264. https://doi.org/10.1128/AEM.01226-14

Chapter 6
Interdisciplinary Uses of Some Edible Species

Beatrice Mofoluwaso Fasogbon

Abstract As the global population has been predicted to reach 9 billion by 2050, there is a need to meet the growing demand for food. The current food production needs to be doubled in size by finding more sustainable food production methods and other sources of nutrition. Edible insects are been investigated to offer a potential solution to this problem. The consumption of insects is now of public interest mainly because of its availability and cheap sources of nutrition. With further work on research, edible insects might become an important and alternative source of food, oil, protein, and bioactive components for human consumption and industrial utilization in the nearest future.

Keywords Interdisciplinary · Edible insects · Nutritional composition

6.1 Introduction

The consumption rate of animal products is expected to increase by approximately 60–70% in the year 2050, due to the shifts in dietary habits, not leaving the developing countries behind (Ayieko et al. 2012; Makkar et al. 2014). The Food and Agriculture Organization of the United Nations (UN) emphasized this in their report (OECD-FAO 2018) that the global production of meat will increases by 15% in the year 2027 and developing countries are projected to account for the massive bulk for the total increase. This may be because of the population growth, and especially in the developing countries. This increase, along with the limited available land area, may pose a challenge for the meat industry worldwide, and alternative protein sources may be needed to feed the growing population (van Huis 2016).

Edible insects like other animal-source food (meat, fish, eggs, milk, etc.) can contribute essentially to the nutritional diet of humans. They have been considered

B. M. Fasogbon (✉)
Department of Food Technology, Federal Institute of Industrial Research, Lagos, Nigeria

© Springer Nature Switzerland AG 2020
A. Adam Mariod (ed.), *African Edible Insects As Alternative Source of Food, Oil, Protein and Bioactive Components*, https://doi.org/10.1007/978-3-030-32952-5_6

to give satisfactorily energy, good protein, amino acid, fatty acid, and micronutrients (minerals, vitamins, and in some cases folic acid) content (Capinera 2004; Johnson 2010; Rumpold and Schlüter 2013a). More than 1900 of the edible insect species have been documented in the literature, and common to the tropical countries. Among the commonly eaten ones are the beetles, caterpillars, ants, bees, grasshoppers, locusts, crickets, true bugs, termites, and flies (Pal and Roy 2014). They have been recognized worldwide as a nutritious food which could improve the nutrition status (Nowak et al. 2016), and potential contributors to food security. Edible insects have been recognized to complementing efforts to feed the increasing global population (FAO 2012; Van Huis et al. 2012).

6.2   African Edible Insects

Insects can be found in a great quantity all over the African continent. During staple food scarcity, insects become important sources of food, and they therefore are essential in the area of food security. Edible insects are consumed in most communities in sub-Saharan Africa. A large proportion of these insects are captured in the wild (Rumpold and Schlüter 2013a) using trap technology such as nets, water-filled bowls, and lights, while many are domesticated including bee honey and silkworms. (Raubenheimer and Rothman 2013). Some of those insects widely consumed and marketed in Africa, including caterpillar, cricket, grasshopper, and termites. The fat content of these insects usually ranges between 10 and 30% fat based on their fresh weight (Durst et al. 2010) with the highest content found in the larval and pupal stages than at the adult stage (Chen et al. 2009). Average protein contents of edible insects amount up to 50–82% (dry weight) (Schabel 2010).

Caterpillars (*Cirina forda*) are particularly prevalent during the rainy season but vary within the country (Vantomme, Göhler, and N'Deckere-Ziangba 2004). They have been recommended as a major edible insect for maximum meat and/or protein yield since they are coldblooded, wingless, and transform plant biomass to animal biomass 10 times more efficient than cattle and on much less land (Schabel 2010). They sell for about twice the price of beef in Nigeria (DeFoliart 1999). Termites are rich in fat and proteins, and are the second most eaten insect all over the world (Chung 2010). The regularly consumed termite species are the large *Macrotermes* species, the winged termites are usually harvested when they emerge from the holes near termite nests after the first rains fall at the end of the dry season (Anankware et al. 2016). Crickets contain little protein but high in fat content Agbidye et al. (2009). Grasshoppers (Acridids) have higher protein and micronutrients than other protein sources like fish and soybean meal usually used as a feed for poultry (Anand et al. 2008). Literature reported that the protein of the house cricket (*Acheta domesticus*) was higher to soy protein (Finke et al. 1989) and the removal of the chitin improves the protein quality a little further. The most popular edible beetle palm weevil (*Rhynchophorus* spp.), locust, and pentatomid bugs are eaten widely throughout sub-Saharan Africa. All these properties of edible insect commonly consumed in Africa may be connected to the report of Van Huis (2003) where it was reported that edible insects in substantial amount could alleviate hunger. Smith and Pryor (2013) and van Huis et al. (2013) submitted that greater micronutrient contents can be provided by consuming an entire insect than only the insect parts.

In Africa, researchers reported that ants, termites, beetle grubs, caterpillars and many species of moths and few species of butterflies and grasshoppers are widely popular among the inhabitants, and are considered an important source of nutrition (Srivastava et al. 2009). The level of nutrition varies across and within the insect species based on the insect feeds, their stage of development, sex and environmental factors (Ramos-Elorduy et al. 2002; Finke and Oonincx 2014). In the study of Belluco et al. (2013), juvenile rats demonstrated that crickets (*Acheta domesticus*) offer a superior source of protein when compared to soy protein from the plant.

Several products derived from edible insects are used as food ingredients (food coloring) or consumed for medicinal purposes (pharmaceuticals) (Schabel 2010). Shea tree caterpillar (*Cirina butyrospermi*) is the most commonly consumed insect in Burkina Faso (Anvo et al. 2016). In Kenya, termites and lake flies are processed (baked, boiled, and cooked) into conventional products such as crackers, muffins, sausages, and meatloaf with an increased shelf life in order to encourage entomophagy (Ayieko et al. 2010). The history of trado-medicine in Africa has successfully revealed the use of insects in the treatment of various diseases in human beings and animals, including diseases such as fever, scabies, epilepsy, migraine, and hemorrhage. Insects are also used to increase milk flow in lactating women (Tango 1994).

A huge variety of insect species including those in Africa have been proven to have remarkable commercial and pharmaceutical values. Products like honey and silk from bees and silkworm are been sold in local and international markets (FAO 2009), likewise carmine, a red dye from scale insects (Hemiptera) used in foods, textiles, and pharmaceuticals. Many researchers have exploited the use of edible insect to produce antibacterial and anticancer drugs. Pierisin and Cecropin, for example, have been reported to exhibits cytotoxic effects against human gastric cancer and mammalian lymphoma and leukemia cells respectively (Srivastava and Gupta 2009). Resilin, on the other hand, a rubber-like protein that allows insects to jump, and Elvin et al. (2005) reported that it has been used in medicine to repair arteries.

6.3   Nutritional Composition of African Edible Insects

Scientific research has described the nutritional quality of insects as high enough to meet human requirements (Wang et al. 2006; Chen et al. 2008; Zhou et al. 2009; Li et al. 2010), therefore, insects are potentially good sources of protein for human food (DeFoliart 1997) and animal feed. Rumpold and Schlüter (2013a) however reported that the nutritional composition of edible insects varies between and within

Table 6.1 Average content of protein, fat, and energy of specific insect orders

Insect orders	Protein (% dry matter)	Fat (% dry matter)	Fiber (%)	Energy (kcal/100 g)
Coleoptera (adult beetles, larva)	40.69	33.40	10.74	490.30
Hemiptera (true bugs)	48.33	30.26	12.40	478.99
Hymenoptera (ants, bees)	46.47	25.09	5.71	484.45
Isoptera (termites)	35.34	32.74	5.06	–
Lepidoptera (butterflies, months)	45.38	27.66	6.60	508.89
Odonata (dragonflies, damselflies)	55.23	19.83	11.79	431.33
Orthoptera (crickets, grasshoppers, locusts)	61.32	13.41	9.55	426.25

Source of data: Rumpold and Schlüter (2013b)

species based on the metamorphic stage, habitat and diet of the insects. Table 6.1 shows the average content of protein, fat and energy of specific insect orders.

Generally, insects are rich in fat most especially termites, caterpillars, and palm weevil larvae, According to Chakravarthy et al. (2016), edible insects are an excellent source of protein, vitamins, and minerals and tend to be low in carbohydrates. Insect diet has outstanding advantages, with many species that provide adequate calorie, protein, and amino acid content, high monounsaturated and/or polyunsaturated fats, and rich in micronutrients such as copper, zinc, iron, magnesium, as well as folic acid, riboflavin, pantothenic acid, and biotin. Edible insects can meet up the required nutrients and dietary energy needed for the human body. In the report of Santos Oliveira et al. (1976), 100 g of caterpillars provides the daily individual's requirement of 76% and over 100% vitamins and minerals. Scientific research has also shown that some insects and their extracts contain anti-fatigue and antioxidant compounds, and they also function in immune regulation (Liu and Wei 2002).

A typical summary of the nutrient and fatty acid composition, vitamin and mineral contents of five insect species common in Africa is extracted from the report of Rumpold and Schlüter (2013b) as shown in Table 6.2. One hundred gram of insect has similar energy content when compared to 100 g of fresh weight meat (Sirimungkararat et al. 2010), except for pork which may be because of its extreme fat content. Agbidye et al. (2009) stated that consuming 100 g of caterpillars, for example, can provide about 76% of the daily amount of protein and almost 100% of vitamins required for humans. Eating silkworm, caterpillars have been reported by Gordon (1998) to be sufficient for daily requirements of copper, zinc, iron, thiamine, and riboflavin.

Most insect species are similar to fish in high fatty acids; they are therefore important food supplement for undernourished children (Chakravarthy et al. 2016). The unsaturated omega-3 and omega-6 fatty acid compositions in mealworms are also comparable with that in fish (FAO 2013).

Table 6.2 Nutrient, fatty acid, vitamin, and mineral composition of some edible insect species common in Africa

Composition (dry matter)	*Rhynchophorus phoenicis* African palm weevil (larvae)	*Cirina forda* Westwood Moth (larvae)	*Acheta domesticus* House cricket (adults)
Nutrient composition			
Protein (%)	10.33–41.69	20.20–74.35	64.38–70.75
Fat (%)	19.50–69.78	5.25–14.30	18.55–22.80
Fiber (%)	2.82–25.14	1.80–9.40	NR
NFE (%)	5.49–48.60	2.36–66.60	2.60
Ash (%)	2.54–5.70	1.50–11.51	3.57–5.10
Energy (kJ/kg)	20,038–20,060	15,030.61	19,057.89
(%) Fatty acids			
Palmitic acid (C16:0)	32.40–36.00	13.00	NR
Stearic acid (C18:0)	0.30–3.10	16.00	NR
Palmitoleic acid (C16:1n7)	3.30–36.00	0.20	NR
Oleic acid (18:1n9)	30.00–41.50	13.90	NR
Linoleic acid (18:2n6)	13.00–26.00	8.10	NR
Linoleic acid (18:3n3/6)	2.00–3.50	45.50	NR
Minerals (mg/100 g)			
Calcium	54.10–208.00	7.00–37.20	132.14–210.00
Potassium	1025.00–2206.00	47.60–2130.00	1126.62
Magnesium	33.60–131.80	1.87–69.89	80.00–109.42
Phosphorus	352.00–685.00	45.90–1090.00	780.00–957.79
Sodium	44.80–52.00	44.40–210.00	435.06
Iron	14.70–30.80	1.30–64.00	6.27–11.23
Zinc	26.50–15.80	4.27–24.20	18.64–21.79
Manganese	0.80–3.50	7.00–10,163.10	2.97–3.73
Copper	1.60	NR	0.85–2.01
Selenium	1.60	NR	0.60
Vitamins			
Retinol (µg/100 g)	11.25	2.99	24.33
A-Tocopherol [IU/kg]	NR	NR	63.96–81.00
Ascorbic acid (mg/100 g)	4.25	1.96	9.74
Thiamin (mg/100 g)	3.38	NR	0.13
Riboflavin (mg/100 g)	2.21–2.51	2.21	11.07
Niacin (mg/100 g)	3.36	NR	12.59
Pantothenic acid (mg/100 g)	NR	NR	7.47
Biotic (µg/100 g)	NR	NR	55.19
Folic acid (mg/100 g)	NR	NR	0.49

NFE nitrogen n-free extract (i.e., Carbohydrate), *NR* not reported. *Sources*: Rumpold and Schlüter (2013b)

Fig. 6.1 (**a**) Palm weevil delicacy (Source: @tenazfood_n_grillz). (**b**) Stick palm weevil hawked on Nigeria street (Source: @kudoshealthyliving)

6.4 Uses of Edible Insects

It is no longer news that the global population will reach 9 billion by 2050, it is, therefore, essential to meet the growing demand for food by doubling the current food production, and find more sustainable methods and sources of nutrition. This has required deep research on how edible insects could offer a potential solution to the problem. Insects can be processed into different forms for food, feed, oil, protein, and other healthy products as described in Fig. 6.1.

6.4.1 Edible Insects as Foods

Insect-based foods have been reported to be a healthy choice for human consumption due to the insect's excellent source of protein, fat, energy, and fiber, which varies based on the species (Belluco et al. 2013; Rumpold and Schlüter 2013b). In the study of Kipkoech et al. (2017), nutrients available in farmed crickets are an indication that the farmed crickets can be used in child food ingredients to improve child nutrition. Tao and Li (2018) also reported that the inclusion of cricket and locust flour in an extruded rice product has an acceptable shelf-life and sensory characteristics.

In the developed countries, edible insects have incorporated even in some simple food product and people are getting used to its consumption. This includes the use of cricket flour in the production of cookies, chips, and pasta from Thailand, mealworm inclusion in the production of insect burger from Belgium and Netherlands, cricket bread, termite sauce, and so on (van Huis 2016).

In Africa, some insects such as termites are commonly eaten raw, baked or fried. Edible insects commonly eaten in Angola are *Macrotermes subhyalinus*, *Rhynchophorus phoenicis*, and *Usta terpsichore*. Over 60 of edible insects have been reported to be food in Zambia, including honeybees, a saturniid caterpillar is known locally as mumpa (DeFoliart 1999). In sub-Saharan Africa edible insects have been used by humans, they are been consumed directly as food or used as food ingredients. Insects are sometimes served alongside staple foods to serve as the main source of protein in the meal. This is common in West Africa, where insects such as shea caterpillar (*Cirina* spp.) are served in a tomato sauce with rice or maize meal, or as a sandwich filling (Payne and Itterbeeck 2018).

In Zimbabwe, the consumption of insect has been in existence for a long time. According to the report of Chavunduka (1975), winged insects and giant crickets (*Brachytrupes membranaceus*) are frequently consumed raw, grilled, or fried without additional fat. Caterpillars, termites, locusts, honeybees, and ants are among the favorites and largely consumed insects in the country. In recent reports of Musundire et al. (2016a), these insects are now consumed majorly as a supplement for the nutrient provided when eaten with cereals, nuts, and vegetables. The larvae of moths, wasps, butterflies, beetles, adult grasshoppers, moths, crickets, and so on are popularly eaten boiled or roasted over an open fire in Papua New Guinea. In the Southern part of Africa, mopane worms (*Imbrasia belina*), locusts, bugs, termites, honeybees, and crickets are the most widely consumed insects (Halloran and Vantomme 2013).

Kelemu et al. (2015) also reported the long history of Kenyans consuming insect (especially cricket species variants) as food, with high production systems of an insect for food and feed. Several food products have been formulated and made from common edible insects such as termites, lake flies, and cricket in Kenya. These products including biscuits, crackers, bread made with crickets, have been formulated to school feeding programs, and also for who are not familiar with eating insect (Münke-Svendsen et al. 2016), and they were well received by the consumers. *Serene*, a widely harvested and consumed insect species in Zambia and other regions in Eastern Africa are consumed as a traditional snack (Mmari et al. 2017). In Kenya, termite-based complementary foods were developed to combat child malnutrition, which is affordable, safe and with adequate nutrient density (Kinyuru et al. 2015). In several western states of Nigeria including Delta and Edo States, palm weevil (*Rhynchophorus phoenicis* Fabr.) is eaten and marketed in Nigerian (Fig. 6.1a, b). Termites (*Macrotermes bellicosus*) are also commonly eaten in several parts of western Nigeria (Banjo et al. 2006) in a popular form shown in Fig. 6.2. The grubs of *R. phoenicis* and the coconut rhinoceros beetles (Scarabaeidae) are a common food, usually soaked and cooked before eaten in the Democratic Republic of Congo (DRC) (Onyeike et al. 2005) and Nigeria.

Edible insects such as termite are surplus providing essential nutrient for human and animal, but they are limited by their seasonal availability and are highly perishable. Processes such as boiling, steaming, frying, baking, and drying are traditional preparation methods of edible insects (Rumpold et al. 2014). Processes such as conventional cooking, drying, grinding, and mixing with other ingredients could extend

Fig. 6.2 Termite (Source: @pulsenigeriafood)

their shelf life of the insects considerably and contribute to promoting entomophagy throughout Africa. Edible insects could be milled into powdery or paste forms which are easily incorporated into other foods to make them acceptable nutritionally and appealing in sensory.

Milled flies are been formed into the cake after sun-drying (van Huis 2003) for consumption. Consumers that are not familiar with eating whole insects could find the ground or paste form acceptable. Women from a part of South Africa roast, sundry, and mill caterpillars for future use, which they mix with stewed watermelon when needed (Nonaka 1996). Termites have been processed into ingredients form for fortifying cereal porridge for children (Kinyuru et al. 2013; FAO 2013), wheat buns (Kinyuru et al. 2009), termite and lakefly-based crackers, muffins, meatloaf, and sausages for commercialization (Ayieko et al. 2010). In a recent report by Adeboye et al. (2019), termite powder was used in a nontraditional form to enrich a ready-to-cook vegetable soup powder.

A number of studies have evaluated and reported the effect of processing on the nutritional properties of edible insects in Africa. Degutting according to the report of Madibela et al. (2009) increased the level of crude protein of mopane caterpillars with low levels of ash and condensed tannins. Boiling decreased the crude protein and ash content of mopane caterpillar (Madibela et al. 2007). The energy and protein level of *Hemijana variegata* caterpillars in South Africa is lowed due to the boiling process (Egan et al. 2014). Frying increased the levels of Potassium in Sudanese tree locusts and reduced that of Phosphorus (El Hassan et al. 2008). Frying also increased the lipid content of *Rhynchophorus phoenicis* and *Oryctes monoceros* (Edijala et al. 2009). The toasting of edible termites and grasshoppers has been found to significantly reduce their vitamin contents (Edijala et al. 2009). Madibela et al. (2007) stated that roasting increased the contents of minerals and crude protein of the mopane caterpillars.

Oven-drying has been reported to increase the concentration of minerals and essential and nonessential amino acids in *Sternocera orissa* (Shadung et al. 2012). It has been found that prolonged oven-drying decreased the energy value of H. variegata caterpillars by 9% with no effect on the proximate composition (Egan et al. 2014). Aniebo and Owen (2010) found out that oven-dried maggots contained higher protein and fat content than sun-dried ones, while solar-drying of dried grasshoppers and termites resulted in significant loss of riboflavin, folic acid, niacin, pyridoxine, retinol, ascorbic acid, and α-tocopherol (Kinyuru et al. 2010). Thermal processing of the edible insect can lead to the loss of water-soluble vitamins as well as other beneficial bioactive compounds through leaching in boiling water (Gokoglu et al. 2004; Musundire et al. 2014a; Musundire et al. 2016b) or increase in fat content due to absorption when frying the insects (Gokoglu et al. 2004).

Several products derived from insects are used as food ingredients (bee honey, food coloring) because of its nutritional and functional benefits; or used for medicinal purposes (e.g., bee venom) (Schabel 2010). Because edible insects are also promoted for their protein content, there are reports on the effects of some processing methods on important functional properties (water absorption) of the edible insect flours. Boiled and sundried *C. forda* larvae possessed good water absorption capacity (Osasona and Olaofe 2010). This is similar to the observation of Omotoso (2006), who reported that boiling, oven-drying, and milling improved the water absorption capacity of *C. forda*. Roasting and grilling has been reported by Womeni et al. (2012) to decrease the water absorption capacity of *R. phoenicis* while boiling and smoking did not affect this property Roasting and grilling also enhanced the aggregation of proteins, which makes the insect products unsuitable for formulations requiring high viscosity (Womeni et al. 2012). The oil absorption capacity of a food product is a desirable property in flavor retention and for improved palatability or mouthfeel (Kinsella 1976). El Hassan et al. (2008) reported that Sudanese tree locust flour dried and milled showed higher oil absorption capacity than those boiled and dried. This is related to the report of Omotoso (2006) who also observed high values of oil absorption capacity for freshly harvested, boiled, oven-dried, and milled *C. forda*. The emulsifying activity, emulsion capacity and foaming capacity of flours processed from boiled and sun-dried *C. forda* larvae are good, but with a poor foaming ability which was due to poor dispensability (Osasona and Olaofe 2010).

Termites are low in anti-nutrients (Adepoju and Omotayo 2014) suggesting their use in formulating nutrient-dense complementary foods with nutraceutical benefits. Important nutrients have been reported to increase significantly in maize with the addition of termites (Banjo et al. 2006). The addition of the insect (whole or granular) to traditionally eaten foods is a great way to consume insect and introduce it into a daily diet without having an effect on eating habits. The powder form of insects can be perfectly used in food industry for all kinds of products, and they will still maintain their textural or functional properties.

Research about insects as food has developed promptly over the past decade, including that of animal feed. Insects have gained usefulness as ingredients in the production of animal feeds for conventional livestock (e.g., poultry and swine), fish, and pets. Hwangbo et al. (2009) reported that maggot meal could replace fish meal in the production of broiler chickens. Insects have therefore been researched as innovative ingredients for animal feeds (Huis et al. 2013). The study of Pambo et al. (2016) showed a high acceptance of edible insects for food and as an alternative to conventional meat among consumers. Yen (2010) also noted that using insects as human food can cause a more energy-efficient food production and ease environmental conservation.

6.4.2   Edible Insects as a Protein

The relative increase in the demand for meat production and other sources of protein is pronounced in the developing country, and it is expected to increase as the population increases in the nearest future. It is of prominent importance to find an alternative source for protein. Literature reports showed that proteins can be extracted from the insects (as shown in Fig. 6.3) and applied in food and/or feed as an alternative to soy or meat, in order to increase the protein and/or functional components, or reduce anti-nutritional components of the food.

Protein, a major component of edible insects comprises about 30–65% of the total dry matter. Banjo et al. (2006) reported that some insects (*Analeptestri fasciata, Rhynchophorus phoenicis*, and *Zonocerus variegatus*) found in south-western Nigeria have crude protein content ranging between 26.80% and 29.62%. The evaluation of Xiaoming et al. (2010) showed that the protein content of 100 species of insects ranged between 13% and 77% of dry matter. Variation in the percentage amount of protein is dependent on the feed (vegetables, grains, or waste) and the metamorphosis stage (Ademolu et al. 2010) of the insects. Grasshoppers fed with bran would contain an extraordinary amount of essential fatty acids than those fed on maize. Insect protein has high nutritional benefits, which qualifies it as a meat replacer (Rumpold and Schlüter 2013a).

Amino acids are the building blocks required for the biosynthesis of all proteins through human metabolism (Van Huis et al. 2013). Insects contain a number of essential amino acids (Bukkens 2005) including high phenylalanine and tyrosine. Some of these insects are significantly high in lysine, tryptophan, and threonine, which is deficient in certain cereal proteins. Lysine-rich caterpillars complement lysine-poor staple protein food in the Democratic Republic of the Congo; also eating palm weevil larvae in Papua New Guinea balances the lysine and leucine that is limiting in tubers (rich in tryptophan) (Bukkens 2005). In some African countries like Kenya, Nigeria, and Angola where maize is a staple food, supplementing diets with insects like termite are now been accepted as part of the diets. Rumpold and Schlüter (2013a) has extracted the amino acid spectra of some selected insect order and compared it to the amino acid required for human nutrition as stated by the World Health Organisation. The author reported that the selected insects (except the rich in fat palm weevil larvae) show a high-quality amino acid profile with high contents of phenylalanine and tyrosine and that they generally meet the requirements except for the amino acid methionine.

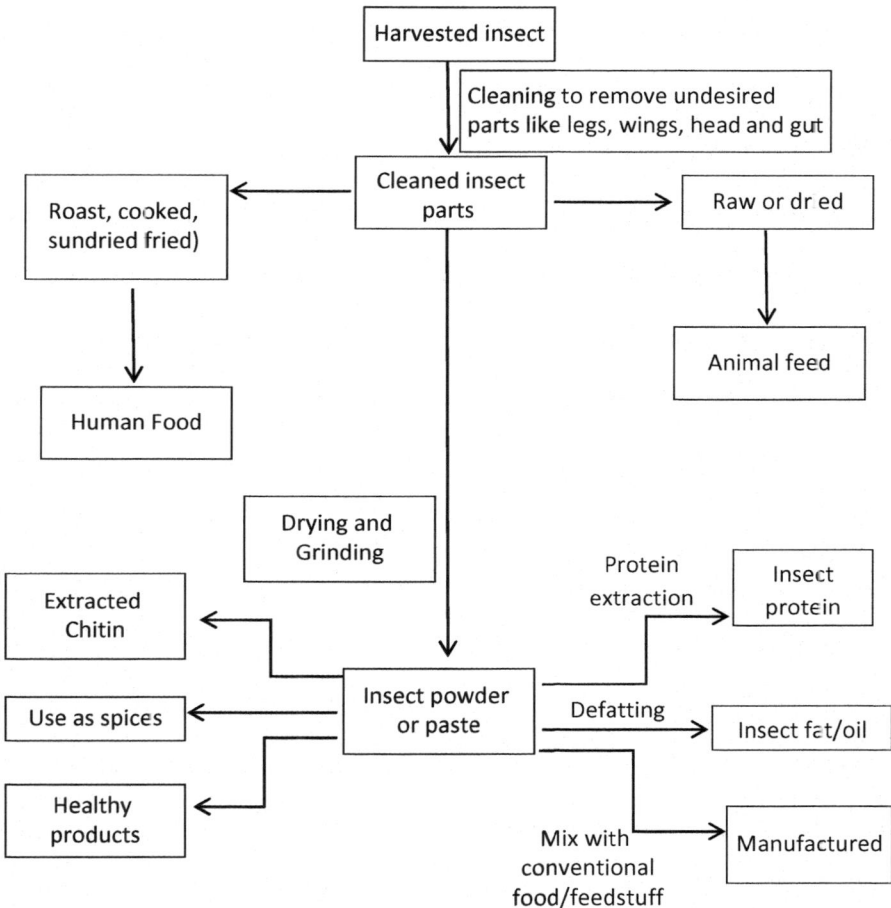

Fig. 6.3 Extraction of protein from edible insects

Extracting insect proteins for human food products could be a useful way of increasing acceptability among cautious consumers (Halloran et al. 2016), and may find usefulness in the diet of individuals with increased protein requirements like sportsmen (Zielińska et al. 2019). The proper extraction and use of the extracted insect proteins can make food with desired properties, without modifying the sensory of products well known to the consumers. Supplementing food products with insect protein is a way of increasing the knowledge of the functional properties, amino acid profile, thermal stability, and so on of the extracted proteins and how these properties vary across the insect species (Williams et al. 2016). Studies have shown that that the cricket's protein (Acheta domesticus) fed to juvenile rats are higher when compared to soy protein (Belluco et al. 2013), but recommendations from human trials have not been ascertained (Williams et al. 2016).

6.4.3   Edible Insects as Oil

Fat is another important component of insects following protein, and like protein, the fat profiles of insects highly rely on their feed (Bukkens 2005). Fat content among insects ranged from 7 to 77 g/100 g (dry weight), with a higher concentration in the larvae than the adults (Ramos-Elorduy et al. 1997). The fat content of wasp larvae has been reported to be 29.01% (Deng et al. 2013), while its pupae and adults are 27.25% and 17.22% respectively. Termites among other soft-bodied adult insects and larvae have the highest levels of fat than those with hard exoskeleton like crickets and grasshoppers (Bukkens 1997). Termite (*Macrotermes bellicosus*) and the caterpillar (*Imbrasia belina*) commonly consumed in Nigeria contain an average of 3.6% cholesterol (Ekpo et al. 2009). Fats contained in the insect can contribute to human nutrition through the provision of energy and essential fatty acids (Ramos-Elorduy 2008).

   In the fatty acid spectra of edible insects, the ratios of saturated fatty acids (SFA) to unsaturated fatty acids range from 0.4 to 0.65 (Rumpold and Schlüter 2013a). The unsaturated fatty acid profile of insects is considerably similar to that of poultry and white fish but significantly higher in polyunsaturated fatty acids (PUFAs) than either poultry or red meat (Rumpold and Schlüter 2013a). The polyunsaturated/saturated fatty acid ratios of most of the insects are greater than 0.20, which is related with low levels of cholesterol thus less risk of some coronary heart diseases (Kinyuru et al. 2013). Insects contain a minute amount of Omega-3fatty acids, eicosapentaenoic acid and docosahexaenoic acid (DHA) but do contain linoleic acid and linolenic acid (Rumpold and Schlüter 2013a; Calder 2017) which are nutritionally important for the healthy development of children and infants (Michaelsen et al. 2009). Palmitic acid contents in insects are comparatively high, and termites are rich in palmitic acid (30.47%) (Womeni et al. 2009). Womeni et al. (2009) also confirm the presence of the essential linoleic and α-linolenic acids in the oils extracted from several insects consumed in Cameroon.

   The presence of unsaturated fatty acids in the processed insect foods will cause the product to go rancid quickly; the fats can, therefore, be extracted making it useful for several other things. Meanwhile, the insect oils have low amounts of polyunsaturated fatty acids (linoleic and linolenic acid) giving them high oxidative stability, which is of high importance than the antioxidants the oil contains. Traditionally, fats (e.g., oil) extracted from some insect species are used extensively for frying meat and other food products. Oil from Pentatomid insects is used in preparing food (van Huis et al. 2013). The oil extracted from *R. phoenicis* grubs contains a high level of unsaturated components hence its use as edible oil (Okaraonye and Ikewuchi 2008). Mariod et al. (2004) reported the use of melon bug (*Aspongopus viduatus F*) and sorghum bug (*Agonoscelis pubescens*) oil used as a famine food and for traditional medicine in the western parts of Sudan. Mustafa et al. (2008) report the high antibacterial activity possessed by the insects, making their oil a useful preservative in meat and meat products to controlling gram-positive bacteria. Mariod et al. (2005) demonstrated improving the oxidative stability of sunflower kernel oil was blending it with edible oil of melon/sorghum bug. Mariod (2013) likewise explored the industrial uses of insect (melon and sorghum bugs) oil and confirmed that the resultant fatty acid esters of the oils are comparable with the DIN 51606 specifications for biodiesel.

Chitin, a structural nitrogen-based carbohydrate found in the exoskeleton of insects is considered an anti-nutrient due to its negative effects on protein digestibility (Belluco et al. 2013). Alternatively, chitosan a material derived from chitin has found usefulness in the production of polymer for food packaging (FAO 2013). Chitin and its derivative chitosan have properties that could improve the immune response (resistance against pathogenic bacteria and viruses) of specific groups of people, and an indication that chitin could reduce allergic reactions to certain individuals (Goodman 1989; Muzzarelli 2010).

6.4.4 Edible Insects as Bioactive Components

Bioactive compounds are nutritional components that occur in foods in small quantities. Insects can provide bioactive compounds that have health benefits beyond simple nutritional values, and reduce health risks and strengthen the immune system (Roos and van Huis 2017). The nutraceutical benefit of insects varies as a result of the wide range of edible species, and the differences in their metamorphic stage, habitat, and diet. Several studies have reported antioxidant activity in insect species (Dutta et al. 2016; Zielińska et al. 2017), and its potential ability to prevent molecular damage in the human body. However, there has been slow progress in developing insect products solely for their bioactive activity, which may be due to the difficulties in identifying the insect species, drug toxicity, or development and production costs (Ratcliffe et al. 2011; Ratcliffe et al. 2014).

Antimicrobial peptides (AMPs) have been produced from insects to protect against microbial infections and environmental changes (Soares and Mello 2004). Yen (2015) detected the presence of alkaloids, flavonoids, anthraquinones, tannins, steroids, triterpenoids, and cyanogenic glycosides inedible stink bug (*Encosternum delegorguei*). In addition to the good quality of silk produced from silkwork, the insect has also been exploited for its different bioactive properties (Islam et al. 2016).

Bioactive substances extracted from insects have been used in alternative medicine, and research shows that insect constitutes an almost inexhaustible source of the active ingredients needed. Potentially bioactive peptides derived from insect proteins and phenolic compounds have had interesting activities (even in small animals) as antifungal, antibacterial, antidiabetic, and antioxidants (Zielińska et al. 2019). The study of Nongonierma and FitzGerald (2017) has shown that the bioactive strength of edible insect protein hydrolysates or peptides is comparable or higher than that of other dietary proteins (plants and animals).

Species of insects such as wax moth (*Galleria mellonella*), the yellow mealworm (*Tenebrio molitor*) and the silkworm (*Bombyx mori*) have been found to contain active levels of ACE inhibitory that are comparable with other food sources (Cito et al. 2017), this means consuming these insects can prevent the risk of cardiovascular diseases. Ryu et al. (2002) confirmed this by reporting the blood-glucose-lowering effects of silkworm powder in mice. Seo et al. (2017) also studied some obese mice and concluded that the daily intake of yellow mealworm larvae powder has a potential of inducing weight loss due to the effect of the bioactive compounds contained in them. Consumption of termites was believed to improve fertility (Kelemu et al. 2015). Edible stink bug *E. delegorguei* in South Africa and Zimbabwe had the potential of curing asthma and enhancing sexual desires (Teffo 2006; Musundire et al. 2014b), as they contain alkaloids, flavonoids, anthraquinones, tannins, steroids, triterpenoids, and cyanogenic glycosides which varies with seasons and processing conditions.

Studies on insect-to-insect bioactivity effects in human are researched; however, Nguyen et al. (2016) reported the reduction in the symptoms of rotenone-induced Parkinson's disease found in *Drosophila* flies feed with boiled and freeze-dried powder of the silkworm (*B. mori*). Some Nigerians consume crickets (*B. membranaceus*) as food to aid mental development and for pre/postnatal care (Banjo et al. 2003). ACE inhibitory peptides present naturally in insects make it a valuable alternative to synthetic drugs against hypertensive treatment. Studies show that the hydrolysates of silkworm protein contain ACE activity inhibitors (Wang et al. 2008), making it a potential source of ACE inhibitor drugs.

6.5   Conclusion

Edible insects are excellent sources of nutritional and functional properties, with novel bioactive compounds which are presently explored to address the enormous global health challenges in Africa among other developing countries. Literature has shown the usefulness of edible insects in Africa as alternative sources of food, oil, protein, and bioactive components, which further need strengthening. Indication of edible insects health benefit on human requires further research to ascertain the entire traditional claim.

References

Adeboye O, Fasogbon BM, Adegbuyi KO (2019) Formulation of a nutritive vegetable soup powder enriched with termite. In: Abstract proceedings of the 5th regional food science and technology summit, NSPRI, Ilorin, pp 430

Ademolu KO, Idowu AB, Olatunde GO (2010) Nutritional value assessment of variegated grasshopper, *Zonocerus variegatus* (L.) (Acridoidea: Pygomorphidae), during post-embryonic development. Afr Ent 18:360–364

Adepoju OT, Omotayo OA (2014) Nutrient Composition and Potential Contribution of Winged Termites (Marcrotermes bellicosus Smeathman) to Micronutrient Intake of Consumers in Nigeria. British J Appl Sci Technol 4(7):1149–1158

Agbidye FS, Ofuya TI, Akindele SO (2009) Marketability and nutritional qualities of some edible forest insects in Benue State, Nigeria. Pak J Nutr 8:917–922

Anand H, Ganguly A, Haldar P (2008) Potential value of Acridids as high protein supplement for poultry feed. Int J Poult Sci 7:722–725

Anankware JP, Osekre EA, Obeng-Ofori D, Khamala CM (2016) Identification and classification of common edible insects in Ghana. Int J Ent Res 1(5):33–39

Aniebo A, Owen O (2010) Effects of age and method of drying on the proximate composition of housefly larvae (Musca domestica Linnaeus) meal (HFLM). Pak J Nutr 9:485–487

Anvo M, Toguyen A, Otchoumou A (2016) Nutritional qualities of edible caterpillars Cirinabutyrospermi in southwestern of Burkina Faso. Int J Innov Appl Stud 18:639

Ayieko MA, Oriamo V, Nyambuga IA (2010) Processed products of termites and lake flies: improving entomophagy for food security within the Lake Victoria region. Afri J Food Agric Nutr Dev 10(2):2085–2098

Ayieko MA, Kinyuru JN, Ndong'a MF, Kenji GM (2012) Nutritional value and consumption of black ants (*Carebaravidua* Smith) from the Lake Victoria region in Kenya. Adv J Food Sci Technol 41:39–45

Banjo AD, Lawal OA, Owolana OA, Olubanjo OA, Ashidi JS, Dedeke GA, Owa DA, Sobowale OA (2003) An ethno-zoological survey of insects and their allies among the remos (Ogun State) South Western Nigeria. Indilinga: Afr J Indigenous Knowl Syst 2(1):61–68

Banjo AD, Lawal OA, Songonuga EA (2006) The nutritional value of fourteen species of edible insects in southwestern Nigeria. Afr J Biotechnol 5(3):298–301

Belluco S, Losasso C, Maggioletti M (2013) Edible insects in a food safety and nutritional perspective: a critical review. Comp Rev Food Sci Food Saf 12:296–313

Bukkens SGF (2005) Insects in the human diet: nutritional aspects. In: Paoletti MG (ed) Ecological implications of mini livestock, role of rodents, frogs, snails, and insects for sustainable development. Science Publishers, New Hampshire, pp 545–577

Calder PC (2017) Omega-3: the good oil. Nutr Bull 42(2):132–140. https://doi.org/10.1111/nbu.12261

Capinera JL (2004) Encyclopedia of entomology. Kluwer Academic Publishers, Dordrecht. ISBN:0-7923-8670-1

Chakravarthy AK, Jayasimha GT, Rachana RR, Rohini G (2016) Insects as human food. In: Chakravarthy A, Sridhara S (eds) Economic and ecological significance of arthropods in diversified ecosystems. Springer, Singapore, pp 133–146. https://doi.org/10.1007/978-981-10-1524-3_7

Chavunduka DM (1975) Insects as a Source of Protein to the Africa. Rhodesia Sci News 9:217–220

Chen XM, Feng Y, Zhang H (2008) Review of the nutritive value of edible insects. In: Durst PB, Johnson DV, Leslie RN, Shono K (eds) Forest insects as food: humans bite back. RAP Publication, Chiang Mai, pp 85–92

Chen X, Feng Y, Chen Z (2009) Common edible insects and their utilization in China. Entomol Res 39(5):299–303

Chung AYC (2010) Edible insects and entomophagy in Borneo. In: Durst PB, Johnson DV, Leslie RN, Shono K (eds) Forest insects as food: humans bite back. Proceedings of a workshop on Asia-Pacific resources and their potential for development, Chiang Mai, Thailand, pp 141–150

Cito A, Botta M, Francardi V, Dreassi E (2017) Insects as source of angiotensin converting enzyme inhibitory peptides. J Insects Food Feed 3:231–240

DeFoliart GR (1997) An overview of the role of edible insects in preserving biodiversity. Ecol Food Nut 36(2–4):109–132

DeFoliart G (1999) Insects as food: why the Western attitude is important. Annu Rev Entomol 44:21–50

Deng ZB, Yang W, Yang CP et al (2013). Nutrition analysis and evaluation from wasps. Acta Nutrimenta Sinica 35(5):514–515

Durst PB, Johnson DV, Leslie RL, Shono K (2010) Forest insects as food: humans bite back, proceedings of a workshop on Asia-Pacific resources and their potential for development. FAO, Regional Office for Asia and the Pacific, Bangkok, pp 1–64

Dutta P, Dey T, Manna P, Kalita J, Dewanjee S, (2016) Antioxidant potential of Vespa affinis L., a traditional edible insect species of North East India. PLoS One 11(5):e0156107

Edijala J, Egbogbo O, Anigboro A (2009) Proximate composition and cholesterolconcentrations of *Rhynchophorusphoenicis* and *Oryctesmonoceros* larvae subjected todifferent heat treatments. Afr J Biotechnol 8:2346–2348

Egan BA, Toms R, Minter L, Olivier PA (2014) Nutritional significance of the edible insect, *Hemijanavariegata* Rothschild (Lepidoptera: Eupterotidae) of the Blouberg region, Limpopo, South Africa. Afr Entomol 22:15–23

Ekpo KE, Onigbinde AO, Asia IO (2009) Pharmaceutical potentials of the oils of somepopular insects consumed in southern Nigeria. Afr J Pharm Pharmacol 3:51–57

El Hassan NM, Hamed SY, Hassan AB, Eltayeb MM, Babiker EE (2008) Nutritional evaluation and physiochemical properties of boiled and fried tree locust. Pak J Nutr 7:325–329

Elvin CM, Carr AG, Huson MG, Maxwell JM, Pearson RD, Vuocolo T, Liyou NE, Wong DCC, Meritt DJ, Dixon NE (2005) Synthesis and properties of crosslinked recombinant proresilin. Nature 437:999–1002. https://doi.org/10.1038/nature04085.

FAO (2009) The state of food and agriculture: livestock in the balance. Food and Agriculture Organization of the United Nations (FAO), Rome

FAO (2012) Assessing the potential of insects as food and feed in assuring food security. Food and Agriculture Organization of the United Nations (FAO), Summary Report, Rome, Italy, vol 1, pp 1–38

FAO (2013) Edible insects: future prospects for food and feed security. Forestry paper no. 171. Food and Agriculture Organization of the United Nations, (FAO), Rome, Italy

Finke MD, Oonincx D (2014) Insects as food for insectivores. In: Morales-Ramos JA, Rojas MG, Shapiro-Ilan DI (eds) Mass production of beneficial organisms. Elsevier, Amsterdam, pp 583–616

Finke MD, Defoliart G, Benevenga NJ (1989) Use of a four-parameter logistic model to evaluate the quality of the protein from three insect species when fed to rats. J Nutr 119:864–871

Gokoglu N, Yerlikaya P, Cengiz E (2004) Effects of cooking methods on the proximate composition and mineral contents of rainbow trout (Oncorhynchus mykiss). Food Chem 84:19–22

Goodman WG (1989) Chitin: a magic bullet? Food Insects Newslett 2(3):1, 6–7

Gordon DG (1998) The eat-a-bug cookbook. Ten Speed Press, Berkeley, CA, p 103

Halloran A, Vantomme P (2013) The contribution of insects to food security, livelihoods and the environment. FAO, Rome

Halloran A, Roos N, Flore R, Hanboonsong Y (2016) The development of the edible cricket industry in Thailand. J Insect Food Feed 2(2):91–100. https://doi.org/10.3920/JIFF2015.0091

Hwangbo J, Hong EC, Jang A, Kang HK, Oh JS, Kim BW, Park BS (2009) Utilization of house fly-maggots, a feed supplement in the production of broiler chickens. J Environ Biol 30(4):609–614

Islam SB, Bezbaruah S, Kalita J (2016) A review on antimicrobial peptides from Bombyxmori L and their application in plant and animal disease control. J Adv Bio Biotechnol 9(3):2394–1081. https://doi.org/10.9734/JABB/2016/27539

Johnson DV (2010) Proceedings of a workshop on resources and their potential for development. FAO Regional Office for Asia and the Pacific, Bangkok, pp 5–22

Kelemu S, Niassy S, Torto B, Fiaboe K, Affognon H, Tonnang H, Maniania NK, Ekesi S (2015) African edible insects for food and feed: inventory, diversity, commonalitiesand contribution to food security. J Insects Food Feed 1:103–119

Kinsella J (1976) Functional properties of food proteins: a review. Crit Rev Food Sci Nutr 7:219–280

Kinyuru JN, Kenji GM, Njoroge MS (2009) Process development, nutrition and sensory qualities of wheat buns enriched with edible termites (Macrotermessubhylanus) from Lake Victoria region, Kenya. Afr J Food Agric Nutr Dev 9:1739–1750

Kinyuru JN, Kenji GM, Muhoho SN (2010) Nutritional potential of longhorn grasshopper (*Ruspoliadifferens*) consumed in Siaya District, Kenya. J Agric Sci Technol 12(1):1–24

Kinyuru JN, Konyole S, Roos N, Onyango C, Owino V, Owuor B, Estambale B, Friis H, Aagaard-Hansen J, Kenji G (2013) Nutrient composition of four species of winged termites consumed in Western Kenya. J Food Comp Anal 30(2):120–124

Kinyuru JN, Konyole SO, Onyango-Omolo SA, Kenji GM, Onyango CA, Owino VO, Roos N (2015) Nutrients, functional properties, storage stability and costing of complementary foods enriched with either termites and fish or commercial micronutrients. J Insects Food Feed 1(2):149–158

Kipkoech C, Kinyuru JN, Samuel I, Nanna R (2017) Use of house cricket to address food security in Kenya: Nutrient and chitin composition of farmed crickets as influenced by age. Afr J Agric Res 12(44):3189–3197

Li ZR, Liu J, Wang CK, Zhou Q (2010) Effect of *Periplanetaamericana* on meat quality traits of broilers. Fujian J Agric Sci 25:14–17

Liu GQ, Wei MC (2002) The review on functional factors in insects and exploitation prospect of functional food. Food Sci Technol 27:21–25

Madibela OR, Seitiso TK, Thema TF, Letso M (2007) Effect of traditional processing methods on chemical composition and in vitro true dry matter digestibility of the Mophane worm (Imbrasia belina). J Arid Environ 68(3):492–500

Madibela O, Mokwena K, Nsoso S, Thema T (2009) Chemical composition of Mopaneworm sampled at three sites in Botswana and subjected to different processing. Trop Anim Health Prod 41:935–942

Makkar HP, Tran G, Heuzé V, Ankers P (2014) State-of-the-art on use of insects as animal feed. Anim Feed Sci Technol 197:1–33

Mariod AA (2013) Insect oil and protein: biochemistry, food and other uses: review. Agric Sci 4:76–80

Mariod AA, Matthaus B, Eichner K (2004) Fatty acid, tocopherol and sterol composition as well asoxidative stability of three unusual Sudanese oils. J Food Lipids 11:179–189

Mariod A, Matthäus B, Eichner K, Hussein IH (2005) Improving the oxidative stability of sunflower oil by blending with *Sclerocarya birrea* and *Aspongopu sviduatus* oils. J Food Lipids 1(2):150–158

Michaelsen KF, Hoppe C, Roos N, Kaestel P, Stougaard M, Lauritzen L, Mølgaard C (2009) Choice of foods and ingredients for moderately malnourished children 6 months to 5 years of age. Food Nutr Bull 30(3):343–404

Mmari MW, Kinyuru JN, Laswai HS, Okoth JK (2017) Traditions, beliefs and indigenous technologies in connection with the edible longhorn grasshopper Ruspolia differens (Serville 1838) in Tanzania. J Ethnobiol Ethnomed 13:60. https://doi.org/10.1186/s13002-017-0191-6

Münke-Svendsen C, Kinyuru J, Ayieko M, Makkar H (2016) Technical brief #1: insects as food and feed in Kenya – past, current and future perspectives. GREEiNSECT - insect for green economy. https://www.researchgate.net/publication/295918503

Mustafa NEM, Mariod AA, Matthäus B (2008) Antibacterial activity of the Aspongopus viduatus (melon bug) oil. J Food Safety 28:577–586

Musundire R, Zvidzai CJ, Chidewe C, Samende BK, Manditsera FA (2014a) Nutrient and anti-nutrient composition of *Henicuswhellani* (Orthoptera: Stenopelmatidae), an edible ground cricket, in south-eastern Zimbabwe. Int J Trop Insect Sci 34:223–231

Musundire R, Zvidzai CJ, Chidewe C (2014b) Bio-active compounds composition in edible stink-bugs consumed in South-Eastern districts of Zimbabwe. Int J Biol 6:36–45

Musundire R, Osuga I, Cheseto X, Irungu J, Torto B (2016a) Aflatoxin contamination detected in nutrient and anti-oxidant rich edible stinkbug stored in recycled grain containers. PLoS One 11:1–16

Musundire R, Zvidzai CJ, Chidewe C, Ngadze RT, Macheka L, Manditsera FA, Mubaiwa J, Masheka A (2016b) Nutritional and bioactive compounds composition of Eulepida mashona, an edible beetle in Zimbabwe. J Insects Food Feed 2:179–187

Muzzarelli RAA (2010) Chitins and chitosans as immunoadjuvants and non-allergenic drug carriers. Mar Drugs 8(2):292–312

Nguyen P, Kim KY, Kim AY, Kim NS, Kweon HY (2016) Increased health span and resistance to Parkinson's disease in Drosophila by boiled and freeze-dried mature silk worm larval powder. J Asia Pac Entomol 19:551–561

Nonaka K (1996) Ethnoentomology of the central Kalahari san. African Study Monographs 22:29–46

Nongonierma AB, FitzGerald RJ (2017) Unlocking the biological potential of proteins from edible insects through enzymatic hydrolysis: A review. Innovative Food Sci Emerg Technol 43:239–252

Nowak V, Persijn D, Rittenschober D, Charrondiere UR (2016) Review of food composition data for edible insects. Food Chem 193:39–46

OECD-FAO (2018) In: Chapter 6 Meat agricultural outlook 2018-2027, Paris, pp 150. http://www.fao.org/3/i9166e/i9166e_Chapter6_Meat.pdf

Okaraonye CC, Ikewuchi JC (2008) *Rhynchophorus phoenicis* (F) larva meal: nutritional value and health implications. J Biol Sci 8:1221–1225

Oliveira JFS, Passos de Carvalho J, Bruno de Sousa RFX, Madalena Simão M (1976) The nutritional value of four species of insects consumed in Angola. Ecol Food Nutr 5(2):91–97

Omotoso OT (2006) Nutritional quality, functional properties and anti-nutrient compositions of the larva of *Cirina forda* (Westwood) (Lepidoptera: Saturniidae). J Zhejiang Univ Sci B 7(1):51–55

Onyeike EN, Ayalogu EO, Okaraonye CC (2005) Nutritive value of the larvae of raphia palm beetle (Oryctes rhinoceros) and weevil (*Rhynchophorus phoenicis*). J Sci Food Agric 85:1822–1828

Osasona AI, Olaofe O (2010) Nutritional and functional properties of *Cirina forda* larva from Ado-Ekiti, Nigeria. Afr J Food Sci 4:775–777

Pal P, Roy S (2014) Edible insects: future of human food - a review. Int Lett Nat Sci 21:1–11

Pambo KO, Okello JJ, Mbeche R and Kinyuru JN (2016) Consumer acceptance of edible insects for non-meat protein in Western Kenya. In: Paper prepared for presentation at the 5th African Association of Agricultural Economists (AAAE) conference at the United Nations conference center, Addis Ababa-Ethiopia, 23–26 Sept 2016.

Payne CLR, Itterbeeck JV (2018) Ecosystem services from edible insects in agricultural systems: a review. Insects 8:24. https://doi.org/10.3390/insects8010024

Ramos-Elorduy JB (2008) Energy supplied by edible insects from Mexico and their nutritional and ecological importance. Ecol Food Nutr 47(3)

Ramos-Elorduy J, Moreno J, Prado E (1997) Nutritional value of edible insects from the state of Oaxaca, Mexico. J Food Comp Anal 10:142–157

Ramos-Elorduy J, Gonzalez E, Hernandez A (2002) Use of Tenebrio molitor (Coleoptera: Tenebrionidae) to recycle organic wastes and as feed for broiler chickens. J Econ Entomol 95:214–220

Ratcliffe NA, Mello CB, Garcia ES, Butt TM, Azambuja P (2011) Insect natural products and processes: new treatments for human disease. Insect Biochem Mol Biol 41(10):747–769

Ratcliffe N, Azambuja P, Mello CB (2014) Recent advances in developing insect natural products as potential modern day medicines. J Evid Based Complement Alternat Med 2014:904958

Raubenheimer D, Rothman JM (2013) Nutritional ecology of entomophagy in humans and other primates. Annu Rev Entomol 58(1):141–160. https://doi.org/10.1146/annurev-ento-120710-100713. PMID:23039342

Roos N, van Huis A (2017) Consuming insects: are there health benefits? J Insects Food Feed 3(4):225–229

Rumpold BA, Schlüter OK (2013a) Potential and challenges of insects as an innovative source for food and feed production. Inn Food Sci Emerg Technol 17:1–11. https://doi.org/10.1016/j.ifset.2012.11.005

Rumpold BA, Schlüter OK (2013b) Nutritional composition and safety aspects of edible insects. Mol Nut Food Res 57(3):802–823

Rumpold BA, Fröhling A, Reineke K, Knorr D, Boguslawski S, Ehlbeck J, Schlüter O (2014) Comparison of volumetric and surface decontamination techniques for innovative processing of mealworm larvae (Tenebrio molitor). Innovative Food Sci Emerg Technol 26:232–241

Ryu KS, Lee HS, Kim IS (2002) Effects and mechanisms of silkworm powder as a blood glucose-lowering agent. Int J Indust Entomol 4:93–100

Schabel HG (2010) Forest insects as food: a global review. In: Durst PB, Johnson DV, Leslie RN, Shono K (eds) Forest insects as food: humans bite back. FAO, Bangkok, Thailand, pp 37–64

Seo M, Goo TW, Chung M, Baek M, Hwang JS, Kim MA, Yun EY (2017) Tenebrio molitor larvae inhibit adipogenesis through AMPK and MAPKs signaling in 3t3-l1 adipocytes and obesity in high-fat diet-induced obese mice. Int J Mol Sci 18:518

Shadung KG, Mphosi MS, Mashela PW (2012) Influence of drying method and location on amino acids and mineral elements of *Sternocera orissa* Buguet 1836 (Coleoptera: Buprestidae) in South Africa. Afr J Agric Res 7:6130–6135

Sirimungkararat S, Saksirirat W, Nopparat T, Natongkham A (2010) Edible products from eri and mulberry silkworms in Thailand. In: Durst PB, Johnson DV, Leslie RL, Shono K (eds) Forest insects as food: humans bite back, proceedings of a workshop on Asia-Pacific resources and their potential for development. FAO, Regional Office for Asia and the Pacific, Bangkok, pp 189–200

Smith R, Pryor R (2013) Work Package 5: Pro-Insect Platform in Europe. http://www.proteinsect. eu/fileadmin/user_upload/deliverables/D5.1t-FINAL.pdf

Soares JW, Mello CM (2004) Antimicrobial peptides: a review of how peptide structure impacts antimicrobial activity. Proc SPIE 5271:20

Srivastava JK, Gupta S (2009) Health promoting benefits of chamomile in the elderly population. In: Watson RR (ed), Complementary and alternative therapies in the aging population. Elsevier Inc, Academic Press

Srivastava SK, Babu N, Pandey H (2009) Traditional insect bioprospecting-as human food and medicine. Indian J Trad Knowl 8:485–494

Tango M (1994) Insect as human food. Food Insects Newslett 7(3):3–4

Tao J, Li YO (2018) Edible insects as a means to address global malnutrition and food insecurity issues. Food Quality Saf 2:17–26. https://doi.org/10.1093/fqsafe/fyy001

Teffo LS (2006) Nutritional and medicinal value of the edible stinkbug, Encosternum delegorguei Spinola consumed in the Limpopo Province of South Africa and its host plant Dodoneaeviscosa Jacq. var. angustifolia. Doctoral Thesis, University of Pretoria, South Africa

van Huis A (2003) Insects as Food in sub-Saharan Africa. Int J Trop Insect Sci 23(03):163–185

van Huis A (2016) Edible insects are the future? Proc Nutr Soc 75(03):294–305. https://doi. org/10.1017/S0029665116000069

van Huis A, van Gurp H, Dicke M (2012) Het insectenvkookboek. Atlas, Amsterdam, the Netherlands

van Huis A, Van Itterbeeck J, Klunder H, Mertens E, Halloran A, Muir G, Vantomme P (2013) Edible insects future prospects for food and feed security. FAO Forestry, paper 171

Vantomme P, Göhler D, N'Deckere-Ziangba F (2004) Contribution of forest insects to food security and forest conservation: the example of caterpillars in Central Africa. Odi Wildlife Policy Briefing, 3:1–4

Wang D, Zhang Q, Zhai SW (2006) Advances in the application of insect protein, chitosan and fatty acids to animal nutrition and feed. J Northwest Forestry Univ 21:135–138

Wang W, Shen S, Chen Q, Tanga B, He G, Ruan H, Das UN (2008) Hydrolyzates of silkworm pupae (Bombyxmori) protein is a new source of angiotensin I-converting enzyme inhibitory peptides. Curr Pharm Biotechnol 9:307–314

Williams JP, Williams JR, Kirabo A, Chester D, Peterson M (2016) Nutrient content and health benefits of insects. Chapter 3. In: Insects as sustainable food ingredients, pp 61–84. https://doi. org/10.1016/B978-0-12-802856-8.00003-X

Womeni HM, Tiencheu B, Linder M, Nabayo EMC, Tenyang N, Mbiapo P, Villeneuve P, Fanni J, Panmentier M (2012) Nutritional value and effect of cooking, drying and storage process on some functional properties of *Rhynchophorus phoenicis*. Int J Life Sci Pharm Rev 2:203–219

Xiaoming C, Ying F, Hong Z, Zhiyong C. (2010) Review of the nutritive value of edible insects. In Durst PB, Johnson DV, Leslie RL, Shono K (eds) Forest insects as food: humans bite insect physiology and ecology250 back, proceedings of a workshop on Asia-Pacific resources and their potential for development. FAO Regional Office for Asia and the Pacific, Bangkok

Yen AL (2010) Edible insects and other invertebrates in Australia: future prospects. In P.B. Durst, D.V. Johnson, R.L. Leslie. & K. Shono, eds. Forest insects as food: humans bite back, proceedings of a workshop on Asia-Pacific resources and their potential for development. Bangkok, FAO Regional Office for Asia and the Pacific, Bangkok, pp 65–84

Yen AL (2015) Insects as food and feed in the Asia Pacific region: current perspectives and future directions. J Insects Food Feed 1(1):33–55

Zhou Q, Li ZR, Liu J, Lin Q, Wang CK, Jiang X (2009) Effect of *Periplaneta americana* meal on immunity and antioxidation of broliers. J Fujian Agric Forestry Univ 38:175–180

Zielińska E, Karaś M, Jakubczyk A (2017) Antioxidant activity of predigested protein obtained from a range of farmed edible insects. Int J Food Sci Technol 52(2):306–312

Zielińska E, Karaś M, Jakubczyk A, Zieliński D, Baraniak B (2019) Edible insects as source of proteins. In: Mérillon JM, Ramawat KG (eds) Bioactive molecules in food, Reference series in phytochemistry. Cham, Springer Nature, pp 389–441. https://doi.org/10.1007/978-3-319-78030-6_67

Chapter 7
Sensory Quality of Edible Insects

Marwa Yagoub Farag Koko and Abdalbasit Adam Mariod

Abstract Insects are considered as a protein-rich food mainly for many rural communities and ethnic groups, especially in Africa, Australia, Thailand, and some tropical countries. From a nourishment perspective, The high protein, vitamin, mineral, carbohydrate, and fatty acid content of bugs varies depending on the insect feed, environment, type, and improvement phase of the insect. Market consumption size was over 55$ million USD in 2017, and the business hopes to observe noteworthy additions at over 43.5% by 2024. Eatable insects show extraordinary potential as an environmentally friendly decision for future food frameworks. The utilization of insects along these lines contributes emphatically to the environment, food and nourishing security, and a sound life for present and coming generations. Recently, there is an increasing concern about insects as human food, and their role in food and agricultural industries and research institution laboratories and different practices. This chapter discusses the nutritional composition of edible insects considering the safe consumption in human nutrition.

Keywords Insects · Nutrition values · Protein · Protein source · Sensory quality

Abbreviations

EFSA   European Food Safety Authority
FAO   Food and Agriculture organization

M. Y. F. Koko (✉)
National Center of Food Research, Khartoum North, Sudan

A. A. Mariod
Indigenous Knowledge Center, Ghibaish College of Science and Technology, Ghibaish, Sudan

© Springer Nature Switzerland AG 2020
A. Adam Mariod (ed.), *African Edible Insects As Alternative Source of Food, Oil, Protein and Bioactive Components*, https://doi.org/10.1007/978-3-030-32952-5_7

7.1 Edible Insects and Their Global Consumption

Currently, more than 2000 insect species are known to be edible. Universally, the most common consumable species included are caterpillars, honeybees, wasps, and ants. They are trailed by grasshoppers, crickets, cicadas, leaf hoppers and bugs, termites, dragonflies, flies, and a different species (Jongema 2015). The biggest utilization of edible insects is located in Africa, Asia, and Latin America (Jongema 2015). In most European countries, the human utilization of the edible insects is low and frequently socially improper or even forbidden. The nutritional benefit of edible insects is similar to regularly eat meats. Thinking about the developing populace on the planet and the expanding interest for a generation of customary hamburger, pork and chicken products, palatable creepy crawlies ought to be genuinely considered as a wellspring of insect protein (Dreon and Paoletti 2009).

As far as cultivating conditions, the accompanying bug species could be reared and expended in Europe: house cricket (*Acheta domesticus*), Jamaican field cricket (*Gryllus assimilis*), African transient grasshopper (*Locusta migratoria*), desert insect (*Schistocerca gregaria*), yellow mealworm scarab (*Tenebrio molitor*), superworm (*Zophobas morio*), lesser mealworm (*Alphitobius diaperinus*), western bumble bee (*Apis mellifera*), and wax moth (*Galleria mellonella*) (EFSA 2015).

Expanding populace development on the planet builds interest in protein sources yet the measure of the accessible farmland is restricted. In 2050 the total populace is evaluated at in excess of nine billion individuals, bringing about an extra requirement for nourishment of a large portion of the present needs. Traditional protein sources might be lacking and we should concentrate on elective sources (Godfray et al. 2010), which might be eatable insects (Luan et al. 2013). In Africa, Southeast Asia, and the northern part of Latin America, this huge gathering of insects is a mainstream delicacy and an intriguing collection of nourishment advancement. For instance, in Mexico, chapulines (grasshoppers of the class Sphenarium) are a continuous national dish together with meat and beans (Cerritos and Cano-Santana 2008).

7.2 Nutritional Assessment of Edible Insects

The dietary benefit of palatable edible insects is extremely different primarily on account of the substantial number and fluctuation of species. Healthy benefits can differ extensively even inside a gathering of edible insects relying upon the phase of transformation, source of the bug, and its eating regimen (Finke and Oonincx 2014). Correspondingly, the healthy benefit changes as per the arrangement and handling before utilization (drying, cooking, searing, and so forth) (van Huis et al. 2013). As per Payne et al. (2016) insect nutritional composition demonstrated high decent variation between species. The Nutrient Value Score of crickets, palm weevil hatchings, and mealworm was essentially more advantageous than on account of hamburger and chicken and none of six tried edible insects were measurably less solid than meat. Most consumable bugs give adequate vitality and proteins allow in the human eating routine, and also meeting the amino acid necessities. Bugs additionally have a high content of mono- and polyunsaturated fats; they are rich in components such as copper, magnesium, manganese, phosphorus, selenium, and zinc, and vitamins like riboflavin, pantothenic acid, biotin, and folic acid (Rumpold and Schlüter 2013).

The health benefits of the edible insects are an exceptional factor, not slightest on account of the wide assortment of species. Indeed, even in a similar gathering of eatable bug species, qualities may contrast contingent upon the transformative phase of the edible insects, specifically, for species with a total transformation known as holometabolous species, for example, ants, honey bees, and insects), and their natural surroundings and diet. As with most foods, drying, boiling, or frying as a preparation and processing method before consumption will also influence the nutritional composition of edible insects. Only a limited number of specific studies were analyzed for the nutritional composition of edible insects; in any case, these information are not constantly similar because of the previously mentioned differences among insects and due to the shifting techniques utilized to break down compounds. In addition, where normally consumed, insects contain just a part of local diets. For instance, in certain African societies insects comprise nearly 5–10% of the protein used (Ayieko and Oriaro 2008).

The collected nutritional compositions for 236 edible insects was reported and distributed depending on the dry weight. Albeit critical difference was seen in the gathered information, numerous eatable insects give attractive measures of energy and protein, meet amino acid necessities for people, are high in monounsaturated or potentially polyunsaturated fats, and are rich in micronutrients, for example, copper, iron, magnesium, manganese, phosphorous, selenium, and zinc, just as riboflavin, pantothenic acid, biotin, and, in some cases, folic acid (Rumpold and Schlüter 2013)

7.2.1 Dietary Energy

Ramos-Elorduy et al. (1997) reported the chemical analysis of 78 insect species of Oaxaca state, Mexico, and the results of the calorie content was about 293–762 kcal per 100 g of dry matter. For example, the uncultivated energy was in the normal range but still higher than metabolized energy) of the migratory locust (*Locusta migratoria*) was in the range of 598–816 kJ per 100 g fresh weight (recalculated from dry matter), depending on the insect's diet (Oonincx and van der Poel 2011).

7.2.2 Proteins and Amino Acids

Proteins are natural mixes comprising amino acids. They are significant components of food nutrition yet in addition add to its physical and sensory properties. The nutritive esteem relies upon a few variables: protein content, which differs broadly among all foods; protein quality, which relies upon the sort of amino acids present (fundamental or trivial) and whether the quality agrees to human needs; and protein digestibility, which alludes to the edibility of the amino acids present in the food. Amino acids are the building blocks required for the biosynthesis of all proteins through human digestion to guarantee appropriate development, advancement, and support. Basic amino acids are fundamental on the grounds that the body cannot synthesize them and thus should acquire them through food. Eight amino acids are named essential: phenylalanine, valine, threonine, tryptophan, isoleucine, methionine, leucine, and lysine.

Grain proteins are key staples in weight control plans worldwide and are frequently low in lysine and, sometimes, come up short on tryptophan (e.g., maize) and threonine. In some insect species, these amino acids are very much represented (Bukkens 2005). For instance, a few caterpillars of the Saturniidae family, palm weevil larvae and aquatic insects have amino acid scores for lysine higher than 100 mg amino acid for each 100 g rough protein. However, so as to make proposals in regards to the utilization of palatable insects as food advancements in diet, it is imperative to take a gander at customary eating regimens completely, and specifically at staple nourishments, and to think about their wholesome quality against that of eatable insects locally accessible in the district. In the Democratic Republic of the Congo, for instance, lysine-rich caterpillars complement lysine-poor staple proteins.

Similarly, individuals in Papua New Guinea eat tubers that are poor in lysine and leucine; however, they adjust for this dietary gap by eating palm weevil larvae. The tubers provide tryptophan and aromatic amino acids, which are restricted in palm weevils (Bukkens 2005).

In nations in Africa, where maize is a staple food—for example, Angola, Kenya, Nigeria, and Zimbabwe—there is very so often broad tryptophan and lysine deficiency; supplementing diets with termite species like *Macrotermes bellicosus* (Angola) ought to be a moderately simple advance, as they have already form accepted pieces of traditional diets. Not all termite species are appropriate, nonetheless: *Macrotermes subhyalinus*, for instance, is not rich in these amino acids (Sogbesan and Ugwumba 2008).

7.2.3   Fatty Acid Composition

Consumable insects are a significant wellspring of fat. Womeni et al. (2009) studied the content and arrangement of oils extracted from a few insects. Their oils are rich in polyunsaturated fats and regularly contain the fundamental linoleic and α-linolenic acids. The wholesome significance of these two basic unsaturated fats is well established, mostly for the sound improvement of kids and newborn children (Michaelsen et al. 2009). More noteworthy consideration has been given to the potential lack in intake of these omega-3 and omega-6 unsaturated fatty acids in recent times, and insects could assume a significant role, specifically in landlocked nations with lower access to fish food sources, by providing these fundamental fatty acids through available diets. The fatty acids of insects have all the earmarks of being affected by the plants on which they feed (Bukkens 2005). The nearness of fatty acids will likewise offer ascent to fast oxidation of insect food items when preparing, thus making them go rancid rapidly.

7.2.4 Micronutrients

Micronutrients—including minerals and vitamins—assume a significant role in the dietary benefit of food. Micronutrient insufficiency, which is typical in many developing nations, can have major unfriendly well-being results, adding to impairments in development, safe capacity, mental and physical advancement, and regenerative results that cannot generally be turned around by nutrition intercessions (FAO 2011). In insects, transformative stage and diet, exceptionally impact dietary benefit, owning widely inclusive expressions about the micronutrient content of insect types of little value. Additionally, the mineral and vitamin content of palatable insects depicted in the literature are exceedingly variable crosswise over species and orders. Utilization of the whole insect body raises nutritional content. An investigation on little fish, for instance, recommended that devouring the entire organism—including all tissues—is a superior wellspring of minerals and vitamins than the utilization of fish fillets. Similarly, consuming the whole insect is relied upon to give higher micronutrient content than eating singular insect parts.

7.2.5 Minerals

Minerals have a significant impact in biological procedures. The prescribed dietary recompense and sufficient intake are commonly used to measure recommended day by day intake of minerals. Obviously the mopane caterpillar—in the same way as other insect is a superb source of iron. Most eatable insects gloat equivalent or higher iron content than hamburger (Bukkens 2005). The hamburger has an iron content of 6 mg for every 100 g of dry weight, while the iron content of the mopane caterpillar, for instance, is 31–77 mg for each 100 g. The iron content of grasshoppers (*Locusta migratoria*) shifts somewhere in the range of 8 and 20 mg for each 100 g of dry weight, contingent upon their diet (Oonincx et al. 2010).

Edible insects are rich source of iron and their inclusion in the daily diet could improve iron status help forestall anemia in some nations. WHO has mentioned iron insufficiency as the world's most normal and boundless nutritional issue. In developing nations, one of every two pregnant ladies and around 40% of preschool kids are accepted to be anemic. Well-being results incorporate poor pregnancy results, debilitated physical and intellectual advancement, expanded danger of death in kids and decreased work profitability in adults. Anemia is a preventable deficiency yet adds to 20% of every maternal death. Given the high iron content of several insect groups, further assessment of increasingly palatable insect species is justified. Zinc insufficiency is another general medical issue, particularly for youngster and maternal health. Zinc insufficiency can prompt development hindrance, deferred sexual and bone development, skin lesions, diarrhea and expanded powerlessness to contaminations intervened by means of imperfections in the immune system. In general, most insects are accepted to be great sources of zinc. Zinc in beef averages 12.5 mg per 100 g of dry weight, while the palm weevil larvae (*Rhynchophorus phoenicis*), for instance, contains 26.5 mg per 100 g (Bukkens 2005).

7.2.6   Vitamins

Vitamins are fundamental for stimulating metabolic procedures and upgrading immune system, capacities are available in most eatable insects. Bukkens (2005) reported that thiamine (vitamin B1) content ranges from 0.1 mg to 4 mg for every 100 g of dry issue. Riboflavin (vitamin B2) content went from 0.11 to 8.9 mg per 100 mg. By correlation, entire meal bread gives 0.16 mg and 0.19 mg per 100 g of B1 and B2, respectively. Vitamin B12 is found just in food of animal origin and is very much represented in mealworm larvae, *Tenebrio molitor* (0.47 µg per 100 g) and house crickets, *Acheta domesticus* (5.4 µg per 100 g in adults and 8.7 µg per 100 g in nymphs). Numerous species have exceptionally low amounts of vitamin B12, which is the reason more research is expected to recognize edible insects rich in B vitamins (Bukkens 2005; Finke 2002).

Vitamin A amounts in *Ichthyodes truncata* and *I. epimethea* ranged from 32 µg to 48 µg per 100 g and 6.8 µg to 8.2 µg per 100 g of dry issue for retinol and β-carotene, separately. The level of these vitamins were not as much as the levels found in yellow mealworm hatchlings, super worms, and house crickets (Finke 2002; Bukkens 2005; Ooninicx and van der Poel 2011). For the most part, insects are not the best source of vitamin A. α-tocopherol and β + γ tocopherol (nutrient E) levels were 35 mg and 9 mg for every 100 g found in the palm weevil hatchlings, separately; the daily prescribed intake is 15 mg (Bukkens 2005). The vitamin E content in silkworm powder (*Bombyx mori*) is additionally moderately high, at 9.65 mg per 100 g (Tong et al. 2011).

7.2.7   Fiber Content

Fibers considered one of the major components of the nourishing part of eatable insects which contain significant amounts of fiber. The fiber content is determined using destructive acids and neutral reagents. The most commonly occurring fiber in consumable insects is chitin. A great deal of data is available on the fiber content of insects. Finke (2007) assessed the chitin content of insect species raised industrially as food for insectivores, and discovered it to extend from 2.7 mg to 49.8 mg per kg (fresh) and from 11.6 mg to 137.2 mg per kg (dry issue). Some contend that chitin functions like any other dietary fiber (Muzzarelli et al. 2001), and this could suggest a high fiber content in eatable insects, particularly species with hard exoskeletons (Bukkens 2005).

References

Ayieko, MA, Oriaro V (2008) Consumption, indigeneous knowledge and cultural values of the lakefly species within the Lake Victoria region. Afr J Environ Sci Technol 2(10):282–286

Bukkens SGF (2005) Insects in the human diet: nutritional aspects. In: Paoletti MG (ed) Ecological implications of minilivestock; role of rodents, frogs, snails, and insects for sustainable development. Science Publishers, Enfield, NH, pp 545–577

Cerritos R, Cano-Santana Z (2008) Harvesting grasshoppers Sphenarium purpurascens in Mexico for human consumption: a comparison with insecticidal control for managing pest outbreaks. Crop Prot 27(3–5):473–480

Dreon AL, Paoletti MG (2009) The wild food (plants and insects) in Western Friuli local knowledge (Friuli-Venezia Giulia, North Eastern Italy). Contrib Nat Hist 12:461–488

EFSA (2015) Risk profile related to production and consumption of insects as food and feed. EFSA J 13:4257

FAO (2011) State of food and agriculture 2010-2011. Women in agriculture: closing the gender gap for development. Rome

Finke MD (2002) Complete nutrient composition of commercially raised invertebrates used as food for insectivores. Zoo Biol 21:269–285

Finke MD (2007) Estimate of chitin in raw whole insects. Zoo Biol 26(2):105–115

Finke MD, Oonincx DD (2014) Insects as food for insectivores. In: Morales-Ramos J, Rojas G, Shapiro-Ilan DI (eds) Mass production of beneficial organisms: invertebrates and entomopathogens. Elsevier, New York, pp 583–616

Godfray HCJ, Crute IR, Haddad L, Lawrence D, Muir JF, Nisbett N, Pretty J, Robinson S, Toulmin C, Whiteley R (2010) The future of the global food system. Philos Trans R Soc B Biol Sci 365:2769–2777

van Huis A, van Itterbeeck J, Klunder H, Mertens E, Halloran A, Muir G, Vantomme P (2013) Edible insects. Future prospects for food and feed security. FAO, Rome, p 201

Jongema Y (2015) World list of edible insects. Laboratory of Entomology, Wageningen University, Wageningen

Luan Y, Cui X, Ferrat M (2013) Historical trends of food self-sufficiency in Africa. Food Secur 5:393–405

Michaelsen KF, Hoppe C, Roos N, Kaestel P, Stougaard M, Lauritzen L, Mølgaard C, Girma T, Friis H (2009) Choice of foods and ingredients for moderately malnourished children 6 months to 5 years of age. Food Nutr Bull 30:S344–405

Muzzarelli RAA, Terbojevich M, Muzzarelli C, Miliani M, Francescangeli O (2001) Partial depolymerization of chitosan with the aid of papain. In: Muzzarelli RAA (ed) Chitin enzymology. Atec, Italy, pp 405–414

Oonincx DGAB, van der Poel AFB (2011) Effects of diet on the chemical composition of migratory locusts (Locustamigratoria). Zoo Biol 30:9–16

Oonincx DGAB, van Itterbeeck J, Heetkamp MJW, van den Brand H, van Loon J, van Huis A (2010) An exploration on greenhouse gas and ammonia production by insect species suitable for animal or human consumption. PLoS One 5:e14445

Payne CLR, Scarborough P, Rayner M, Nonaka K (2016) Are edible insects more or less 'healthy' than commonly consumed meats? A comparison using two nutrient profiling models developed to combat over- and undernutrition. Eur J Clin Nutr 70:285–291

Ramos-Elorduy J, Moreno JMP, Prado EE, Perez MA, Otero JL, Guevara OL (1997) Nutritional value of edible insects from the State of Oaxaca, Mexico. J Food Compos Anal 10(2):142–157

Rumpold BA, Schlüter OK (2013) Nutritional composition and safety aspects of edible insects. Mol Nutr Food Res 57:802–823

Sogbesan A, Ugwumba A (2008) Nutritional evaluation of termite (*Macrotermessubhyalinus*) meal as animal protein supplements in the diets of *Heterobranchuslongifilis*. Turk J Fish Aquat Sci 8:149–157

Tong L, Yu X, Lui H (2011) Insect food for astronauts: gas exchange in silkworms fed on mulberry and lettuce and the nutritional value of these insects for human consumption during deep space flights. Bull Entomol Res 101:613–622

Womeni HM, Linder M, Tiencheu B, Mbiapo FT, Villeneuve P, Fanni J, Parmentier M (2009) Oils of insects and larvae consumed in Africa: potential sources of polyunsaturated fatty acids. J Oleo Sci 16:230–235

Chapter 8
Automation of Insect Mass Rearing and Processing Technologies of Mealworms (*Tenebrio molitor*)

Nina Kröncke, Andreas Baur, Verena Böschen, Sebastian Demtröder, Rainer Benning, and Antonio Delgado

Abstract Automating the production of insects for rearing, harvesting and process-ing is an important step in making insect mass production more attractive, cost effective and competitive. In this study, we present a functional model for the auto-mated separation and monitoring of the breeding process of mealworms. Data have shown that it is possible to separate mealworm larvae from the substrate and faeces effectively with a zig-zag air separator and make handling and separating processes easier by automation. In addition, an integrated camera system with a neural net-work can assist in assessing the constitution of mealworms by using the segments of the larvae for evaluation. Regarding the processing of mealworm larvae, it has been shown that duration of drying and temperature have a significant influence on the nutritional value and quality of the larvae. In addition, the oil of the dried larvae was extracted with a screw press and the protein-rich press cake was processed with a rolling mill to insect meal, which is comparable to fishmeal and suitable for use in pet food and livestock feed.

Keywords *Tenebrio molitor* · Drying · Automation · Air classification · Image processing

N. Kröncke (✉) · S. Demtröder · R. Benning
University of Applied Sciences Bremerhaven, Bremerhaven, Germany
e-mail: nkroencke@hs-bremerhaven.de; sdemtroeder@hs-bremerhaven.de;
rbenning@hs-bremerhaven.de

A. Baur
Friedrich-Alexander Universität Erlangen-Nürnberg, Erlangen, Germany
e-mail: andreas.baur@fau.de

V. Böschen
Research Institute of Food Technology of the International Research Association of Feed Technology e. V., Braunschweig-Thune, Germany
e-mail: v.boeschen@iff-braunschweig.de

A. Delgado
Friedrich-Alexander-Universität Erlangen-Nürnberg, Erlangen, Germany
e-mail: antonio.delgado@fau.de

© Springer Nature Switzerland AG 2020
A. Adam Mariod (ed.), *African Edible Insects As Alternative Source of Food, Oil, Protein and Bioactive Components*, https://doi.org/10.1007/978-3-030-32952-5_8

Abbreviations

ANN   Artificial Neural Network
DM   Dry matter
d_p Particle equivalent diameter
F_A Buoyancy force
F_G Weight force
F_R Frictional force
g Gravitational constant
V Volume
ρ_f Density of the fluid
ρ_s Density of the particle
v_s Descent rate of a particle

8.1   Automation

On the FAO technical consultation meeting in 2012, mass production of edible insects has been defined as the production of 1 t/day (Vantomme et al. 2012). In order to be able to market edible insects on an industrial scale, it is desirable to set up safe and cost-effective farming systems (Rumpold and Schlüter 2013). In Europe, the mass production of edible insects has so far been expensive and the price comparable to meat, due to the great need for manual labor and little use of automation (Rumpold and Schlüter 2013; Van Huis 2012). Therefore, it is essential to develop a production with automated techniques for rearing, breeding and harvesting, in order to make mass production more attractive and competitive in respect of meat and reduce costs in the production process (Rumpold and Schlüter 2013).

For industrial cultivation of *Tenebrio molitor* larvae automation of the breeding and the processing is indispensable. For the successful industrial mass rearing of mealworms several main issues are very important: climate control (temperature and humidity) to ensure efficient and optimized breeding; separation of frass, food leftovers, larvae and eggs; monitoring of the growth performance of the larvae to choose the right time for feeding and harvesting. Figure 8.1 shows a schematic representation of a functional model for a rearing system of *Tenebrio molitor* with automated parts for separation, which can represent an efficient and timesaving method for industrial mass rearing. Since in a scale-up, the surface grows proportionally to the square-, but the volume is proportional to the cube of the multiplier, a rearing container cannot be enlarged infinitely. A oversized container causes several problems such as a buildup of heat inside the insect rearing container, more difficulties to distribute the substrate homogeneously and impair adequate ventilation. In order to avoid these problems, a so-called "numbering-up" is carried out,

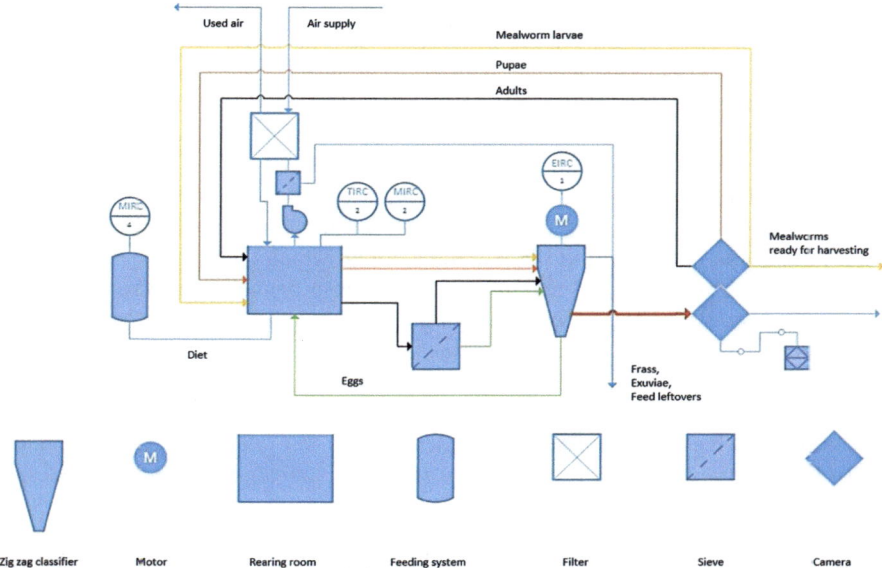

Fig. 8.1 Schematic representation of a functional model for a rearing system and automated processing of mealworms

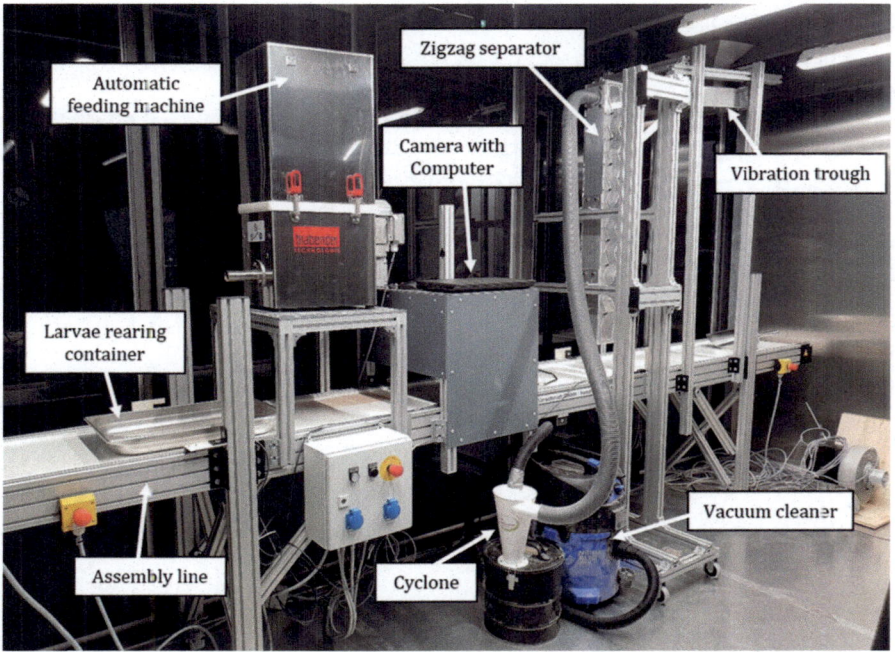

Fig. 8.2 Overview of a functional model (automation) for the mass rearing of mealworms

where the dimensions of the containers are maintained and only the overall amount is increased in order to achieve the required product quantity.

Figure 8.2 shows the functional pattern of the installation, which is placed in a climate room at the University of Applied Sciences Bremerhaven. This arrangement consists of a 4.3 m long assembly line on which the larvae are transported in their rearing containers (stainless steel containers). At its beginning, a vibrating trough is installed, which is used to feed the downstream zig-zag separator (for details, see Figs. 8.3 and 8.4). The entire content of the breeding container (larvae, food residues, frass) can be added into the vibrating trough which then produces a constant feed. The zig-zag separator separates the larvae from the residues. The larvae pass the separation, are collected with a rearing container and can be transported to the next station. The remains of the substrate along with the frass are sucked up and separated via cyclonic separation. A down stream camera system with an evaluation module provides information about the harvest status of the larvae. Subsequently, the automatic feeding machine (Brabender Group) can refill the rearing container with the substrate and in a final step; the rearing container can then be transferred back to the rearing room.

An important step for a simplified handling of mealworm rearing is the separation of the solids. Therefore, in the following section, the classification of solids by a zig-zag classifier is presented in more detail.

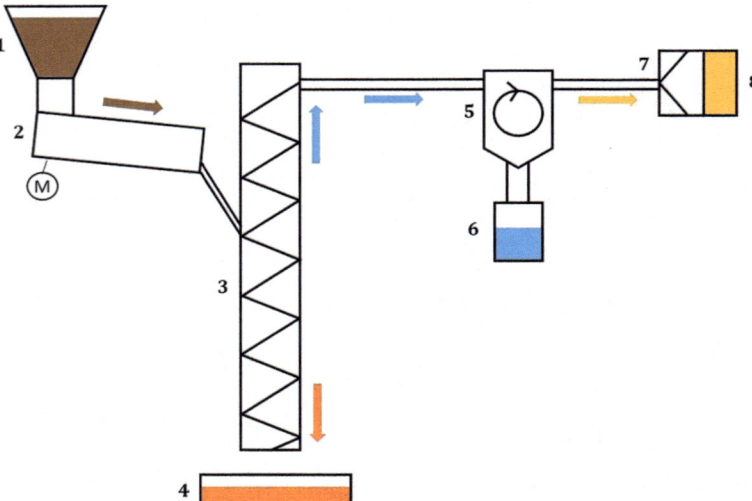

Fig. 8.3 Schematic depiction of a zig-zag air separator. (1) feed material tank, (2) vibrating trough with motor, (3) zig-zag separator, (4) coarse material (larvae, eggs), (5) cyclone separator, (6) fine material tank (frass, exuviae, feed leftovers), (7) filter, (8) vacuum cleaner with fine material residue

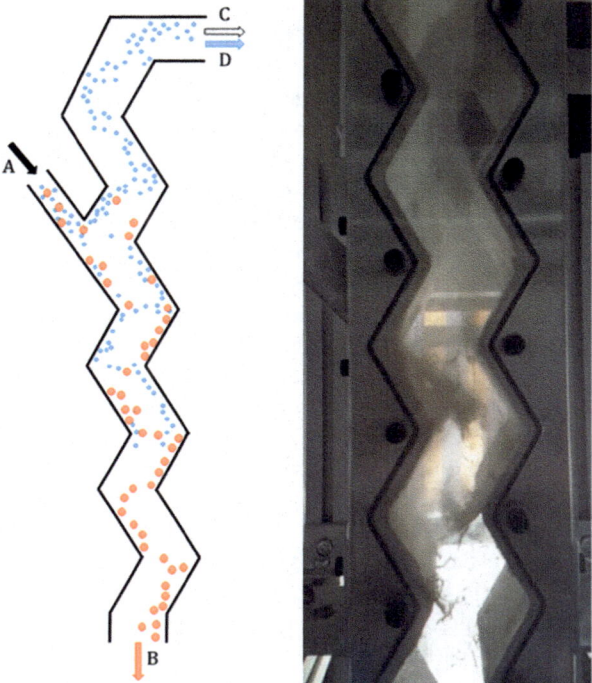

Fig. 8.4 Fundamental principle of zig-zag airseparation (left). (**a**) feed material (substrate, frass, larvae and/or eggs), (**b**) coarse material (larvae or eggs), (**c**) air extraction, (**d**) fine material (substrate or frass). Gradual separation of the feed material into coarse and fine material (right)

8.1.1 Air Separator

Air separation is a process of separating particles into two or more groups depending on their sizes, shapes and specific gravities. This separation technique is being utilized by several industries (i.e., food, cement, or coal) (Ribeiro et al. 2019). Different types of air separators (e.g., dynamic and static separators) have been developed (Altun et al. 2016). Air separators are already used in the food industry for grain processing, whereas the remaining dust and bran particles are separated by air separation (Bond 2004; Arendt and Zannini 2013). The combined use of separators with cyclones can significantly increase the efficiency of separation the fine particles (Caliskan et al. 2019).

An important process step in insect breeding is the separation of the frass and larvae during the fattening as well as the separation of the eggs from the laying substrate. Mealworms produce high amounts of frass, which must be removed at regular intervals from the rearing containers of the larvae in order to achieve a higher yield (Osimani et al. 2018). Conventional methods such as manual screening or sieving are time-consuming and lead to increased dust pollution. To avoid this, a zig-zag separator was tested to separate the larve from frass and unused substrate leftovers. The advantage of a zig-zag separator is the high selectivity that is achieved due to the many screening steps and easy maintaining of hygienic standards. Due to the design of the air separator in this model, the separation stages (kinks) are able to separate both the fine faeces particles from the mealworm larvae, as well as eggs from the laying substrate of the adults, where the sinking rates of the respective components are even more similar. The intake of the feed material was set to the fifth stage from the top, whereby a higher selectivity of the coarse particles should be achieved in comparison to the fine particles.

Figure 8.3 shows a schematic depiction of the zig-zag separator which was designed for this module. First, a vibrating trough feeds the mixture of larvae, frass and feed leftovers to the intake of the zig-zag separator. In the vertical channel, an air flow moves upwards. Depending on the geometry and density of the falling particles, they are taken along with the airflow or fall down due to gravity. At each kink of the zig-zag channel, the heterogeneous mixture passes the air stream again. This corresponds to a screening level. In this way, a cross-current sighting takes place at each kink. The design of the separator with many such stages leads to very sharp separations. The first runs were carried out with an adjustable vacuum cleaner to keep ambient dust levels as low as possible. The vacuum cleaner was replaced by a radial fan later in order to better implement industrial convenience and automation.

Figure 8.4 shows the gradual separation of the feed material into coarse and fine material. In this case, a rearing container with mealworm larvae, frass and exuviae was added as ingredients. With an increasing number of separation stages (every kink is a separation step), the proportion of fine particles will become smaller and smaller.

For the design of the air separator, each object's settling velocity was the crucial factor, and was calculated in advance. The settling velocity (Fig. 8.5) is the speed at

Fig. 8.5 Force balancing for the calculation of settling velocity

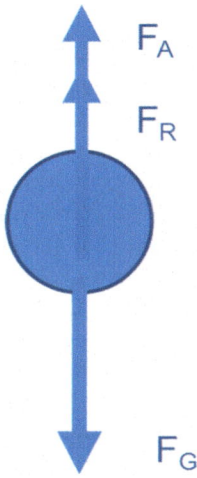

which the gravitational force, the buoyancy force and the friction force increasing with the velocity compensate for each other (Kurzweil et al. 2009):

$F_G = F_R + F_A$

F_G = Weight force

F_R = Frictional force

F_A = Buoyancy force

From this, the formula for the descent rate of a particle (v_s) can be derived (Dietrich 1982):

$$v_s = \sqrt{\frac{3d_p g (\rho_s - \rho_f)}{\rho_f}}$$

d_p = particle equivalent diameter

g = gravitational constant

ρ_s = density of the particle

ρ_f = density of the fluid

For calculating the settling velocity, an equivalent diameter is needed because the formula is valid only for spherical objects. For a mealworm, a cylindrical body was assumed. The following formula is used to calculate the particle equivalent diameter (d_p) at which a spherical object has the same properties as a cylindrical object (Figura 2004):

$$d_p = \sqrt[3]{\frac{6}{\pi} \cdot V}$$

V = Volume of the cylindrical object

The variables for calculating the settling velocity of mealworm larvae, eggs, exuviae, faeces and substrate were determined experimentally by direct measurements of bulk density and gravimetry (Table 8.1).

The larvae can easily be separated from the substrate and the faeces due to the high ratio/differences of/between settling velocity and density. The success of separate the eggs from the laying substrate depends on the attributes of the substrate. In case of small wheat brans, the separation is not possible. By using flour or other products with different attributes, this problem can be bypassed.

Table 8.1 Parameters of *T. molitor* larvae, eggs, exuviae, frass and the feeding substrate (wheat bran)

	Larvae	Eggs	Exuviae	Frass	Substrate (wheat bran)
Size (mm)	15.0–25.0	1.4–1.9	7.5–15.0	0.20–0.8	1.5–4.1
Weight (mg)	100.0 ± 23.0	0.3 ± 0.1	1.2 ± 0.3	0.01 ± 0.01	0.2 ± 0.1
Bulk density (kg/m³)	1358	1136	81	440	1050
Descent rate (m/s)	13.22	1.82	2.25	0.74	1.39

Larvae were about 16 weeks old

8.1.2 Noninvasive Monitoring Systems for the Industrial Rearing of Insects

In the food industry, the quality control is very important at every stage of production. Artificial neural network classifiers (ANN classifiers) have already been successfully implemented for different classification tasks and quality checks on various food products (Delgado et al. 2016; McAllister et al. 2018; Wang et al. 2017). ANN are good classifiers because of their ability to learn not linearly separable patterns and concepts dealing with background noise, uncertainty and random events (Dębska and Guzowska-Swider 2011). In mass rearing of *Tenebrio molitor*, the condition of larvaeand optimal point of harvest is mainly determined by age and size of the larvae (Rumpold and Schlüter 2013). In this presented work, the ANN used a classification model based on the relevant parameters of mealworm larvae to define the general condition. A major advantage of ANN is that it is non-invasive and mealworm larvae can be fed back unharmed to the rearing system after analysis.

For the optical measuring system, an industrial camera (pictor T303M-ETH, Fa. Dr. Merbach) is installed, which was applied for the non-invasive monitoring of the insect rearing container. The camera is installed in a darkened room over an assembly line to ensure consistent lighting conditions (Fig. 8.6). The control software is based on Matlab (Mathworks) and Python (Python Software Foundation). The larvae were placed in their rearing container on the assembly line and moved under the camera, so that from each container a picture was taken for evaluation.

Fig. 8.6 Darkened room over the assembly line with industrial camera inside

For the detection and evaluation of the larval condition, an approach was developed to extract the required information directly from the box without prior cleaning and separation of the mealworm larvae. The extraction of the required data from an inhomogeneous image without a contrasting background requires different, new approaches in contrast to classical image processing. The living larvae require a different interpretation, because they complicate the image analysis due to their activity, the in homogeneity of their size and the appearance. For this purpose, alternative algorithms for segmentation were developed and initial experiments carried out. In the process, relevant variables such as size, circle similarity, diameter, length and breadth of the larvae are extracted, which allow conclusions regarding the growth progress of the mealworms.

8.1.2.1   Image Processing

The evaluation is based on the *irregular watershed* algorithm and an ANN. Watershed algorithm are already used for the extraction of information from sensor networks (Hammoudeh and Newman 2015). With the Watershed algorithm, it is possible to extract the segments of mealworm larvae from an image and evaluate them. Figure 8.7 shows the data matrix for training the network. Manually classified data sets were used with three classes: "Good segments, bad segments and artefacts". For the creation of the data set 109 images were evaluated. The segments were divided into 2938 bad segments, 4717 good segments, and 2257 artefacts. This achieved an accuracy of the neural network of 95.4%.

For the ANN, a "*feed forward back propagation*" network was used:

$$\text{net}_j\left(t\right) = \sum_{i=1}^{n} W_{ij} X_i \left(t\right)$$

W = weight
X = Output of previous neurons
$\text{net}_j(t)$ = Input for the next neuron

Figure 8.8 shows the original unedited image (A) and the pictures with separated (B) and marked (C) segments of mealworm larvae in their rearing tank. The original picture has an inhomogeneous background, which is not suitable for direct evaluation and therefore has to be processed. For evaluation of the images, preprocessing them by increasing the contrast and adjusting the thresh holding is necessary. After calculating, the Euclidean distances, the so-called "markers" of the individual segments can ultimately be determined in order to extract them. The segments are stored individually and evaluated according to the division into the three classes.

The separated segments can be used for further information retrieval such as determination of width and length of the larvae and thus the current state of the larvae's condition or for other applications like identifying pupae and using the information to sort them out from the larvae rearing container.

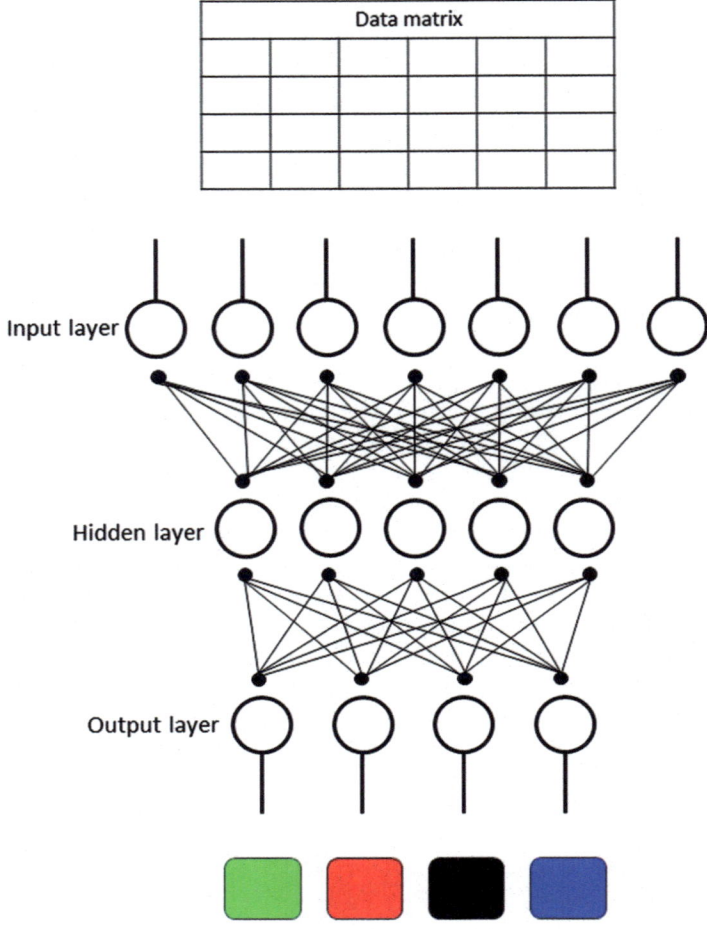

Fig. 8.7 Data matrix of the Artificial Neural Network. Green: Usable segments; Red: Useless segments; Black: Background; Blue: Artefacts and image defects

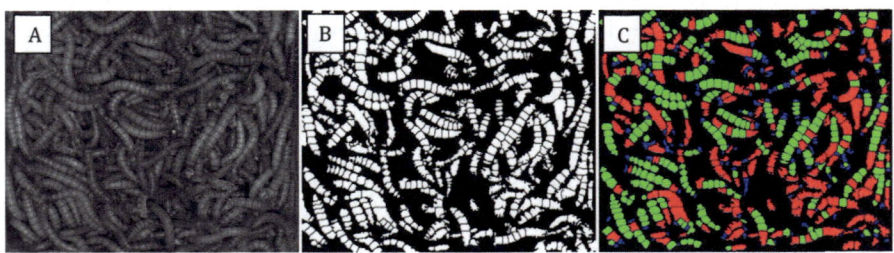

Fig. 8.8 (**a**) Original black and white picture of mealworms; (**b**) Picture with separated important mealworm segments; (**c**) Picture with marked and analyzed mealworm segments: Green: Usable segments; Red: Useless segments; Blue: Artefacts and image defects

8.2   Preparation and Processing of Insects for Animal Feed

In many European countries, such as the Netherlands, France and Germany, an insect-processing industry has developed and uses insects for chips, burgers, bars and flours for animal feed, pet food or human food (FAO 2013; Rumpold and Schlüter 2013; Van Huis 2012). The insect products may have different qualities. Insect flour, for example, can differ significantly in its properties (protein content, residual fat content and chitin) (Le Féon et al. 2019; Thévenot et al. 2018). There are products for the food industry in which the insects or their larvae are dried first and then crushed. The insect material, depending on the insect species, usually has a high fat content from 32% to 45% (Anankware et al. 2014). For numerous recipe design, it is necessary to defat the protein-containing insect material. Similar procedures apply to extraction slices, cakes, legumes and oilseeds. A current problem is the lack of knowledge of separation techniques (protein, fats and chitin) (Vantomme et al. 2012). There are limitations in the application possibilities due to high fat contents (Finke 2002). An advantage resulting from the fat reduction in insect meal is the increasing protein content to levels of up to 70%. It is generally possible to differentiate between wet and dry processes. A variant of the dry processing is to freeze-dry the whole mealworm larvae, but it is a very time- and energy consuming technique (Kröncke et al. 2018; Lenaerts et al. 2018). In order to establish continuous systems for future automation, drying processes from other industrial sectors may be a viable alternative. Depending on the original use of the chosen drying system, possibly only small adjustments have to be made in order to configure a suitable drying technique for insects.

8.2.1   Drying

In this study a belt dryer was used. The aim of the experiment was to dry mealworm larvae and to examine the effects of different drying times and temperatures on nutritional value and microbiological stability. *Tenebrio molitor* larvae were harvested at the age of 16 weeks and frozen at −21 °C in a commercial freezer (HAS 47520, Beko, Germany) before used for the drying process. In this process, various influencing factors had to be determined and adapted beforehand. These are the drying temperature, the thickness of the insect layer on the belt, the air ventilation speed and the duration of the drying process (residence time). Table 8.2 shows the effect of drying time and temperature on the quality of dried mealworms. Moisture content was determined after drying the samples in a drying oven at 103 °C for 4 h. Crude protein, fat and soluble protein content was analyzed as described by the Association of German Agricultural Investigation and Research Institutions (VDLUFA 1976). Total bacterial counts were determined on Plate Count Agar after incubation for 72 h at 30 °C. After a defined residence time in the belt dryer (30 or 70 min respectively), the dried larvae left the process with a final moisture content of between 3.4% and 10.9%.

Table 8.2 Influence of drying time and temperature on the nutritional value and bacterial count of dried mealworm larvae (mean values of three replicates ± standard deviation)

Drying time (min)	70	70	30
Drying temperature (°C)	80	100	120
Moisture (g/100 g)	10.9 ± 0.0^a	3.4 ± 0.0^a	6.0 ± 0.0^a
Protein (g/100 g DM)	53.4 ± 0.4^a	56.4 ± 0.2^a	55.7 ± 0.3^a
Fat (g/100 g DM)	23.3 ± 0.1^a	25.1 ± 0.0^a	24.5 ± 0.1^a
Soluble protein (%)	27.1 ± 0.3^a	20.7 ± 0.1^a	16.9 ± 0.0^a
Bacterial count (cfu/g)	$2.5 \times 10^{5, a}$	$1.8 \times 10^{5, a}$	$2.0 \times 10^{4, a}$

Mealworms were dried with a belt dryer: drying area 250 cm^2 (50 cm × 50 cm), layer thickness 4.5 cm, air velocity 0.52 m/s, starting bacterial count 8.2 × 10^6 cfu/g
DM mass of dried mealworm larvae.
[a]Data per treatment in the same row with a different exponent vary significantly ($p < 0.05$)

Protein solubility is an important attribute and usually determined during development and testing of new protein-rich ingredients such as mealworms as feed or food (Bußler et al. 2016). Protein solubility of dried mealworms was highest at a drying temperature of 80 °C and differed significantly ($p < 0.05$) to the higher temperatures of 100 °C and 120 °C. As the temperature increases, the protein solubility decreases and the denaturation effects increase. High-temperature conditions can greatly reduce the protein solubility because of protein denaturation (Azahog et al. 2016; Bußler et al. 2016). Reduced protein availability of feed for livestock such as poultry or pigs may lead to a potential reduction in growth performance of the animals (Clarke and Wiseman 2007; Dänicke et al. 1998; Fernandez et al. 1994).

The differences in protein and fat content of the larvae are small and do not differ significantly, but were according to literature (Finke 2002; Ghaly and Alkoaik 2009). Furthermore, the final moisture content of the material is of crucial importance. With all drying processes the required moisture content below 14% (according to Feed Regulation (EC) No. 767/2009) was achieved. However, microbial spoilage of the product can still occur during storage (Vandeweyer et al. 2017). The hygienic state of the dried product can be greatly affected by thermal treatment. Based on the starting material with a total bacterial count of 8.2 × 10^6 cfu/g, a germ reduction and thus an improvement of the hygienic condition was achieved with each thermal treatment.

8.2.2 Defatting

The dried larvae material can be defatted in various ways. A schematic overview of the processing of insect material with defatting is shown in Fig. 8.9 (Bußler et al. 2016).

Figure 8.10 shows an overview of the processing with drying step and pressing. Before drying, the larvae were inactivated about a short time temperature process (boiling water). Mealworms were dried as described previously with high temperature and short drying time (see Sect. 8.2.1).

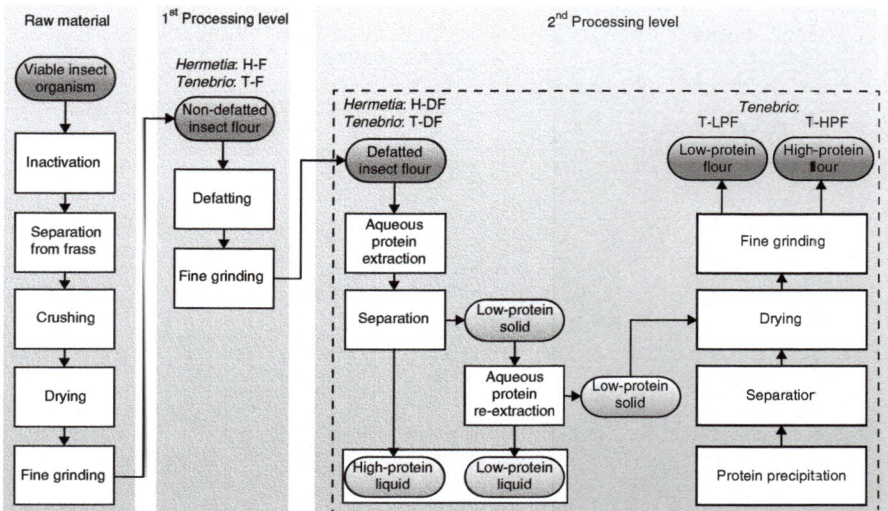

Fig. 8.9 Schematic representation of processing and fractionation of *T. molitor* and *H. illucens* larvae (Bußler et al. 2016)

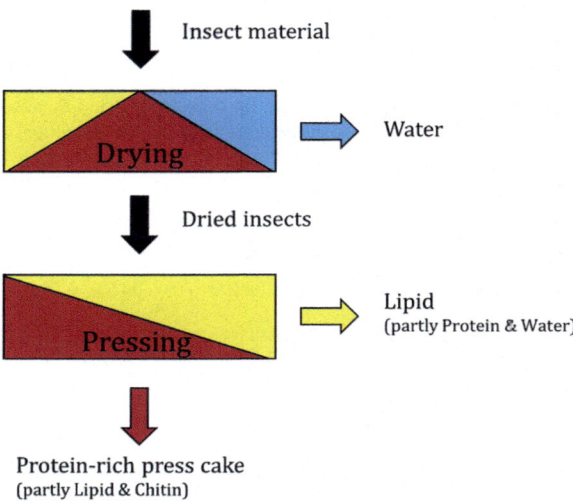

Fig. 8.10 Schematic layout for defatting of insect material (yellow: lipid, blue: water, brown: protein)

In this study, a screw press from Maschinenfabrik Reinartz was used, which is a solvent-free method to extract the insect oil. This process is already established in the oilseed processing industry, for defatting soybeans and rapeseeds (Müller and Relitz 1990). Figure 8.11 shows the screw press producing a protein-rich press cake and oil from mealworms. The dried larvae can be placed in the funnel and separated by pressing which produces oil and the so-called press cake.

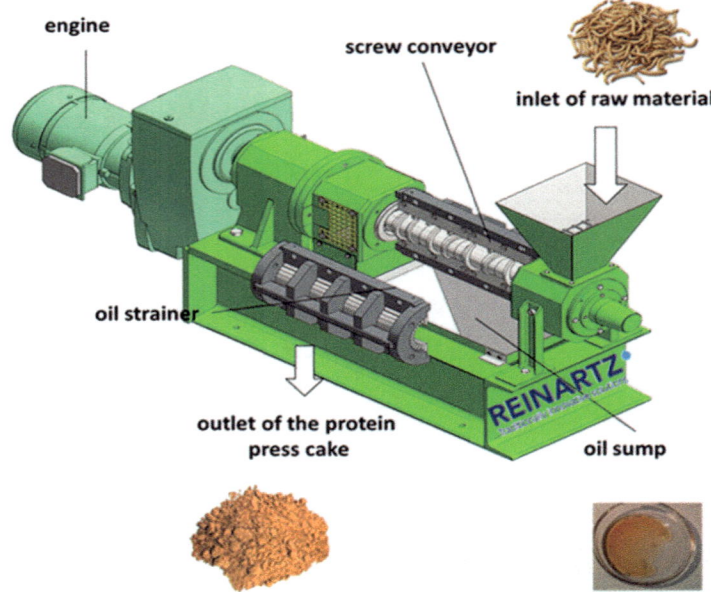

Fig. 8.11 Schematic representation of protein meal and oil extraction from mealworms (*Tenebrio molitor*) using a screw press from Maschinenfabrik Reinartz

Table 8.3 Moisture, protein and fat content of dried mealworm larvae and the press cake after preparation with the screw press compared with commercially available fishmeal (mean values of three replicates ± standard deviation)

Material	Fishmeal	Dried mealworm	Mealworm press cake
Moisture (g/100 g)	5.7 ± 0.0[a]	5.4 ± 0.0[a]	3.9 ± 0.0[a]
Protein (g/100 g DM)	70.8 ± 0.4[a]	48.6 ± 0.1[a]	70.1 ± 0.1[a]
Fat (g/100 g DM)	10.1 ± 0.0[a]	34.9 ± 0.1[a]	6.5 ± 0.0[a]

DM dry matter.
[a]Data per treatment in the same row with a different exponent vary significantly ($p < 0.05$)

The press process is combined with an increase of temperature in press cake and the extracted lipid. In this case the press cake has a temperature of 100 °C and the lipid 60 °C. At the end of the pressing process, two products are created: a protein-rich solid (press cake) and the insect fat. This process can be influenced by various parameters such as bulk density and the water content after drying. These parameters affect not only the process but also the quality of the protein meal (Sitzmann et al. 1992). Table 8.3 shows the moisture, protein and fat content of dried mealworm larvae and the mealworm press cake compared with commercially available fishmeal from herring. Dried mealworm larvae have a high fat content of up to 34.9%. The protein content of the mealworm press cake is comparable to that of fishmeal, but the fat content of the mealworm press cake is significantly lower and reduced after processing.

Regarding the use of protein meals in animal feed, also the physicochemical properties such as the particle structure are relevant (Vukmirovic et al. 2017). Since the press cake has a very coarse particle structure (Fig. 8.13—A), a roller mill (Fig. 8.12) can be used to adjust the particle size in order to create a smaller structure similar to that of fish meal. Roller mills are already established in the processing industry, for example in the manufacture of flour from grains (Campbell 2007). Therefore, a roller mill (MIAG Mühlenbau) with structured roller surface was used to mill the mealworm press cake to a finer particle size.

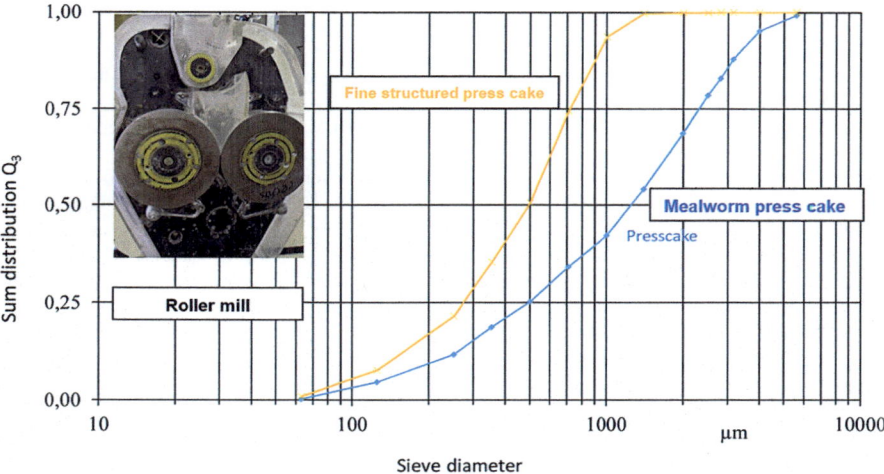

Fig. 8.12 Comparison of the particle size distribution of mealworm press cake (before milling) and the fine structured press cake (after milling), roller mill gap: 0.1 μm

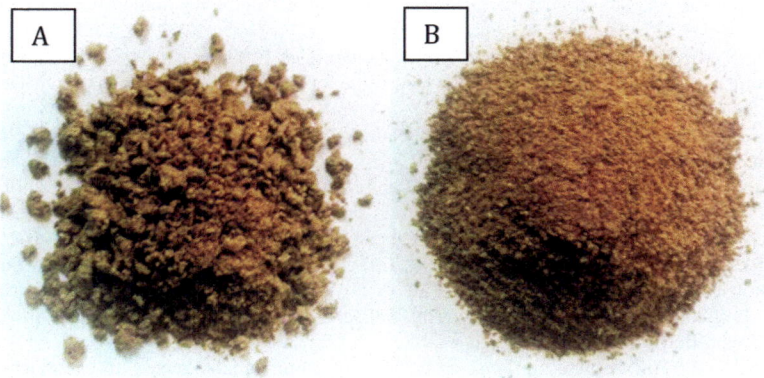

Fig. 8.13 Mealworm press cake before (**a**) and after (**b**) treatment with a roller mill

Figure 8.13 shows the result of milling the press cake by using a roller mill. The mealworm press cake particles was very coarse before the treatment. After milling, a much finer structure was achieved.

Both processes mentioned (wet processing, dry processing) contain several process-specific steps from the inactivation of the larvae to the respective end products. These steps cause the material in the different variants to be thermally and mechanically stressed to varying degrees, and as a result, the product properties (protein meal and fat) are process-dependent in their qualities.

References

Altun O, Toprak A, Benzer H, Darimaz O (2016) Multi component modelling of an air classifier. Miner Eng 93:50–56

Anankware PJ, Fenning KO, Osekre E, Obeng-Ofori D (2014) Insects as food and feed: a review. Int J Agric Res Rev 3:143–151

Arendt A, Zannini E (2013) Cereal grains for the food and beverage industries, Woodhead publishing series in food science, technology and nutrition, vol 248. Woodhead Publishing Limited, Cambrigde

Azahog C, Ducept F, Garcia R, Rakotozafy L, Cuvelier M-E, Keller S, Lewandowski R, Mezdour S (2016) Extraction and physiochemical characterization of Tenebrio molitor proteins. Food Res Int 88:24–31

Bond N (2004) Rice milling. In: Champagne ET (ed) Rice: chemistry and technology. AACC International, Inc., St. Paul, MN

Bußler S, Rumpold BA, Jander E, Rawel HM, Schlüter OK (2016) Recovery and techno-functionality of flours and proteins from two edible insect species: mealworm (Tenebrio molitor) and black soldier fly (Hermetia illucens) larvae. Heliyon 2:1–24

Caliskan ME, Karagoz I, Avci A, Surmen A (2019) An experimental investigation into the particle classification capability of a novel cyclone separator. Sep Purif Technol 209:908–913

Campbell GM (2007) Chapter 7 Roller milling of wheat. In: Salman AD, Ghadiri M, Hounslow MJ (eds) Handbook of powder technology, vol 12. Elsevier, Oxford, pp 383–419. https://doi.org/10.1016/S0167-3785(07)12010-8

Clarke E, Wiseman J (2007) Effects of extrusion conditions on trypsin inhibitor activity of full fat soybeans and subsequent effects on their nutritional value for young broilers. Br Poult Sci 48:703–712

Dänicke S, Kracht W, Jeroch H, Zachmann R, Heidenreich E, Löwe R (1998) Effect of different technical treatments of rapeseed on the feed value for broilers and laying hens. Arch Tierernahr 51:53–62

Dębska B, Guzowska-Swider B (2011) Application of artificial neural network in food classification. Anal Chim Acta 705:283–291

Delgado A, Rauh C, Park J, Kim Y, Groß F, Diez L (2016) Artificial neural networks: applications in food processing. Reference module in food science

Dietrich WE (1982) Settling velocity of natural particles. Water Resour Res 18:1615–1626

Fernandez SR, Zhang Y, Parsons CM (1994) Effect of overheating on the nutritional quality of cottonseed meal. Poult Sci 73:1563–1571

Figura L (2004) Lebensmittelphysik – Physikalische Kenngrößen, Messung und Anwendung. Springer, Berlin

Finke MD (2002) Complete nutrient composition of commercially raised invertebrates used as food for insectivores. Zoo Biol 21:269–285

Food and Agricultural Organization of the United Nations (2013) Edible insects – future prospects for food and feed security, FAO forestry paper 171, ISBN 978–92–5-107595-1

Ghaly AE, Alkoaik FN (2009) The yellow mealworm as a novel source of protein. Am J Agric Biol Sci 4:319–331

Hammoudeh M, Newman R (2015) Information extraction from sensor networks using the Watershed transform algorithm. Inf Fusion 22:39–49

Kröncke N, Böschen V, Woyzichovski J, Demtröder S, Benning R (2018) Comparison of suitable drying processes for mealworms (*Tenebrio molitor*). Innov Food Sci Emerg Technol 50:20–25

Kurzweil P, Frenzel B, Gebhard F (2009) Physik Formelsammlung – Mit Erläuterungen und Beispielen aus der Praxis für Ingenieure und Naturwissenschaftler. Vieweg + Teubner, GWV Fachverlag GmbH, Wiesbaden

Le Féon S, Thévenot A, Maillard F, Macombe C, Forteau L, Aubin J (2019) Life cycle assessment of fish fed with insect meal: case study of mealworm inclusion in trout feed, in France. Aquaculture 500:82–91

Lenaerts S, Van der Borght M, Callens A, Van Campenhout L (2018) Suitability of microwave drying for mealworms (*Tenebrio molitor*) as alternative to freeze drying: impact on nutritional quality and colour. Food Chem 254:129–136

McAllister P, Zheng H, Bond R, Moorhead A (2018) Combining deep residual neural network features with supervised machine learning algorithms to classify diverse food image datasets. Comput Biol Med 95:217–233

Müller S, Relitz H (1990) Erfahrungen mit der automatischen Prozessführung von Hochleistungsschneckenpressen. Fat Sci Technol 92:593

Osimani A, Milanovic V, Cardinali F, Garofalo C, Clementi F et al (2018) The bacterial biota of laboratory-reared edible mealworms (*Tenebrio molitor* L.): from feed to frass. Int J Food Microbiol 272:49–60

Ribeiro PPM, Dos Santos ID, Dutra AJB (2019) Copper and metals concentration from printed circuit boards using a zig-zag classifier. J Mater Res Technol 8(1):513–520

Rumpold BA, Schlüter OK (2013) Potential and challenges of insects as an innovative source for food and feed production. Innov Food Sci Emerg Technol 17:1–11

Sitzmann W, Engel W, Weber K (1992) Die Bedeutung der On-line Feuchtemessung bei der Ölsaatenverarbeitung. Eur J Lipid Sci Technol 94:561–566

Thévenot A, Rivera JL, Wilfart A, Maillard F, Hassouna M, Senga-Kiesse T, Le Féon S, Aubin J (2018) Mealworm meal for animal feed: environmental assessment and sensitivity analysis to guide future prospects. J Clean Prod 170:1260–1267

Van Huis A (2012) Potential of insects as food and feed in assuring food security. Annu Rev Entomol 58:563–583

Vandeweyer D, Lenaerts S, Callens A, Van Campenhout L (2017) Effect of blanching followed by refrigerated storage or industrial microwave drying on the microbial load of yellow mealworm larvae (*Tenebrio molitor*). Food Control 71:311–314

Vantomme P, Mertens E, Van Huis A, Klunder HC (2012) Summary report of the technical consultation meeting "Assessing the Potential of Insects as Food and Feed in assuring Food Security". In: FAO (ed) Rome

VDLUFA (1976) VDLUFA methodenbuch III. In: VDLUFA-Verlag (vol ed) Band III Die chemische Untersuchung von Futtermitteln. vol 3, Band III Die chemische Untersuchung von Futtermitteln. VDLUFA-Verlag, Bonn, pp 2190

Vukmirovic D, Colovic R, Rakita S, Brlek T, Duragic O, Sola-Oriol D (2017) Importance of feed structure (particle size) and feed form (mash vs. pellets) in pig nutrition. A review. Anim Feed Sci Technol 233:133–144

Wang J, Yue H, Zhou Z (2017) Am improved traceability system for food quality assurance and evaluation based on fuzzy classification and neural network. Food Control 79:363–370

Chapter 9
The Legislative Status of Edible Insects in the World

Abdalbasit Adam Mariod

Abstract Edible insects are highly nutritious with high fat, protein, and mineral contents that represent a good alternative food and feed source. Legal frameworks ought to enable the utilization of insects as ingredients for food and feed. Many countries have legislation that refers to insects in food as impurities which prescribes most permissible levels. It seems, therefore, that specific legislation to manage the utilization of insects in food and feed is to be developed.

Keywords Legislation · Edible insects

9.1 Introduction

Before the most important introduction of insect food products into the market, the small-scale production and trade of edible insects was not thought of vital enough to be enclosed in legislation. Since rearing scaled up and bug food products became a phase in food retail, queries relating to correct regulation of this organic phenomenon emerged (Halloran et al. 2018). Legal frameworks ought to enable the utilization of insects as ingredients for food and feed. The utilization of organic waste as a feed ingredient for insects is another legislative concern. In the short term, legislation does not afford the utilization of macromolecule extracts originating from insects. The transmissible *Spongiform Encephalopathies* regulation at the EU level additionally blocks the introduction of insects as an ingredient for feed (van Huis et al. 2013). FAoLEx may be a comprehensive and up-to-date processed legislative info that constitutes one in every of the world's largest electronic collections of national laws and rules on food, agriculture, and renewable natural resources. Laws and rules on cultivation and silkworm-raising are well developed in many countries with prominent honey and silk industries. Many countries have legislation that

A. A. Mariod (✉)
Indigenous Knowledge and Heritage Centre, Ghibaish College of Science and Technology, Ghibaish, Sudan

refers to insects in food as impurities which prescribes most permissible levels. It seems, therefore, that specific legislation to manage the utilization of insects in food and feed is nonetheless to be developed (van Huis et al. 2013). In 2011, the EU Commission issued asking for reports on the present utilization of insects as food, with the promise that reports from every international organization member state would serve to tell legislative proposals for the new method for novel foods (Byrne 2011).

9.2   Legal Provisions on the Utilization of Insects for Food and Feed Production

Specific legal provisions on the utilization of insects for food and feed production would serve to manage and regulate the utilization of insects by business processors and would guarantee shopper access to info. For this purpose, regulators would wish to assess the potential risks related to the utilization of insects, in terms of each species and quantities. Legal frameworks that protect shopper interests may additionally target displays on food packaging and therefore the information provided there for shoppers on the results of risk assessments on the consequences of insects on human health. In 2010, the government of the Lao Peoples Democratic Republic projected to the FAO/WHO Codex Coordinating Committee for Asia that standards for regional trade and food safety for house crickets be developed. This proposal was not accepted, however, as information indicated that there was no verifiable level of trade in insects to warrant such action (http://www.fao.org). There is an increasing lobby within the feed sector in the event of specific legislation on the utilization of insects as feed. Lobbying is being conducted at a national level (including non-public sector-led actions by USA-based corporations to get FDA approval on the utilization of insects in feed) and at an EU level. The farming of insects and edible arthropods for human food, observed as mini-livestock, is rising as an ecologically sound type of agriculture. Recently, as a result of robust lobbying by the feed sector, initiatives have begun to emerge to form a sanctioning setting for the event of rules and standards on the utilization of insects for cultivation feed and to a lesser degree as human food. At the EU level, as an example, the standard and safety criteria for insect-based feed are presently undergoing review. The assembly and consumption of insects ought to even be analyzed from the point of view of their potential impact on health and multifariousness and therefore the potential environmental hazards related to insect production and unleash, together with the accidental unleash of insect species. Risk assessments and containment measures ought to address potential outbreaks of illness which will be harmful to human or animal health and to plant protection. Alternative target areas of legislation may embrace regulation of the trade between countries of living insects as breeding stock. This might embrace the event of standards, codes of practices/standards and product quality metrics to garner believability. The thought of a novel food is guiding the event of rules and standards for insects as human food. The term novel food refers to food products that do not have a history of human consumption within the region or country in question. Samples of definitions of novel foods contained in national legislation are food that does not have a history of human consumption in Australia or New Zealand.

The term might embrace edible insects, oils, berries, and foodstuffs that are the merchandise of biotechnology (including genetically modified foods). Foods that are the merchandise of biotechnology is also thought of as novel globally; however, foodstuffs derived from natural products, though novel in some countries, might represent a substantial portion of the traditional dietary intake in alternative countries. It has been recommended that an extended history of human use implies that insects by choice harvested for human use or consumption do not cause a major risk (Banjo et al. 2006), because the international reference standards for food and feed, a Codex Alimentarius commonplace on the utilization of insects as food and feed ingredients may function a reference to national legislation on insect production and use as food and feed, from each safety and quality viewpoints. However, the Codex Alimentarius does not contain specific standards on recent or processed insects to be used as food and feed (www.codexalimentarius.org).

9.3 Legislation Relating to Edible Insects as Food

Before the most important introduction of insect food product into the market, the small-scale production and trade of edible insects was not thought of vital enough to be enclosed in legislation. Since rearing has been scaled up and bug food product became a phase in food retail, queries relating to correct regulation of this organic phenomenon emerged (Halloran et al. 2018). A transparent legislative framework was thus necessary. Worldwide, the Codex Alimentarius provides a group of tips, standards, and codes of observing to contribute to food safety and quality in an internationally uniform approach. Whereas the Codex features a major influence on food legislation, it is not wrongfully binding (Vandeweyer 2018). In Europe, human food and animal feed are strictly regulated at European level, whereas going away area for specific regulation on a national level. The EU Food Safety Authority (EFSA) and therefore Federal Agency for the safety of the food chain (FASFC) guard the food safety risks in Europe and Belgium, severally. Since the introduction of insects on the market in many European countries, EFSA and therefore the national restrictive agencies have taken action to begin constructing specific rules relating to insects for human and animal consumption. These rules ought to target microbiological criteria applicable on edible insects and therefore the standard of insects as a novel food. To boot, general European (Regulations) and Belgian (Royal Decrees) legislation applicable for animal-based food and feedstuffs, covering as an example aspects of traceability, hygiene, packaging, and Hazard Analysis and demanding management Points (HACCP) is additionally valid for edible insects (FASFC 2016).

9.4 Position of Edible Insects Worldwide

From a point of reading, there are three legal trends regarding edible insects. First, there are the Anglo-Saxon countries United Kingdom of Great Britain and Northern Ireland, USA, Canada, New Zealand and Australia, for whom edible insects do not represent a unique food, and therefore the food agencies have approved import and sales. Then there are the non-English-speaking Western countries, and therefore the international organization, above all, that have felt the necessity to possess rules and supply approvals before permitting any promotion. Non-Western countries comprise the remaining trend there, insects are usually a conventional food, however seldom prepackaged and exported or imported. In these countries, customs and therefore the FDA had never found themselves facing prepackaged product containing insects, as insects were typically found within the native market, unpackaged. And within the absence of rules these agencies have generally shown inconsistent reactions. There are cases wherever the promoting of edible insects is legal, however the import or export is not. Additionally, there is the matter of food legislation, that is commonly lacking business standards for insect foods. Above all, insects do not seem to be enclosed within the FAO/WHO/Codex (http://www.fao.org/FAO-WHO-codexalimentarius), that contains a world guideline for food safety. Customs offices, additionally, usually have problems to find reference points. Consonant system codes set internationally by the globe Customs Organization for the language of products do not contain any definition that refers to insects as food. The creation of recent codes is requested by a member state; however, the method can take most likely years.

9.5 Position of Edible Insects in Some Countries

9.5.1 Canada

Crickets do not seem to be thought of as a unique food, and today the most important breeder in North America is found in country Canada and serves some native startups, including One Hop kitchen. If, however, an insect lacks a history of safe consumption, it might fall back into the novel food class pending an analysis by the Bureau of Microbial Hazards in Food board of directors.

9.5.2 USA

There is no specific set of standards for edible insects in the USA. The FDA has made public its opinion, that is, the current legal basis for the market. To be allowed in market, insects should be bred for human consumption. A product containing insects should, after all, follow the standards required by the FDA together with bacteriological tests and good manufacturing practice certification. The label on the merchandise should mention the common name and therefore the insect's scientific name, and mention the potential risks of hypersensitivity reaction. Import from other countries is allowed, and therefore the USA FDA has already updated its Import Prior Notice with a list of edible insect products.

9.5.3 Australia and New Zealand

These countries share office for the upkeep of food safety called food standards. This agency has self-addressed some cases just like the super larva (*Zophobas morio*), the domestic cricket (*Acheta domesticus*) and therefore the lepidopteran (*Tenebrio molitor*), deciding that they are not novel foods, even supposing they cannot be thought of ancient foods either. Above all, they need nonetheless to encounter food safety issues and consequently have not been placed to the consumption limits or import (http://www.foodstandards.gov.au).

9.5.4 European Union

In 2015, the EU Parliament set that insects constitute the novel foods class, and consequently are subject to extended approval processes. From Gregorian calendar month 2018, the new Novel Food law is in situ, and therefore the application is meant to be simplified. In some countries there is a particular degree of tolerance (e.g., France). In others, like European nation, the tolerance is near to zero. Germany has been come terribly tolerant in 2018 and a few bug products are in supermarkets since then. Five countries did not settle for the 2015 call of the EU and since then has permissible and in some cases regulate the promotion and consumption of insects. These are Belgium, Britain, the Netherlands, Kingdom of Denmark, and Finland, and their position is delineated below.

9.5.5 Belgium

The office for the protection of the organic phenomenon has made a selected regulation for edible insects that made Belgium one in every of the foremost advanced nations in terms of entomophagy, though no insects bred outside of the EU Union are accepted. They updated their regulation in 2018 in keeping with the EU transmutation amount and that they are expected to stay with the EU rules from 2019 (http://www.afsca.be).

9.5.6   Netherlands

The Netherlands is home to some larva and cricket farms designed to breed insects for human consumption. These inlcude the leader Protifarm (and its subsidiary Kreca) and further some startups active in the promotion and production of edible insects. Its legal basis is not clear, though, and therefore the public body for food safety has refused to comment.

9.5.7   Kingdom of Denmark

The Danish Veterinary and Food Administration believes that whole insects (including flour, if made from whole insects) do not comprise the EU novel food legislation. As a result, imports from non-EU countries are feasible for those insects falling beneath the transmutation amount (e.g., mealworm and house crickets). Kingdom of Denmark is jumping ahead with edible insect initiatives (https://www.bugsolutely.com/denmark).

9.5.8   Finland

Finland has followed the Danish example, in 2017, releasing rules for import and sales of edible insects. As for the opposite countries that allowed edible insect before 2018, in 2018 they were within the transmutation amount. It is not clear what is going to happen in 2019.

9.5.9   Germany

The management of food in Germany may be a task for the 16 federal states. The Department of the Federal Government of Consumer Protection and Food Safety (BVL) fulfills just some coordination functions; thus its position is not wrongfully binding, and it is aligned with the EU commission decision: insects as novel food and cannot be sold in Germany till a procedure for novel food approval has been finalized. However in March 2018 underground cluster declared the launch of a larva alimentary paste, thus they got an area or federal inexperienced lightweight for it.

9.5.10 Norway

Norway is not an EU member, however belongs to the EU Economic space and so follows variety of European rules. Still, their interpretation of edible insects is that once they are whole (as hostile elements or isolates of insects), they are doing not comprise the novel food law. Import would be accepted if custom is cleared in an EU country.

9.5.11 Great Britain

For years, the Food Safety Agency has shown a good position in relation to the sale, consumption, and import of edible insects. Owing to Brexit, what would happen in the long run is unsure; however, presumably insects are going to be allowed to being sold on the market.

9.5.12 Switzerland

On December 2016, the council passed an edible insect law, permitting the sale and consumption of three species, crickets (*Acheta domesticus*), *Locusta migratoria*, and mealworm. Among the requirements, the insects should bred for human consumption and once slaughter should be treated in keeping with the factors of food security. The rules released by the food agency are terribly strict and complex. Within the case of import from non-EU countries, they need the insect to be whole, shipped solely by plane to Zurich or Geneva, and accompanied by an unbelievably long list of lab check and certificates.

9.5.13 Non-Western Countries

Southeast Asian countries have a practice of entomophagy, however, do not have rules with reference to the breeding, sale and export of insects. Thailand, the world's largest stock of crickets, has relaxed the rules for cricket farming (GAP Good Agricultural Practice) in 2017. Even in China, insects are a standard cooking ingredient in several regions, however, there is still no mention of this in food law. An exception, though, is silkworm pupae, that was enclosed in 2014 within the list of foods allowed by the Ministry of Health. The South Korea government launched a method to allow some edible insects in 2011. On the list there are larva, crickets (not the usual *Acheta domesticus*, however the *Gryllus bimaculatus* species) and a few larvae. Following this preliminary method, in 2016, the Korean Food and Drug Administration classified crickets and mealworms as traditional foods, while not restrictions.

9.6 Suggestions to Improve the Legislative Framework

The current lack of a well-elaborated legislative framework is a burden for the edible insect sector. In Belgium, a few specific microbiological requirements for edible insects are included in the FASFC circular of 2016 and some additional food safety and process hygiene recommendations are mentioned in a list of action limits composed by the FASFC. However, for other food sectors, a European legislative framework exists in the form of specific microbiological criteria embedded in Regulation (EG) N° 2073/2005. Likewise, it is suggested to compose specific criteria for edible insects as well (Vandeweyer 2018).

References

Banjo AD, Lawal OA, Songonuga EA (2006) The nutritional value of fourteen species of edible insects in southwestern Nigeria. Afr J Biotechnol 5:298–301

Byrne J (2011) FSA flags up potential of purified insect protein. Food Navigator. http://www.fao.org/3/i3253e/i3253e14.pdf. Accessed 30 July 2019

FASFC (2016) Circular concerning the breeding and marketing of insects and insect-based food for human consumption. Brussels, Belgium. http://www.afsca.be/levensmiddelen/omzendbrieven/_documents/2016-04-26_circob_NL_insecten_V2_clean.pdf

Halloran A, Flore R, Vantomme P, Roos N (2018) Edible insects in sustainable food systems. Springer International Publishing, Cham. https://doi.org/10.1007/978-3-319-740119

van Huis A, Van Itterbeeck J, Klunder H, Mertens E, Halloran A, Muir G, Vantomme P (2013) Edible insects: future prospects for food and feed security FAO forestry paper. FAO, Rome, p 171

Vandeweyer D (2018) Microbiological quality of raw edible insects and impact of processing and preservation. PhD. thesis, KU Leuven, Belgium.

Further Reading

http://www.afsca.be/foodstuffs/insects/
http://www.codexalimentarius.org
http://www.foodstandards.gov.au
https://www.bugsolutely.com/denmark
http://www.fao.org/FAO-WHO-codexalimentarius/about-codex/en/#c453333

Chapter 10
Sorghum Bug (*Agonoscelis pubescens*) as a Source of Edible Oil, Protein, and Gelatin

Abdalbasit Adam Mariod

Abstract The chemical composition, amino acids, and fatty acids of the sorghum bug indicate good nutritional value. The quality of its protein showed the sorghum bug to be a good and suitable source of edible protein. Analysis of dried adults showed high levels of crude protein and fat. The bug protein contained 16 known amino acids, including all the essential amino acids. Oleic, palmitic, linoleic, and linolenic acids were the predominant fatty acids in oil from the sorghum bug. The extracted insect oil was suitable for frying and biodeisel production.

Keywords *Agonoscelis pubescens* · Sorghum bug · Oil · Protein · Frying · Biodeisel · Stability

10.1 Introduction

Hundreds of insect species have been utilized as human food, including grasshoppers, caterpillars, beetles, termites, bees, and wasps. Grubs of the palm weevil are fried and eaten in several parts of western Nigeria, where the fried grubs are actively marketed. The caterpillar stage of *Imbrasia belina*, known as the mopane worm, is a popular item of diet in Botswana, Northern South Africa, Zimbabwe, and Namibia (Banjo et al. 2006).

In the course of the past few years there has been a new upsurge of interest in bugs as food. One possible responsible aspect is the increasing realization in the Western world that insects are traditionally and nutritionally essential meals for many non-European (non-EU) cultures (Foliart 1992). It has been postulated that bugs might be an important supply of proteins of suitable quality and excellent digestibility (Zhou and Han 2006).

A. A. Mariod (✉)
Indigenous Knowledge and Heritage Centre, Ghibaish College of Science & Technology, Ghibaish, Sudan

© Springer Nature Switzerland AG 2020   149
A. Adam Mariod (ed.), *African Edible Insects As Alternative Source of Food, Oil, Protein and Bioactive Components*, https://doi.org/10.1007/978-3-030-32952-5_10

The sorghum bug, *Agonoscelis pubescens*, belongs to the order Hemiptera (family Pentatomidae), commonly known in Sudan as dura andat. The mature insect is shield fashioned, approximately 11–13 mm in length and 6–7 mm wide; the top and underside of the frame have a pleasant silvery pubescence. *A. pubescens* is found in a number of African nations south of the Sahara (Sudan, Uganda, Tanzania, Somalia, Malawi, South Africa, Nigeria, Ghana, and Madagascar). In Sudan, the dura andat has a huge distribution in the central, eastern, western, and southern components of the country. The insects attack, in the population or nymph stages, some of the vegetation including lucerne (*Medicago sativa*), sunflower, wheat, and sesame. Especially, the adults infest sorghum for the duration of the milky phase. The insects feed on the developing grains, which become atrophied. The adults find safe haven throughout the dry season in clusters, which may include many heaps or hundreds of thousands of individuals, for about 9 months on the stems and branches of bushes and trees such as *Balanites aegyptiaca* (Heglig), *Zizyphus spina-christi* (Sidr), and *Phoenix dactylifera* (Nakhla) (Schmutterer 1969).

The physiological conditions of the inner organs of the bugs, specially of fat bodies, depend totally on the intensity of insect feeding on ripening sorghum grains during October and November. Adequate feeding all through this time span cumulates in sizable accumulations of metabolic compounds, especially fat. Inadequate feeding, however, results in an atrophied midgut and meager fat deposition. The fat content of resting insect pupae has been determined to be maximum at the beginning of the resting period when the bugs first arrive at the resting sites; females already contain more fat than the males. In fact, the very best fat content material is to be expected at the end of feeding in late November, shortly before the migration flight itself (Razig 1978).

10.2 Biochemistry and Nutrition of *Agonoscelis pubescens* (Sorghum Bug)

The chemical composition, amino acids, and dietary fulfillment of the protein of *Agonoscelis pubescens* have been investigated. The approximate evaluation of the mature *Agonoscelis pubescens* insect showed 7.6% moisture, 28.2% crude protein, 57.3% fat, and about 5% ash on a dry matter basis, respectively. The essential fatty acids of the sorghum bug oil (SBO) were palmitic, stearic, oleic, and linoleic acids. The oil of the sorghum bug showed a lower quantity of saturated fatty acids and better content of overall unsaturated fatty acids. The bug protein contained 16 known amino acids, including all the essential amino acids. In comparison with the amino acid profile advocated by FAO/WHO, the bug protein became of medium first class because of its content of critical amino acids. The mineral evaluation indicated excess P and K content. The two materials (crude protein and fat) are excessive on a dry matter basis, which makes the insect a desirable supply of protein and oil that can be a supplement to fat and protein foods (Mariod et al. 2011).

10.3   Sorghum Bug (*Agonoscelis pubescens*) as a Source of Edible Oil

Certain insects can be pests and at the same time be consumed by humans. One of the most well known pests in Sudan for sorghum in rainfed areas is the sorghum bug. In Western Sudan, sorghum bug adults are gathered and eaten after frying, even as in some areas of Sudan the accumulated insects are extracted via pressure and the acquired oil is utilized for cooking and a few medicinal uses. In Botana, Sudan, nomads use tar derived from suitably heated bugs for their camels in treatment of dermatological infections (Mariod 2005).

The oil content material of the sorghum bug (Fig. 10.1) could be very high compared with that of common oil seeds as well as groundnut, soybean, sunflower, or rapeseed with oil content between 20% and 45%. The oil content of dried ground sorghum adults is 60% (on dry count basis). The recovered oil from sorghum bugs is confirmed by the yellow coloration.

10.3.1   Sorghum Bug Oil Fatty Acids Composition

Sorghum bug oil (SBO) has been analyzed for its content, fatty acid, tocopherol, and sterol composition in addition to oxidative balance. Little information is available about this oil, although it has been extensively utilized in Sudan and some other African areas for generations, now not most effective as a meal but additionally as

Fig. 10.1 Sorghum bug adult insect (*SOR*) and its oil

Table 10.1 Fatty acid
composition, chemical, and
physical properties of
sorghum bug oil (SBO)

Components	SBO[a]
Fatty acid (%)[b]	
16:0	12.2 ± 0.2
16:1 n−7	1.0 ± 0.1
17:0	0.1 ± 0.2
18:0	7.3 ± 0.2
18:1 n−9	40.9 ± 0.2
18:2 n−6	34.5 ± 0.2
SAFA	20.8
MUFA	42.2
PUFA	35.6
Ratio UFA/SAFA	3.7
Other parameters	
Free fatty acids (%)	10.5 ± 0.1
Peroxide value (mEq O_2/kg)	5.1 ± 0.2
Kinematic viscosity at 40 °C (mm²/s)	27 ± 0.1
Total phenolics (mg/100 g oil)	0.95 ± 0.1
Induction period at 120 °C (h)	5.1 ± 0.3
Reduction of 50% DPPH	0.8 ± 0.10

SAFA saturated fatty acids, *MUFA* monounsaturated
fatty acids, *PUFA* polyunsaturated fatty acids, *UFA*
sum of MUFA and PUFA

[a]Data are reported as mean values of triple determinations ± standard deviation (SD) Sorghum bug oil (SBO) are codes for the two oils

[b]Other fatty acids (12:0, 14:0, 18:3, 20:0, 22:0, and 24:0) were present in amounts <1%

a treatment for sickness. The fatty acid composition is important for the use of the oil in one-of-a-kind packages. The predominant fatty acids of SBO (Table 10.1) are palmitic, stearic, oleic, and linoleic acids. Compared with the vegetable oils of Sudan (cottonseed, peanut, sesame, and sunflower oils), the SBO as investigated showed comparable quantities of saturated and unsaturated fatty acids. The SBO showed decreased amounts of saturated fatty acids (20.8%) and better content of overall unsaturated fatty acids (77.8%) (Table 10.1). The most important fraction in the oil is triacyglycerol (TAG), representing 71.6–85.3% of the total lipid (Table 10.2). Lipids from the insect oil contained a lower percentage of TAG (71.6%) and higher percentages of polar lipids (PL) (11.7%) (Table 10.2).

10.3.2 Sorghum Bug Oil Tocopherol and Sterol Content

The tocopherol content of meals is crucial to protect food lipids against autoxidation and thereby to increase their storage life and their value as wholesome foods (Kamal-Eldin and Appelqvist 1996). The tocopherol content of the SBO is given

Table 10.2 Lipid classes (%), tocopherol (mg/100 g oil), sterol (mg/100 g) of sorghum bug oil (SBO)

Lipid class	Percent (%)	Sterol content	mg/100 g oil
Triacylglycerol (%)	71.6	Cholesterol	2.2
Diacylglycerol (%)	13.6	Campesterol	11.6
Free fatty acids (%)	2.3	Stigmasterol	25.4
Polar lipids (%)	11.7	β-Sitosterol	268.8
Steryl esters (%)	0.8	Δ5-Avenasterol	16.3
Type of tocopherol	*mg/100 g oil*	Δ7-Avenasterol	1.6
α-T	0.88	Δ7-Stigmasterol	2.8
β-T	0.00	Others[a]	121.2
γ-T	32.16	Total	449.9
P8	0.21		
δ-T	0.78		
Total	34.03		

Data are means of triplicate results
TAG triacylglycerol, *DAG* diacylglycerol, *FFA* free fatty acids, *PL* polar lipids, *SE* steryl esters
[a]Others include 24-methylcholesterol, campestanol, chlerosterol, sitostanol, 5,24-stigmastadienol

in Table 10.2. The SBO had better quantities of tocopherols, 34.0 mg/100 g, in comparison with other common vegetable oils, including sesame oil, groundnut oil, or sunflower oil, in which the tocopherol content is between 27.9 and 97.6 mg/100 g.

The quantity of sterols in SBO is 449.9 mg/100 g (Table 10.2). The principal sterol of the oil is changed to β-sitosterol. In comparison with different oils commonly used in human food (DIN EN 14108 1997; Gunstone et al. 1994), SBO contained better amounts of overall sterols.

The oxidative balance of the oils is expressed as the induction duration determined by the Rancimat approach at 120 °C (see Table 10.1). The induction duration of SBO is 5.1 h (Table 10.1). This value is somewhat higher as compared with those of other safe-to-eat oils. The oxidative stability of such generally used oils ranges about 0.3 h for linseed oil and approximately 10 h for groundnut oil. The excessive quantities of tocopherols within SBO notwithstanding, the large part of polyunsaturated fatty acids in this oil appears to be particularly chargeable for its induction period. Consequently, in this situation the fatty acid composition has a far higher impact on the stability of the oil than the high amounts of antioxidants present.

10.3.3 Sorghum Bug Oil Phenolic Compounds

The quantity of overall phenolic compounds in the oil extracts from SBO was determined by means of the Folin-Ciocalteau assay (Silvia et al. 1984). The outcomes of this colorimetric approach, expressed as gallic acid equivalents, are shown in Table 10.1. The SBO extract contained 0.95 mg/100 g oil.

10.3.3.1 Antioxidant Activity of Sorghum Bug Oil Extracts

The antioxidant interest of phenolic compounds may additionally result from the neutralization of free radicals with beginning oxidation strategies, or from the termination of radical chain reactions. Because of this, two exclusive strategies were used for the dedication of the antioxidant interest of insect oil extracts: inhibition of β-carotene co-oxidation in a linoleate version system, and DPPH free radical scavenging: the effects showed lower absorbance of β-carotene in the presence of various methanolic extracts in the coupled system of β-carotene and linoleic acid. It became clear that the manipulated sample oxidized with maximum rapidity. A susceptible correlation among extract attention and antioxidant interest was determined: 0.2 mg of SBO extract had considerably higher ($p > 0.05$) antioxidant properties than 0.1 and 0.6 mg extracts. The methanolic extracts of the SBO have been markedly effective in inhibiting the oxidation of linoleic acid, and next bleaching of β-carotene, in contrast with the control. In Table 10.1, the lower absorbance of the DPPH radical (DPPH') resulting from its decrease with the aid of insect oil extracts is illustrated. The 50% loss in the DPPH radical because of its decrease by the insect oil antioxidants (SBO extracts) was expanded in the SBO. We want the most effective 0.8 mg extract of SBO that can attain 50% loss in DPPH.

10.3.4 Sorghum Bug Oil Volatile Compounds

Volatile compounds formed during frying and during different processing steps (degumming, neutralization, bleaching, deodorization) of SBO were adsorbed onto Tenax and analyzed by gas chromatography (GC) after thermal desorption. During different SBO processing steps, hexanal was the most prominent volatile compound. The total amount of volatiles in SBO seems to be very high, as can be clearly shown by the number and thickness of the peaks. The total amount of volatiles as well as the amount of hexanal decreased during different processing steps in all samples: the total amount decreased by more than 98%. The least total amount of volatiles was found in the deodorization step, as volatiles do not exist in larger amounts in deodorized oil because of the effect of the high-vacuum and -temperature treatments (Mariod et al. 2007). During frying using SBO, the total amount of volatiles decreased with frying time and the following aldhydes were identified in SBO: pentanal, hexanal, heptanal, octanal, nonanal, *trans*-2-hexenal, *trans*-2-heptenal, *trans*-2-octenal, *trans*-2-nonenal, *trans*-2-decenal, *trans*-2-undecenal, and decadienal, with nonenal and *trans*-2-nonenal as the most prominent volatile compounds (Mariod et al. 2007).

10.3.5 Steps for Processing Sorghum Bug Oil

In a refining test, on a laboratory scale crude sorghum oil (SBO) was processed by using alkali refining. Excellent modifications had been characterized with the dedication of free fatty acids, peroxide value, tocopherols, sterols, phosphatides, and balance toward oxidation (Rancimat test). Similarly, the fatty acid composition was determined. It is clear that the content material of phosphatides, peroxides, tocopherols, and sterols as well as oxidative stability were decreased at some point of processing and free fatty acids were almost completely eliminated. The content material of phosphorus was decreased in SBO, by using 19%, at the same time as whole oil processing eliminated 99% of free fatty acids in the crude oil. The level of overall tocopherols decreased at some stage in processing by as much as 83.8%. The coloration decreased through the processing steps up to bleaching; then in the deodorization step it darkened sharply. No change in the fatty acid composition was observed. The order of oxidation stability became deodorized > crude > degummed > neutralized = bleached. Overall sterols (367.1 mg/100 g) were reduced by 42% in the processed (deodorized) sorghum worm oil as compared with the crude oil (Mariod et al. 2006a).

10.3.6 Oxidative Stability of Sorghum Bug Oil

Mariod et al. (2008) maintained (autoxidized) SBO in the dark at 30° ± 2 °C for 24 months. Oil aliquots were been withdrawn every 2–4 months for analyses of modifications in four indices, specifically, fatty acid composition, tocopherol content, peroxide value, and oxidative balance index by using Rancimat After 24 months of storage, the fatty acid composition of the oil confirmed no change although the tocopherols were reduced. The storage oil confirmed a periodic growth in the peroxide value and had much less balance as measured by the Rancimat test.

10.3.7 Behavior of Crude Sorghum Bug Oil During Deep-Frying

The conduct of crude sorghum bug oil at some stage in deep-frying of par-fried potatoes was studied with reference to chemical, physical, and sensory parameters, such as the content material of FFA, tocopherols, polar compounds, oligomer triglycerides (TG), volatile compounds, oxidative balance, and total oxidation (TOTOX) price. The outcomes showed that the oil was appropriate for deep-frying of potatoes. The use of sorghum worm oil was not satisfactory after 6–12 h, considering the sensory assessment. Searching at the chemical and physical parameters, sorghum oil passed the limits for the amount of polar compounds, oligomer triglycerides, and free fatty acids, respectively, proposed by the German Society of Fat Sciences (DGF) as standards for the rejection of used frying oils, at 18 h at the latest. The outcomes showed that SBO was appropriate for deep frying of potatoes, but remarkable differences in palatable product were found (Mariod et al. 2006b).

10.3.8   Transesterification of Sorghum Bug Oil

The transesterification of vegetable oils constitutes a green method to offer a fuel with chemical residences that closely resemble those of diesel fuel. The percentage of reagents therefore influences the method in terms of conversion performance (Freedman et al. 1984), and this factor differs in line with the vegetable oil. Numerous researchers have diagnosed the maximum crucial variables that have an effect on the transesterification response, specifically, the response temperature, the type and quantity of catalyst, the ratio of alcohol to vegetable oil, the stirring rate, the reaction time, etc. (Dorado et al. 2004). The acquired biodiesel characteristics were studied according to the DIN EN 14214 specifications for biodiesel. Most of the biodiesel characteristics met the DIN specifications (water content, iodine number, phosphorus content). Mariod et al. (2006) stated that it was feasible to combine the methyl and ethyl esters catalyzed by means of H_2SO_4 from the SBO. They carried out initial assessments of the use of each methanol and ethanol under identical situations (catalyst, response time, temperature), and separated the ester layer from the glycerin layer without difficuly. Transesterification, the usage of ethanol, gave better yield than the usage of methanol in SBO samples. The insect oil supplies excessive biodiesel yield. The kinematic viscosity values of SBO samples have been higher than those of biodiesel standard limits (Mariod et al. 2006).

10.4   Sorghum Bug as a Source of Protein and Gelatin

In agreement with Mariod et al. (2011), *Agonoscelis pubescens* includes a high concentration of protein (28.2%). The overall amount of amino acids changed to 268.8 mg/g crude protein, less than the 864.2 mg/g crude protein in eggs suggested as a primary protein source in the human eating regimen. The proportion of sulfur-containing amino acids (methionine and cystine) in *A. pubescens* became 7.2% mg/g crude protein, which was considered to be an affordable quantity for its important characteristic in cellular approaches in oxidation and reduction (cystine); it acts as a methyl donor (methionien) in metabolism (Heimann 1980). The overall quantity of crucial amino acids determined in *A. pubescens* became 119.3 mg/g crude protein, which was better than the requirement of the FAO/WHO (113 g protein for adults). The stages of essential amino acids (EAAs) were similar to those of the FAO/WHO amino acid reference requirement (FAO/WHO 1991) set up for human adults. In keeping with the amino acid scores, the primary restricting amino acid of sorghum bug oil was lysine and the second was methionine + cysteine, followed by leucine (Mariod et al. 2011). In looking for new assets of protein and gelatin, researchers have investigated many wild plant species. According to Sudanese indigenous information, many insects have food and medicinal uses. Gelatin fit for human consumption was extracted from the sorghum worm insect using hot water, with moderate acid and distilled water. Sodium dodecyl sulfate–polyacrylamine gel electrophoresis (SDS-PAGE) of the insect gelatins showed very low molecular weight chains, and the two gelatins contained 40 kDa as the main aspect. Differential scanning calorimetry results showed a difference among extraction strategies regarding extracted gelatin quality. Fourier transform infrared spectroscopy (FTIR) spectra of sorghum bug gelatin absorption were located in six bands. Microstructures of the insect gelatin tested with the scanning electron microscope confirmed that the sorghum bug exhibited a satisfactory gelatin network with very small voids (Mariod 2013). When ice cream made by using using 0.5% insect gelatin was compared with that made using 0.5% commercial gelatin as a stabilizing agent, use of the insect gelatin was seen to be appropriate by the panelists, and no large variations among ice cream made using insect gelatin as compared with that made with commercial gelatin were seen in their choices (Mariod and Fadul 2015).

10.5 Conclusions

Insects are conventional meals in Sudanese cultures, particularly in western (Kordofan and Darfour) and Blue Nile states, wherein they have an essential role in human nutrition and can provide many nutrients. *Agonoscelis pubescens* (the sorghum bug), a pest on sorghum plants, also has very high nutritional quality and is certainly an excellent supply of protein, fatty acids, and various nutrients (minerals) that could be regularly provided in the growing regions. The consumption of this insect, consequently, has to be endorsed, and it might be ideal if this species could be expanded into cultivated as well as accumulated yield, which would change the food supply from being seasonal to a meal object that could available uninterruptedly. Also, advanced processing techniques are needed to provide readily available merchandise that is primarily insect based with a protracted shelf-life.

References

Banjo AD, Lawall OA, Songonuga EA (2006) The nutritional value of fourteen species of edible insects in southwestern Nigeria. Afr J Biotechnol 5:298–301

Dorado MP, Ballesteros E, Lopez FJ, Mittelbach M (2004) Optimization of alkali-catalyzed transesterification of *Brassica carinata* oil for biodiesel production. J Energy Fuels 18:77–83

FAO/WHO (1991) Protein quality evaluation. Report of joint FAO/WHO expert consultation. FAO food and nutrition paper 51. FAO/WHO, Rome

Foliart GD (1992) Insects as human food. Crop Prot 11:395–399

Freedman B, Pryde EH, Mounts TL (1984) Variables affecting the yields of fatty esters from transesterified vegetable oils. J Am Oil Chem Soc 61:1638–1643

Gunstone FD, Harwood JL, Padley FB (1994) The lipid handbook 2nd edn. Chapman and Hall, UK

Heimann W (1980) Fundamentals of food chemistry, 2nd edn. Horwood, Chichester

Kamal-Eldin A, Appelqvist L (1996) The chemistry and antioxidant properties of tocopherols and tocotrienols. Lipids 31:671–701

Mariod A (2005) Investigations on the oxidative stability of some unconventional Sudanese oils traditionally used in human nutrition. PhD thesis, Westfälische Wilhelms-University, Muenster-Germany

Mariod AA (2013) Insect oil and protein: biochemistry, food and other uses. Review Agric Sci 4(9B):76–80

Mariod AA, Fadul H (2015) Extraction and characterization of gelatin from two edible Sudanese insects and its applications in ice cream making. Food Sci Technol Int 21(5):380–391

Mariod AA, Klupsch S, Hussein IH, Ondruschka B (2006) Synthesis of alkyl esters from three unconventional Sudanese oils for the use as biodiesel. J Energy Fuels 20(5):2249–2252

Mariod AA, Matthäus B, Eichner K, Hussein IH (2006a) Effects of processing steps on the quality and stability of three unconventional Sudanese oils. Eur J Lipid Sci Technol 108(4):298–308

Mariod AA, Matthäus B, Eichner K, Hussein IH (2006b) Frying quality and oxidative stability of two unconventional oils. J Am Oil Chem Soc 83:529–538

Mariod AA, Matthäus B, Eichner K, Hussein IH (2007) Volatile compounds emissions from three unconventional Sudanese oils during alkali-refining and frying. Sudan J Agric Sci 8:157–167

Mariod AA, Matthäus B, Eichner K, Hussein IH (2008) Long-term storage of three unconventional oils. Grasas Aceites 1:16–22

Mariod AA, Bushra M, Abdel-Wahab SI, Siddig I, Nooraini MA (2011) Proximate, amino acid, fatty acid and mineral composition of two Sudanese edible insects. Int J Trop Insect Sci 31(3):145–153

Razig AA (1978) Control strategy against the millet bug, Agonoscelis pubescens (Hemiptera, Pentatomidae) based on a forecast system. In: Crop pest management in the Sudan. Proceedings of a symposium, July 1978, Khartoum, Sudan, pp 251–256

Schmutterer H (1969) Pest of crops in northeast and central Africa. Gustav Fisher, Stuttgart/ Portland

Silvia MT, Miller EE, Pratt D, Chia E (1984) Seeds as a source of natural lipid antioxidants. J Am Oil Chem Soc 61:928–931

Zhou J, Han D (2006) Proximate, amino acid and mineral composition of pupae of the silkworm Antheraea pernyi in China. J Food Compos Anal 19:850–853

Chapter 11
Watermelon Bug (*Aspongopus viduatus*) as a Source of Edible Oil, Protein, and Gelatin

Abdalbasit Adam Mariod

Abstract The amino acids, fatty acids, and nutritional quality of the watermelon bug (*Aspongopus viduatus*) show it to be a good and suitable source of edible oil and protein. The approximate analysis of *Aspongopus viduatus* adults showed 8.3% moisture, 27.0% crude protein, 54.2% fat, and 3.5% ash, respectively, on a dry matter basis. The bug protein contained 16 known amino acids, including all the essential amino acids. The predominant fatty acids in the melon bug oil were oleic, palmitic, linoleic, and linolenic acids. The extracted insect oil was suitable for cooking and biodeisel production.

Keywords *Aspongopus viduatus* · Watermelon bug · Oil · Protein · Biodiesel · Stability.

11.1 Introduction

Hundreds of insect species have been used as human food. Some of the most vital groups encompass grasshoppers, caterpillars, beetles, winged termites, bees, wasps, and a variety of aquatic insects. Grubs of the palm weevil are fried and eaten in several parts of western Nigeria, where the fried grubs are actively marketed. The caterpillar stage of *Imbrasia belina*, known as the mopane worm, is a popular item of the diet in Botswana, northern South Africa, Zimbabwe, and Namibia (Banjo et al. 2006).

During the past few years, there has been a new upsurge of interest in insects as meal items. One notable aspect is the growing focus within the Western world that insects are historically and nutritionally essential foods for many non-European cultures (De Foliart 1992). It has been postulated that insects might be an essential supply of proteins of proper quality and high digestibility (Zhou and Han 2006).

A. A. Mariod (✉)
Indigenous Knowledge and Heritage Centre, Ghibaish College of Science & Technology, Ghibaish, Sudan

© Springer Nature Switzerland AG 2020
A. Adam Mariod (ed.), *African Edible Insects As Alternative Source of Food, Oil, Protein and Bioactive Components*, https://doi.org/10.1007/978-3-030-32952-5_11

159

Watermelon (*Citrullus vulgaris*) is a crop cultivated in rainfed locations in West Kordofan and the Darfor area in Sudan. The crop is attacked by watermelon insects that suck juice from the fruit and seed. *Aspongopus viduatus* (the melon bug) is a worm 20 mm in height, belonging to the order Hemiptera, whose members are referred to as "genuine bugs." These insects have unique front wings, referred to as hemelytra, wherein one half is leathery and the apical half is membranous (http://www.Science.mcmaster.ca). Melon bugs are extensively dispersed in Sudan, especially within the western areas (Kordofan and Darfor states), in which field watermelons are considered as one of the most important crops in conventional rainfed agriculture. For small farmers of those states, watermelons are strategic vegetation because of their multiple uses, from their utility as a major source for consuming water at some stage in summertime months in a few remote regions to the utility of extraction residues of the fruit in animal nutrients. The adult bug is considered to be the primary pest of watermelons. The nymphs pierce leaves, stems, and suck the sap, resulting in wilting, fruit drop, and plant death. In the faraway territories of Sudan, oil from the bug *Aspongopus viduatus* (Pentatomidae) is used as a sweet oil; there are few statistics regarding the poisonous impact of this oil. This bug oil corresponds in its essential contents with maximum values of animal oils (Tauscher et al. 1981). An incorporated pest control application was designed with the aid of the Elobied Agricultural Studies Station (North Kordofan state of Sudan) with enhancement of network participation in a so-called "Handpicking of melon bug adults" in four unique regions of Northern Kordofan state for two seasons. At some point in those seasons, a total of 15 t of melon bug adults were gathered within the first season and 226 lots in the second (Bashir et al. 2002). Melon bugs are suitable for eating, and in the ultimate nymph stage, a very soft level, the insects are cooked and eaten. Many Namibians collect the adults and use them as a delight in, or as a spice in a ground form for, cooking food (http://www.natmus.cul.na). In the Western Kordofan state of Sudan, the bug is thought of domestically as "Um-buga" and is used as food. The oil is extracted by means of boiling the insects in hot water. The oil is utilized in cooking (at some points of famine and shortage of food) and in a few medicinal packages, consisting of a skin lesion remedy (Mariod et al. 2004). In faraway territories of Sudan, oil from these bugs is used as sweet oil. No toxic effect of this oil has been described, and the fatty acid composition corresponds to that of most animal oils (Tauscher et al. 1981).

11.2   Biochemistry and Nutrition of *Aspongobus viduatus* (Melon Bug)

The approximate analysis of dry ground melon bug adults showed that moisture, crude protein, fat, ash, and carbohydrate content were 8.3%, 27.0%, 54.2%, 3.5%, and 7.0%, respectively. The concentrations of major elements such as Mg, Ca, K, P, and Na in the melon bug were 301.1, 1021.2, 200.08, 1234.30, and 402.1 mg/100 g, respectively (Mariod et al. 2011a).

11.3   *Aspongobus viduatus* (Melon Bug) as a Source of Edible Oil

11.3.1   Aspongobus viduatus *(Melon Bug) Oil Fatty Acid Composition*

Little information is available regarding melon bug oil, even though it has been extensively used in Sudan and some other African nations for generations: no longer most effectively as a meal, but also as a remedy for sickness. The fatty acid composition is vital for using the oil in exceptional applications. The major fatty acids of the melon bug oil (Table 11.1) are palmitic, stearic, oleic, and linoleic acids. The melon bug oil (MBO) showed a higher amount of saturated fatty acids (37.4%) and a high content material of unsaturated fatty acids (61.5%) (Mariod et al. 2007).

Table 11.1 Fatty acid composition, chemical and physical properties of melon bug oil (MBO)

Components	MBO[a]
Fatty acid (%)[b]	
16:0	30.9 ± 0.3
16:1n−7	10.7 ± 0.2
17:0	2.4 ± 0.1
18:0	3.5 ± 0.1
18:1 n−9	46.6 ± 0.3
18:2 n−6	3.9 ± 0.1
SAFA	37.4
MUFA	57.5
PUFA	3.9
Ratio UFA/SAFA	1.6
Other parameters	
Free fatty acids (%)	3.0 ± 0.1
Peroxide value (mEq O_2/kg)	2.1 ± 0.2
Kinematic viscosity at 40 °C (mm^2/s)	35 ± 0.1
Total phenolics (mg/100 g oil)	20.66 ± 0.3
Induction period at 120 °C (h)	38 ± 0.4
Reduction of 50% DPPH	0.13 ± 0.04

SAFA saturated fatty acids, *MUFA* monounsaturated fatty acids, *PUFA* polyunsaturated fatty acids, *UFA* sum of MUFA and PUFA
[a]Data are reported as mean values of triplicate determinations ± standard deviation (SD)
[b]Other fatty acids (12:0, 14:0, 18:3, 20:0, 22:0, and 24:0) were present in amounts <1%

11.3.2 Aspongopus viduatus *Oil: Tocopherol and Sterol Content*

The tocopherol content of ingredients is crucial to protect meal lipids from autoxidation and, thereby, to increase their storage existence and their value as healthful ingredients (Kamal-Eldin and Appelqvist 1996). The tocopherol content of the melon bug oil (MBO) is given in Table 11.2. The MBO had a low quantity of tocopherols, 0.3 mg/100 g compared with other not unusual vegetable oils, including sesame oil, groundnut oil, or sunflower oil, wherein the quantity of tocopherols is between 27.9 and 97.6 mg/100 g (Mariod et al. 2007).

The amount of sterols in MBO is 17.5 mg/100 g (Table 11.2). The principal sterol of the oil is β-sitosterol; compared with different oils commonly utilized in human food, lower amounts of total sterols were determined within the MBO.

The oxidative balance of the oils is expressed as the induction length determined by means of the Rancimat technique at 120 °C (Table 11.1). Oil from the watermelon bug showed a remarkably excessive stability (38.0 h). This value could be very high compared with those of other safe-to-eat oils. The oxidative stability of such commonly used oils varies between 0.3 h for linseed oil and about 10 h for groundnut fats. The high oxidative stability of the MBO is perhaps a result of the low amounts of polyunsaturated fatty acids along with linoleic and linolenic acids (Mariod et al. 2007).

Table 11.2 Lipid classes (%), tocopherol (mg/100 g oil), and sterol (mg/100 g oil) content of melon bug oil (MBO)

Lipid class	Percent (%)	Sterol contents	mg/100 g oil
Triacylglycerol (%)	74.9	Cholesterol	1.4
Diacylglycerol (%)	5.8	Campesterol	1.8
Free fatty acids (%)	2.7	Stigmasterol	0.8
Polar lipids (%)	14.4	β-Sitosterol	10.6
Steryl esters (%)	2.2	Δ5-Avenasterol	0.5
Type of tocopherol	*mg/100 g oil*	Δ7-Avenasterol	0.0
α-T	0.17	Δ7-Stigmasterol	0.9
β-T	0.00	Others[a]	1.5
γ-T	0.13	Total	17.5
P8	0.00		
δ-T	0.00		
Total	0.30		

Data are means of triplicate results
TAG triacylglycerol, *DAG* diacylglycerol, *FFA* free fatty acids, *PL* polar lipids, *SE* sterylesters
[a]Others include 24-methylcholesterol, campestanol, chlerosterol, sitostanol, 5,24-stigmastadienol

11.3.3   Melon Bug Oil Phenolic Compounds

Mariod et al. (2007) used a colorimetric method to determine the quantity of total phenolic components in the oil extracts from MBO as gallic acid equivalents. The quantity of total phenolic in the oil extracts from MBO was 20.66 mg/100 g oil, which said to be a high amount of phenolics (Mariod et al. 2007).

11.3.3.1   Antioxidant Activity of Melon Bug Oil Extracts

The antioxidant activity of phenolic chemical compounds may result from the neutralization of free chemical groups initiating oxidation outgrowth, or from the final result of radical chain reactions. For this reason, Mariod et al. (2007) utilized two different methods for the determination of the antioxidant bodily function of melon hemipteran insect pest crude oil extracts: inhibition of β-carotene co-oxidization in a linoleate model system, and DPPH free radical scavenging, revealed the diminution in absorbance of β-carotene in the presence of different methanolic extracts in the coupled system of β-carotene and linoleic acid. It was clear that the dominance sample oxidized most rapidly. A weak correlation between extract concentration and antioxidant activity was noted: 0.2 mg of crude oil extract had significantly higher ($p > 0.05$) antioxidant properties than 0.1 and 0.6 mg extracts. The methanolic extract of the oil changed to be markedly powerful in inhibiting the oxidization of linoleic acid and subsequent bleaching of β-carotene in contrast with the control. The DPPH˙ radical is considered to be a model of a solid lipophilic radical. A sequence response in lipophilic chemical groups was initiated by means of lipid autoxidation. Antioxidants react with DPPH˙, reducing a variety of DPPH˙ molecules identical to the wide variety in their available hydroxyl group. Consequently, the absorption at 517 nm was proportional to the sum of residual DPPH˙ (Xu et al. 2005). In Sect. 11.1, the decrement in absorbance of the DPPH radical (DPPH˙) caused by its decreasing via insect oil extracts is illustrated. The 50% deprivation in DPPH radical because of its reduction by means of insect oil antioxidant (vegetable oil extracts) is multiplied in MBO; we want the best 0.13 mg extract of melon bug oil to reach 50% loss.

11.3.4   Steps for Processing Melon Bug Oil

In a refining experiment, on a laboratory scale crude melon bug oil (MBO) was processed through alkali refining. Modifications have been characterized via the dedication of free fatty acids, peroxide value, tocopherols, sterols, phosphatides, and balance toward oxidation (Rancimat test). In addition the fatty acid composition was determined. It is clear that the content of phosphatides, peroxides, tocopherols, and sterols, in addition to oxidative stability, were decreased for the duration of processing at the same time that free fatty acids had been almost definitely eliminated. The content of phosphorus become decreased in SBO, by 19%, whereas the full oil processing eliminated 99% of free fatty acids in the crude oil. The overall level of tocopherols was reduced by 83.8% at some point of processing. The color was reduced through the processing steps, as much as bleaching, then in the deodorization step it darkened sharply. No alteration in the fatty acid composition was observed. The order of oxidation balance became deodorized > crude > degummed > neutralized = bleached. Total sterols (367.1 mg/100 g) decreased by 42% in the processed (deodorized) sorghum bug oil compared with the crude oil (Mariod et al. 2006a).

11.3.5   Oxidative Stability of Watermelon Bug Oil

The oxidative balance of MBO stated as the induction duration measured by way of the Rancimat technique at 120 °C is without a doubt high (38.0 h) as compared to other edible oils. As an example, the oxidative balance of sesame oil is 1.6 h and that for sunflower oil is 5.4 h. The excessive oxidative balance of MBO can be caused by the low amounts of polyunsaturated fatty acids (PUFA), which include linoleic and linolenic acid (Mariod 2013). Mariod et al. (2008) stored (autoxidized) melon bug oil in the dark at $30° \pm 2$ °C for 24 months. Oil aliquots were withdrawn every 2–4 months for investigation of changes in four quality indices, specifically, fatty acid composition, tocopherol content, peroxide value, and oxidative balance index with the aid of Rancimat. After 24 months of storage the fatty acid composition of the oil confirmed no change even as tocopherols were reduced. The MBO showed the most effective moderate changes in their oxidative balance as indicated by way of the peroxide value and induction duration throughout the 24 months of storage.

Mariod et al. (2005) described the better oxidative stability of sunflower kernel oil (SKO) by mixing with some Sudanese oils. Mixes (9:1, 8:2, 7:3, 6:4 w/w) of sunflower oil with melon bug (*Aspongopus viduatus*) oil (MBO) were investigated with consideration of the fatty acid content, oxidative stability (Rancimat 120 °C), and stability at 70 °C, measured by the increase in peroxide value. With increasing ratio of MBO in SKO, the content of linoleic acid decreased from 46.3% to 30.1% and the measure of oleic acid increased from 41.3% to 43.9%. By mixing SKO with MBO, oxidative stability in the Rancimat test was changed from 5% to 68% compared to the sunflower kernel oil as control with increasing parts of MBO. Storage of the mixes at 70 °C revealed that the increase in peroxide value as a measure for oxidative deterioration was remarkable lower for the mixtures of MBO with SKO than for pure SKO. The study demonstrates a way to improve the stability of sunflower oil by blending with *Aspongopus viduatus* oil.

11.3.6 Transesterification of Watermelon Bug Oil

The transesterification of vegetable oils constitutes a green technique to offer a fuel with chemical properties that might closely resemble those of diesel fuel. The percentage of reagents thus affects the manner in terms of conversion performance (Freedman et al. 1984), and this element differs according to the vegetable oil. Numerous researchers have recognized the maximum essential variables that affect the transesterification response, specifically, response temperature, kind and quantity of catalyst, ratio of alcohol to vegetable oil, stirring rate, reaction time, etc. (Doraco et al. 2004). The obtained biodiesel properties were investigated following the DIN EN 14214 standards for biodiesel. Most of the biodiesel properties fit the DIN standards (water content, iodine number, phosphorus content). Mariod et al. (2006b) reported that the methyl and ethyl esters catalyzed by H_2SO_4 can be made from melon bug oil (MBO). They conducted a preliminary test utilizing both methanol and ethanol under the same conditions (catalyst, reaction time, temperature). These authors separated the ester layer easily from the glycerin layer. Transesterification utilizing ethanol resulted in higher yield than using methanol in MBO samples. The insect oil gave a high biodiesel yield. The kinematic viscosity values of MBO samples were higher than those of biodiesel specification limits. The viscosity values of samples transesterified utilizing ethanol were lower than of those using methanol but all the viscosity values were above the DIN EN 14214 specifications for biodiesel. All the viscosity values were above diesel fuel viscosity (Mariod et al. 2006b).

11.3.7 Quality and Stability of Deodorized MBO

Mariod et al. (2012) deodorized melon bug oil for unique periods and at unique temperatures. They investigated the quality changes and balance in opposition to oxidation (Rancimat test) besides the fatty acid composition. These authors noted a considerable decrease in oil quality with increase of deodorization temperature. Tocopherols were decreased because of accelerated temperature, and can be completely removed at some point of deodorization temperatures. The fatty acid composition of the deodorized oils did not go through any modifications during deodorization, although oxidative stability was affected by temperature and time of deodorization.

11.3.7.1 Antibacterial Activities of Melon Bug Oil

The antibacterial activity of melon bug unrefined oil, silicic acid column-purified oil, and phenolic compound-free oil was investigated by the agar diffusion method against seven bacterial isolates. Four of these are food-related bacterial pathogen species: *Staphylococcus aureus*, *Salmonella enterica* serovar *paratyphi*, *Escherichia coli*, and *Bacillus cereus*; the three other isolates are *Bacillus subtilis*, *Enterococcus faecalis*, and *Pseudomonas aeruginosa*. The main components of this oil were evaluated. The unrefined oil and the phenolic compound-free oil gave high antibacterial activities against some test species, whereas the silicic acid column-purified oil produced no antibacterial activity. The study reports the possibility of utilizing this oil in food preservation (Mustafa et al. 2008).

11.4 Melon Bug as a Source of Protein and Gelatin

According to Mariod et al. (2011a), melon bug meal contains 28.2% of dry weight as protein and 360.5 mg/g of the crude protein as total amino acids; 57.1% of the crude protein is methionine and cystine (sulfur-containing amino acids), amino acids that are important in the oxidation and reduction system (cysteine) and act as a methyl donor (methionien) in metabolism. Of the crude protein, 208.5 mg/g was essential amino acids, which was higher than the recommendation of the FAO/WHO (113 g protein for adults). According to the amino acid scores, the first limiting amino acids of the melon bug protein were lysine and the second was leucine, followed by phenylalanine + tyrosine (Mariod et al. 2011a).

According to Sudanese indigenous knowledge, many insects have food and medicinal uses, and so can be a good source for Halal gelatin. Mariod et al. (2011b) extracted edible gelatin from dried adult melon bug insects using hot water and mild acid and distilled water. Sodium dodecyl sulfate–polyacrylamine gel electrophoresis (SDS-PAGE) patterns of the obtained insect gelatins had very low molecular weight chains, and this gelatin contained 40 kDa as the main component. Differential scanning calorimetry results confirmed the difference between extraction methods concerning the extracted gelatin quality. Fourier transform infrared spectroscopy (FTIR) spectra of melon bug gelatin showed absorption in more than six bands. Microstructures of the insect gelatin examined with the scanning electron microscope showed that melon bug gelatin showed a finer structure with smaller protein strands and voids (Mariod 2013). Ice cream made by using 0.5% insect gelatine was compared with that made using 0.5% commercial gelatine as the stabilizing agent. The properties of the ice cream produced using insect gelatin were found to be acceptable for the panelists, and no significant differences between ice cream made using insect gelatin when compared with that made using commercial gelatin were found in their general preferences (Mariod and Fadul 2015).

11.5 Melon Bug as an Animal Feed

Jadalla et al. (2014) evaluated the impacts of including dried defatted watermelon bug to replace cereals in the diet on broiler chick performance. They fed 200 broiler chicks a pre-starter diet for 1 week and then offered five diets prepared using defatted bugs in amounts of 0%, 15%, 30%, 45%, and 60%. The chicks were weighed weekly during the experimental period. The results indicated that feed intake of broiler chicks increased significantly ($p < 0.05$) with the addition of dried defatted watermelon bug. Weight gains and feed conversion ratio were also significantly improved ($p < 0.05$).

11.6 Conclusions

Insects are unconventional foods in Sudanese cultures, mainly in the west of the country, and in Blue Nile state, where they have an important role in human nutrition and offer many nutrients. The watermelon bug, which is a pest for watermelon, also has high nutritional quality and is indeed a good source of protein, fatty acids, and other nutrients (minerals) that are often in short supply in developing areas. The consumption of this insect, therefore, should be encouraged, and it would be ideal if this species were to be cultivated as well as collected, which would change the food source from being seasonal to a continuously available food item. Also, processing methods are needed to make the insects into readily available insect-based products with a long shelf-life.

References

Banjo AD, Lawall OA, Songonuga EA (2006) The nutritional value of fourteen species of edible insects in southwestern Nigeria. Afr J Biotechnol 5:298–301

Bashir YGA, Ali MK, Ali KM (2002) IPM options for the control of field Watermelon pests in Western Sudan a paper presented in the 67th meeting of the pests and diseases committee Agricultural Research Corporation ARC July 2002 Medani Sudan (unpublished)

De Foliart GR (1992) Insects as human food. Crop Prot 11:395–399

Dorado MP, Ballesteros E, Lopez FJ, Mittelbach M (2004) Optimization of alkali-catalyzed transesterification of *Brassica carinata* oil for biodiesel production. J Energy Fuels 18:77–83

Freedman B, Pryde EH, Mounts TL (1984) Variables affecting the yields of fatty esters from transesterified vegetable oils. J Am Oil Chem Soc 61:1638–1643

Jadalla JB, Mekki DM, Bushara I, Habbani AMH (2014) Effects of inclusion of different levels of watermelon bug meal in broiler diets on feed intake, body weight changes and feed conversion ratio. Global J Anim Sci Res 2(1):18–25

Kamal-Eldin A, Appelqvist L (1996) The chemistry and antioxidant properties of tocopherols and tocotrienols. Lipids 31:671–701

Mariod AA (2013) Insect oil and protein: biochemistry, food and other uses. Rev Agric Sci 4(9B):76–80

Mariod AA, Fadul H (2015) Extraction and characterization of gelatin from two edible Sudanese insects and its applications in ice cream making. Food Sci Technol Int 21(5):380–391

Mariod AA, Matthäus B, Eichner K (2004) Fatty acid, tocopherol and sterol composition as well as oxidative stability of three unusual Sudanese oils. J Food Lipids 11:179–189

Mariod AA, Matthäus B, Eichner K, Hussein IH (2005) Improving the oxidative stability of sunflower oil by blending with *Sclerocarya birrea* and *Aspongopus viduatus* oils. J Food Lipids 12:150–158

Mariod AA, Matthäus B, Eichner K, Hussein IH (2006a) Effects of processing steps on the quality and stability of three unconventional Sudanese oils. Eur Lipid Sci Technol 108(4):298–308

Mariod AA, Klupsch S, Hussein IH, Ondruschka B (2006b) Synthesis of alkyl esters from three unconventional Sudanese oils for the use as biodiesel. J Energy Fuels 20(5):2249–2252

Mariod AA, Matthäus B, Eichner K, Hussein IH (2007) Fatty acids composition, oxidative stability and transesterification of lipids recovered from melon and sorghum bugs. Sudan J Sci Technol 8:16–20

Mariod AA, Matthäus B, Eichner K, Hussein IH (2008) Long-term storage of three unconventional oils. Grasas Aceites 1:16–22

Mariod AA, Bushra M, Abdel-Wahab SI, Ibrahim SI, Nooraini MA (2011a) Proximate, amino acid, fatty acid and mineral composition of two Sudanese edible insects. Int J Trop Insect Sci 31(3):145–153

Mariod AA, Abdelwahab SI, Ibrahim MY, Mohan S, Abd Elgadir M, Ain NM (2011b) Preparation and characterization of gelatins from two Sudanese edible insects. J Food Sci Eng 1:45–55

Mariod AA, Matthäus B, Eichner K, Hussein IH, Abdelwahab SI, Abdul AB (2012) Effects of deodorization on the quality and stability of three unconventional Sudanese oils. J Food 37(4):189–196

Mustafa NEM, Mariod AA, Matthäus B (2008) Antibacterial activity of the *Aspongopus viduatus* (melon bug) oil. J Food Saf 28:577–586

Tauscher B, Müller M, Schildknecht H (1981) Composition and toxicology of oil extracts (edible oil) from *Aspongopus viduatus*. Chem Mikrobiol Technol Lebensm 7:87–92

Xu J, Chen S, Hu Q (2005) Antioxidant activity of brown pigment and extracts from black sesame seed (*Sesamum indicum* L.). Food Chem 91:79–83

Zhou J, Han D (2006) Proximate, amino acid and mineral composition of pupae of the silkworm *Antheraea pernyi* in China. J Food Compos Anal 19:850–853

Further Reading

http://www.natmus.cul.na
http://www.Science.mcmaster.ca

Chapter 12
Nutritional Composition of African Edible Acridians

Sévilor Kekeunou, Alain Simeu-Noutchom, Marcelle Mbadjoun-Nziké, Mercy Bih Achu-Loh, Patrick Akono-Ntonga, Alain Christel Wandji, and Joseph Lebel Tamesse

Abstract The current inadequacy between the accelerated growth of the African population and the availability of nutrients has led us to study the nutritional composition of Acridian. This chapter reviews work on edible Acridians in Africa and assesses their nutritional value and potential impact on malnutrition in Africa. The results show that about 74 Acridians are eaten in Africa, amongst which the nutritional value of *Zonocerus variegatus* (Linnaeus, 1758), *Kraussaria angulifera* (Krauss, 1877), *Acanthacris ruficornis citrina* (Serville, 1839), *Locusta migratoria* (Linnaeus, 1758), *Cyrtacanthacris aeruginosus unicolor* (Stoll, 1813), *Schistocerca gregaria* (Forskål, 1775) and *Anacridium melanorhodon* (Walker, 1870) were studied. These Acridians have a high protein content with good levels of essential amino acids. They are rich in polyunsaturated fatty acids, vitamins, minerals but poor in carbohydrates. Although their nutrients vary from one species to another, depending on the stages of development and the habitat, edible Acridians can play an important role in the fight against malnutrition in Africa, if they are well exploited.

Keywords Africa · Foods · Acridian · Nutritional value · Minerals · Organic elements

S. Kekeunou (✉) · A. Simeu-Noutchom · M. Mbadjoun-Nzike · A. C. Wandji
Laboratory of Zoology, Faculty of Science, University of Yaoundé I, Yaoundé, Cameroon
e-mail: skekeunou@gmail.com

M. B. Achu-Loh
Laboratory for Food Science and Metabolism, Faculty of Science, University of Yaoundé I, Yaoundé, Cameroon

P. Akono-Ntonga
Laboratory of Animal Biology and Physiology, Faculty of Science, University of Douala, Douala, Cameroon

J. L. Tamesse
Laboratory of Zoology, Higher Teachers Training College, University of Yaounde I, Yaounde, Cameroon

© Springer Nature Switzerland AG 2020
A. Adam Mariod (ed.), *African Edible Insects As Alternative Source of Food, Oil, Protein and Bioactive Components*, https://doi.org/10.1007/978-3-030-32952-5_12

12.1   Introduction

The inadequacy that exists today between the accelerated growth of the African population and the availability of nutrients poses the need to seek new food resources
(Courade 1996). In fact, in the twenty-first Century, famine and malnutrition still affect
several countries in South America (Bolivia, Paraguay, Peru), Asia (Mongolia,
Pakistan), but the countries most affected are in Africa. The situation is all the more
difficult as we find ourselves in an African context in which the demographic forecasts
for the next 100 years show that this situation will not improve (United Nations 2011).

For example, people in the Northern parts of Cameroon are facing a severe nutrient deficiency that affects mainly women and children: 25% are affected by protein–
energy malnutrition, 50.5% by protein, 5% have anemia (Ali et al. 2010). Current
vertebrate protein sources are expensive in domestic markets, yet several recent
studies show that the most diverse and numerous animals such as Hexapods can
play an important role in this quest for new sources of protein (Kelemu et al. 2015,
Aman et al. 2016; Kekeunou and Tamesse 2016; Tamesse et al. 2016). In many parts
of the world, edible insects have long played a key role in meeting human nutritional needs (Banjo et al. 2006). Aman et al. (2016) showed that more than 524
insect species are consumed in Africa.

The greatest diversity of edible insect species in Africa can be found in Orders
such as Lepidoptera, Orthoptera, Coleoptera, Isoptera, Hymenoptera, and
Hemiptera. Recent studies indicate that the protein and energy content of insects
may be comparable to that of conventional meat sources (Ramos-Elorduy et al.
2012; Chakravorty et al. 2014). The protein content of these edible insect is higher
than that of other foods of animal or plant origin (Teffo et al. 2007). Their nutritional
composition varies from one species to another and within the same species, depending on the stage of development, habitat, and diet (Rumpold and Schlüter 2013).
The crude protein content of insects ranges from 40 to 75% dry weight (Rumpold
and Schlüter 2013; Verkerk et al. 2007), with a beneficial amino acid profile and
variable fat content (>50% in some species) (Verkerk et al. 2007). Among the
Orthoptera, the most consumed in Africa are crickets and Acridian (Malaisse 2004).

In this article, we have made an inventory of edible acridians in Africa and present their nutritional composition in order to make available to the scientific community, the knowledge acquired on the subject and make recommendations to
African organizations, sub-regions and the States on the possibilities of using
Acridians in the fight against malnutrition in the continent.

12.2   Edible Acridians in Africa

Acridians, grouped within the infraorder Acrididea, have at least 12,000 species
distributed in seven superfamilies: Acridoidae (11 families and 7680 species),
Eumastacoidea (8 families and 1269 species), Pneumoroidea (1 family and 17

species), Pyrgomorphoidea (1 family and 476 species), Tanaoceroidea (1 family and 3 species), Trigonopterygoidea (2 families and 16 species), and Tetrigoidea (Hujon-Song 2010).

About 74 species are consumed in Africa by humans (Table 12.1). All these species belong to Acridoidea and Pyrgomorphoidea. These Acridians are usually caught in plantations (Ali et al. 2010). Among the species consumed (Aman et al. 2016), *Schistocerca gregaria* (Forskål), *Locusta migratoria migratorioides* (Reiche and Fairmaire), *Nomadacris septemfasciata* (Serville), *Locustana pardalina* Walker, and *Anacridium melanorhodon* (Walker) are of continental importance (Kelemu et al. 2015). In Southern Africa, *N. septemfasciata* and *L. pardalina* appear to be the dominant species, particularly in South Africa, Zambia, Botswana, and Lesotho (Table 12.1). *A. melanorhodon melanorhodon* is consumed in Sudan and South Sudan (Hassan et al. 2007). *Zonocerus variegatus* (Linnaeus) is consumed in 19 African countries, both in Central Africa and in West and East Africa (Chiffaud and Mestre 1990; De Visscher 1990; Kekeunou and Tamesse 2016). The eating habits of Acridians influence their chemical composition (Finke et al. 1989). Other factors such as sex, stage and environmental factors (temperature, day length, humidity, light intensity, etc.) can also influence the chemical composition of insects (Finke et al. 1989).

Acridians are eaten cooked: grilled directly on fire or on embers (Barreteau 1999), fried (Idowu and Modder 1996; Adeoye 2014), roasted or fried and consumed in the sauce (Ali et al. 2010). In Nigeria (West Africa), *Z. variegatus* is sundried or roasted with salt and pepper after wings, elytra and head have been removed (De Visscher 1989). In Cameroon, this Grasshopper is fried with a mixture of spices in well-heated red palm oil before consumption. The wings and the contents of the digestive tract are removed before cooking to avoid intoxication (Kekeunou and Tamesse 2016).

12.3   Organic Matter Composition of Edible Acridians

Nutrients in Acridians include proteins, carbohydrates, fats, and vitamins.

12.3.1   Proteins

The study shows that some Acridians contain about 56–77% protein in dry weight (D.W) and this varies according to the species (Badanaro et al. 2015; Aman et al. 2016). *Z. variegatus*, consumed in 19 African countries (Kekeunou and Tamesse 2016), contains 58.0 g/100 g of crude protein (Solomon et al. 2008), with a good amino acid composition. 17 amino acids are present in the proteins of *Z. variegatus* with contents varying between 6.5×10^2 and 133.7×10^2 mg/100 g of protein. Glutamic acid has the highest content, while cysteine has the least (Adeyeye 2005). These amino acids are

Table 12.1 List of Acridian species consumed in Africa

Family/subfamily	Species	Countries and reference
Acrididae	*Acanthacris ruficornis* (Fabricius 1787)	Sahel, Congo, Niger, Zambia, Zimbabwe, DRC, Zambia, South Africa, Cameroon, Congo, CA Republic, Zimbabwe, Burkina Faso, Malawi, Mali, Niger, Togo, Benin (Chavanduka 1976; Nkouka 1987; Bani 1995; Mbata 1995; Van Huis 2003; Kelemu et al. 2015)
	Acanthacris ruficornis citrina (Serville 1838)	Cameroon (Barreteau 1999; Van Huis 2003; Zakari et al. 2015)
	Acanthacris nigrovariegata (Bolivar 1889)	Zambia (Mbata 1995; Van Huis 2003)
	Aiolopus thalassinus thalassinus (Fabricius 1781)	Togo, Burkina Fasso (Van Huis 2003; Tchibozo et al. 2016)
	Acorypha clara (Walker 1870)	Cameroon (Barreteau 1999; Van Huis 2003)
	Acorypha glaucopsis (Walker 1870)	Cameroon (Barreteau 1999; Van Huis 2003)
	Acorypha picta (Krauss 1877)	Cameroon (Barreteau 1999; Van Huis 2003)
	Acrida bicolor (Thunberg 1815)	Cameroon, Zimbabwe (Chavanduka 1976; Barreteau 1999; Van Huis 2003)
	Acrida sp.	Benin, Togo, Niger, Burkina Faso, Mali, Guinea Conakry (Van Huis 2003; Tchibozo et al. 2016)
	A. sulphuripennis (Gerstäcker 1869)	Zambia (Mbata 1995; Van Huis 2003)
	Acrida turrita Linn.	Cameroon (Barreteau 1999; Van Huis 2003)
	Acridoderus strenuus Walker	Niger, Sahel (Van Huis 2003; Tchibozo et al. 2016)
	Acrotylus blondeli (Saussure 1884)	Niger (Van Huis 2003)
	A. longipes (Charpentier 1845)	Niger (Van Huis 2003)
	Afroxyrrhepes procera (Burmeister 1838)	Congo (Nkouka 1987; Van Huis 2003)
	Anacridium burri (Dirsh and Uvarov 1953)	Southern Africa (Malaisse 1997; Van Huis 2003)
	Anacridium melanorhodon (Walker 1870)	Cameroon, Niger, Sahel, Cameroon, Sudan, Niger (Barreteau 1999; Zakari et al. 2015; Van Huis 2003; Kelemu et al. 2015)
	A. wernerellum (Karny 1907)	Niger, Sahel (Van Huis 2003)
	Brachycrotaphus tryxalicerus (Fischer 1853)	Cameroon (Barreteau 1999; Van Huis 2003)

(continued)

Table 12.1 (continued)

Family/subfamily	Species	Countries and reference
	Catantops stramineus (Walker 1870)	Niger (Van Huis 2003; Zakari et al. 20_5)
	Cataloipus fuscocoeruleipus Sjöstedt	Sahel (Van Huis 2003)
	Cataloipus cymbiferus (Krauss 1877)	Mali (Van Huis 2003; Tchibozo et al. 2016)
	C. haemorrhoidalis (Krauss 1877)	Niger (Van Huis 2003)
	C. cymbiferus (Krauss 1877)	Cameroon (Barreteau 1999; Van Huis 2003)
	Chirista compta (Walker 1870)	Congo (Nkouka 1987; Van Huis 2003)
	Cyathosternum spp.	Zimbabwe (Gelfand 1971; Van Huis 2003)
	Cyrtacanthacris aeruginosas (Stoll 1813)	Nigeria, Zambia, Guineay Conakry (Fasoranti and Ajiboye 1993; Banjo et al. 2006; Mbata 1995; Tchibozo et al. 2016)
	C. tatarica Linn.	Botswana, Zambia (Mbata 1995; Nonaka 1996; Van Huis 2003)
	Diabolocatantops axillaris (Thunberg 1815)	Cameroon, Niger, Sahel (Barreteau 1999; Van Huis 2003)
	Exopropacris modica (Karsch 1893)	Cameroon (Barreteau 1999; Van Huis 2003)
	Eurysternacris brevipes (Chopard 1947)	Burkina Faso (Van Huis 2003; Tchibozc et al. 2016)
	Gastrimargus africanus (Saussure 1888)	Cameroon, Congo, Niger, Sahel, Cameroon, Congo, Niger, Lesotho, Liberia (Nkouka 1987; Barreteau 1999; Van Huis 2003; Kelemu et al. 2015)
	G. procerus (Gerstäcker 1889)	Cameroon, Niger (Lévy-Luxereau 1980; Barreteau 1999; Van Huis 2003)
	Harpezocatantops stylifer (Krauss 1877)	Cameroon, Niger (Lévy-Luxereau 1980; Barreteau 1999; Van Huis 2003)
	Heteracris pulchripes guineensis (Krauss 1890)	Congo (Nkouka 1987; Van Huis 2003)
	Hieroglyphus daganensis (Krauss 1877)	Sahel Africa (Van Huis 2003)
	Hieroglyphus africanus (Uvarov 1922)	Benin, Burkina Faso (Tchibozo et al. 2016; Van Huis 2003)
	Homoxyrrhepes punctipennis (Walker 1870)	Cameroon, Togo (Barreteau 1999; Tchibozo et al. 2016)

(continued)

Table 12.1 (continued)

Family/subfamily	Species	Countries and reference
	Humbe tenuicornis (Schaum 1853)	Niger (Van Huis 2003)
	Kraussaria angulifera (Krauss 1877)	Cameroon, Sahel, Benin, Niger, Mali (Barreteau 1999; Zakari et al. 2015; Tchibozo et al. 2016; Van Huis 2003)
	Krausella amabile (Krauss 1877)	Cameroon (Barreteau 1999; Van Huis 2003)
	Lamarckiana cucullata (Stöll 1813)	Botswana (Van Huis 2003)
	Locusta migratoria capito (Sauss.)	Madagascar (Van Huis 2003)
	Locusta migratoria migratoriodis (Reiche and Fairmaire 1849)	Cameroon, Congo, Tanzania, Zambia, Zimbabwe, Benin, Burkina Faso (Nkouka 1987; Harris 1940; Mbata 1995; Chavanduka 1976; Gelfand 1971; Barreteau 1999; Tchibozo et al. 2016; Van Huis 2003)
	Locustana pardalina (Walker 1870)	South Africa, Southern Africa, Zambia, Zambia, South Africa, Zimbabwe, Botswana, Malawi, Libya (Quin 1959; Mbata 1995; Kelemu et al. 2015)
	Mesopsis abbreviatus (Palisot de Beauvois 1806)	Cameroon (Barreteau 1999; Van Huis 2003)
	Morphacris fasciata (Thunberg 1815)	Togo (Van Huis 2003; Tchibozo et al. 2016)
	Nomadacris septemfasciata (Serville 1838)	Congo, South Africa, Tanzania, Zambia, Zimbabwe, Eastern Africa Niger, Zambia, South Africa, Congo, Zimbabwe, Botswana, Nigeria, Tanzania, Malawi, Uganda, Mozambique (Harris 1940; Quin 1959; Nkouka 1987; Van Huis 2003; Kelemu et al. 2015; Mbata 1995; Chavanduka 1976; Gelfand 1971; Zakari et al. 2015)
	Oedaleus nigeriensis (Uvarov 1926)	Cameroon (Barreteau 1999; Van Huis 2003)
	O. nigrofasciatus (De Geer 1773)	Zambia (Mbata 1995; Van Huis 2003)
	O. senegalensis (Krauss 1877)	Niger, Niger (Tchibozo et al. 2016; Van Huis 2003)
	Ornithacris sp.	Zimbabwe (Van Huis 2003)
	Ornithacris turbida cavroisi (Finot 1907)	Niger, Benin, Togo, Niger Bukina fasso, Mali, Guinea Conakry (Zakari et al. 2015; Habou et al. 2015; Tchibozo et al. 2016)
	Orthacanthacris humilicrus (Karsch 1896)	Niger (Van Huis 2003; Tchibozo et al. 2016)

(continued)

Table 12.1 (continued)

Family/subfamily	Species	Countries and reference
	O. cyanea (Stoll 1813)	Zimbabwe (Gelfand 1971; Van Huis 2003)
	O. turbida cavroisi (Finot 1907)	Congo, Niger, Sahel (Bani 1995; Van Huis 2003)
	Orthacanthacris humilicrus (Karsch 1896)	Niger (Lévy-Luxereau 1980; Van Huis 2003)
	Orthochtha venosa (Ramme 1929)	Cameroon (Barreteau 1999; Van Huis 2003)
	Oxycatantops congoensis (Sjöstedt 1929)	Congo (Bani 1995; Nkouka 1987; Van Huis 2003)
	O. spissus (Walker 1970)	Cameroon, Congo, Sahel (Barreteau 1999; Bani 1995; Nkouka 1987; Van Huis 2003)
	Paracinema tricolor (Thunberg 1815)	Cameroon (Barreteau 1999; Van Huis 2003)
	Schistocerca gregaria (Forskål, 1775)	Africa, Congo, Cameroon, Tanzania, Zambia, Mali, Zambia, South Africa, Cameroon, Congo, Botswana, Tanzania, Sudan, Uganda, Ethiopia, Kenya, Sierra Leone, Morocco, Guinea, Lesotho, Mauritania, Somalia, Eritrea, Guinea Bissau (Harris 1940; Barreteau 1999; Nkouka 1987; Kelemu et al. 2015; Mbata 1995; Tchibozo et al. 2016)
	Sherifuria haningtoni (Uvarov 1926)	Cameroon, Malawi, Lesotho (Kelemu et al. 2015; Barreteau 1999; Van Huis 2003)
	Stenocrobylus festivus (Karsch 1891)	Guinea Conakry (Tchibozo et al. 2016; Van Huis 2003)
	Spathosternum pygmaeum (Karsch 1893)	Benin (Van Huis 2003; Tchibozo et al. 2016)
	Truxalis johnstoni (Dirsh 1950)	Cameroon (Barreteau 1999; Van Huis 2003)
	Tuxalis sp.	Benin (Van Huis 2003; Tchibozo et al. 2016)
	Truxaloides constrictus (Schaum 1853)	Zimbabwe (Gelfand 1971; Van Huis 2003)
	Tylotropidius gracilipes (Brancsik 1895)	Cameroon (Barreteau 1999; Van Huis 2003)
Pyrgomorphidae (Pyrgomorphoidea)	*Chrotogonus senegalensis* (Krauss 1877)	Cameroon (Barreteau 1999; Van Huis 2003)
	Phymateus viridipes (Stål 1873)	Congo, Southern Africa, Zambia, South Africa, Congo, Zimbabwe, Botswana, Mozambique, Namibia (Malaisse 1997; Van Huis 2003; Kelemu et al. 2015)
	Pyrgomorpha cognata (Krauss 1877)	Cameroon (Barreteau 1999; Van Huis 2003)

(continued)

Table 12.1 (continued)

Family/subfamily	Species	Countries and reference
	Zonocerus variegatus (Linn.1758)	Central African Republic, Nigeria, Benin, Togo, Burkina Faso, Guinea Conakry, DRC, Cameroon, Congo, CA Republic, Nigeria, Côte d'Ivoire, Sao Tomé, Guinea, Ghana, Liberia, Guinea Bissau, Benin, Burundi, Cameroon, Ivory Coast, Central Africa, Congo, DRC, Ghana, Liberia, Mali, Chad, Niger, Guinea, Guinea Bissau, Tanzania, Sierra Leone, Nigeria, Sao Tome, Togo (De Visscher 1990; Kelemu et al. 2015; Kekeunou et al. 2006; Idowu and Modder 1996; Kekeunou and Tamesse 2016; Barreteau 1999; Fasoranti and Ajiboye 1993; Kelemu et al. 2015; Van Huis 2003)
	Z. elegans (Thunberg 1815)	Mozambique, South Africa (Quin 1959; Van Huis 2003)

histidine (39.2 × 10^2 mg/100 g), isoleucine (36.7 × 10^2 mg/100 g), leucine (50.6 × 10^2 mg/100 g), aspartic acid (81.9 × 10^2 mg/100 g), serine (46.7 × 10^2 mg/100 g), lysine (48.4 × 10^2 mg/100 g), threonine (30.7 × 10^2 mg/100 g), glutamic acid (133.7 × 10^2 mg/100 g), glycine (44.9 × 10^2 mg/100 g), arginine (60.6 × 10^2 mg/100 g), proline (43.0 × 10^2 mg/100 g), alanine (36.6 × 10^2 mg/100 g), methionine (18.9 × 10^2 mg/100 g), cysteine (6.5 × 10^2 mg/100 g), valine (35.4 × 10^2 mg/100 g), tyrosine (25.3 × 10^2 mg/100 g), and phenylalanine (30.5 × 10^2 mg/100 g crude protein). Among these amino acids, eight are essential (isoleucine, leucine, lysine, methionine, phenylalanine, threonine, tryptophan, and valine) (Adeyeye 2005). Essential amino acids are the first limiting factor in the synthesis of body proteins (Tome 2009). The protein level of *Z. variegatus* can meet the recommended dietary intake of proteins for school children and adults (13.5–56 g protein/day) (FAO 1973). This would make a substantial contribution to infant growth and well-being, hence could help prevent protein deficiency. Thus, *Z. variegatus* can serve as a source of essential amino acids and is a valuable food and could usefully contribute to improving nutrition in Africa. In *Kraussaria angulifera*, the protein content (64.69 ± 0.08 g/100 g) is significantly higher than that of *Z. variegatus* (26.28–58.0 g/100 g), *Acanthacris ruficornis citrina* (58.32 ± 0.47 g/100 g) (Badanaro et al. 2015), and the migratory locust, *Locusta migratoria* (42.16–58.62 g/100 g), with an average of 50.42 g/100 g (Mohamed 2015). Despite the high protein content noted in many Acridians, some species have a very low content, such as *Cyrtacanthacris aeruginosus* unicolor with a crude protein content of 12.1 g/100 g (Length 2006). The protein content can be influenced by sex, age, or mode of cooking of the locust: in the desert locust (*Schistocerca gregaria*), males contain less protein (54.50 g/100 g than females (86 g/100 g) (Habou et al. 2015). Boiled and fried *Anacridium melanorhodon* contain 66.24 g/100 g and 67.75 g/100 g proteins respectively (Nafisa Mohamed 2002). The adult of *Z. variegatus* are higher in protein (21.38 ± 0.02 g/100 g) than the larvae [18.31 ± 0.00 for stage I larvae, 14.4 ± 1.80 for stage II larvae, 16.81 ± 0.00 for stage III larvae, 15.52 ± 2.0 for stage IV larvae, 14.61 ± 2.0 for stage V larvae and 16.1 ± 1.9 g/100 g for stage VI larvae] (Ademolu et al. 2010).The amino acid composition of *Z. variegatus* is similar to that of oleaginous seeds, melon, pumpkin and squash seeds (Olaofe et al. 1994) and yellow bean meal (*Sphenostylis stenocarpa*) (Adeyeye 2005). Amino acids are the elements of growth and development (Kekeunou and Tamesse 2016).

The protein levels of some grasshoppers are higher than those of some African conventional foods as sources of protein. *K. angulifera* and *A. ruficornis citrina* have been shown to be better sources of protein (64.69 ± 0.08 and 58.32 ± 0.47 g/100 g) (Badanaro et al. 2015) than some vegetable proteins (soybean) and animal protein sources (egg, poultry, mutton, pork, beef, fish) with protein contents ranging from 12.6 to 32 g/100 g (Stadlmayr et al. 2012). This highlights the role that edible Acridians can play in the fight against protein malnutrition in Africa. In fact, the nutritional value of foods depends largely on the quality of the nutrients they contain; that of a food protein depends on its amino acid composition especially on essential amino acids with respect to their ability to ensure the proper synthesis of the proteins of the body (Tome 2009; Alamu et al. 2013).

This study shows that some Acridians such as *Z. variegatus*, *A. melanorhodon*, *K. angulifera*, *A. ruficornis citrina*, *L. migratoria*, and *S. gregaria* have high levels of protein that can help combat malnutrition which persists in Africa. The average nutritional requirement of humans for protein, which is 0.066 g/100 g/day (Tome 2009), can be met by consuming sufficient quantities of Acridians. In fact, proteins are at the basis of all activities of the body. They constitute many important biomolecules such as enzymes, hormones, and hemoglobin. Proteins are important components of antibodies and play multiple roles such as enhancing the immune function of the body. They are the only material that produces nitrogen for the maintenance of acid and alkaline balance, the transmission of genetic information, and the transport of important materials in the human body. They provide energy (Bukkens 1997) and also play a major role in the renewal of muscle tissue, nails, hair, bone matrix, and skin (Fortin 2005).

Rice has low levels of lysine, an essential amino acid that can be found in Grasshoppers. Consumption of these Acridians may help to supplement protein levels in some foods such as cereals (maize, rice, ...), which are staple foods worldwide (Van Huis et al. 2014). Edible Grasshoppers can be used as sources of animal protein in African protein malnutrition programs because they are richer in protein than conventional African sources. However, they are less rich in protein than dry and salted beef.

12.3.2 Lipids

Edible Acridians are an important source of fat (Womeni et al. 2009): 18.9 and 20.34 g/100 g, with an average of 19.62 g/100 g in *Locusta migratoria* (Mohamed 2015), a significantly higher fat content than those of *Kraussaria angulifera* (11.71 ± 0.27 g/100 g), *Acanthacris ruficornis* (9.00 ± 0.22 g/100 g), and

Zonocerus variegatus (0.85 ± 0.05 to 4.77 ± 0.00 g/100 g) (Ademolu et al. 2010). These fat contents depend on the stage of development: higher in gregarious *Z. variegatus* larvae (4.30 ± 0.2 in stage I larvae, 4.77 ± 0.00 in stage II larvae, 2.86 ± 0.1 in stage III larvae) than in solitarious larvae (0.71 ± 0.02 in stage IV larvae, 1.07 ± 0.03 in stage V larvae, 0.91 ± 0.00 in stage VI larvae) and adults (0.85 ± 0.05 g/100 g) (Ademolu et al. 2010). Acridian lipids are generally composed of almost 8 fatty acids: palmitoleic acid, oleic acid, linoleic acid, linolenic acid (α and γ-linolenic), lauric acid, palmitic acid, stearic acid, and myristic acid (Womeni et al. 2009). This composition is disproportionate from one species to another. Thus *Acanthacris ruficornis* consists of: lauric acid ($0.64 \times 10^3 \pm 0.01 \times 10^3$); palmitic acid ($39.95 \times 10^3 \pm 0.02 \times 10^3$); stearic acid ($10.21 \times 10^3 \pm 0.03 \times 10^3$); oleic acid ($40.84 \times 10^3 \pm 0.03 \times 10^3$); linoleic acid ($4.49 \times 10^3 \pm 0.05 \times 10^3$); and linolenic acid ($2.18 \times 10^3 \pm 0.00 \times 10^3$ mg/100 g). With the exception of myristic acid ($0.96 \ 10^3 \pm 0.00 \times 10^3$ mg/100 g), found in trace amounts in *A. ruficornis*. *Kraussaria angulifera* is composed of the same fatty acids but at slightly different proportions lauric acid ($0.58 \times 10^3 \pm 0.01 \times 10^3$); palmitic acid ($30.35 \times 10^3 \pm 0.01 \times 10^3$); stearic acid ($9.16 \times 10^3 \pm 0.01 \times 10^3$); oleic acid ($47.06 \times 10^3 \pm 0.02 \times 10^3$); linoleic acid ($5.07 \times 10^3 \pm 0.00 \times 10^3$) and linolenic acid ($6.35 \times 10^3 \pm 0.01 \times 10^3$ mg/100 g) (Badanaro et al. 2015). *Z. variegatus* contains approximately 15.5 g/100 g of lipids (Solomon et al. 2008; Womeni et al. 2009): palmitoleic acid (24×10^3 mg/100 g), α-linolenic acid (15×10^3 mg/100 g), γ-linolenic acid (15×10^3 mg/100 g), oleic acid (29.3×10^3 mg/100 g) and a lower level of stearic acid (5.7×10^3 mg/100 g) (Adeyeye 2011). The roles of linoleic and α-linoleic essential acids are well known, mainly for the development of infants and young children (Michaelsen et al. 2009; Womeni et al. 2009). The lipids of *A. ruficornis* and *K. angulifera* are characterized by a high oleic acid content respectively $40.84 \pm 0.03 \times 10^3$ mg/100 g and $47.06 \pm 0.02 \times 10^3$ mg/100 g, a monounsaturated fatty acid that is involved in the prevention of type 2 diabetes (Ros 2003) and the metabolic syndrome (Gillingham et al. 2011). Linoleic acid (an essential fatty acid that can only be got from food) is involved in the protection of the cardiovascular system (Bang et al. 1971) and in the prevention of cancer (Ha et al. 1987) and atherosclerosis (Kritchevsky et al. 2000; Badanaro et al. 2015). Linolenic acid (also an essential fatty acid acts in the regulation of cholesterol (Demaison and Moreau 2002). Fatty acids are reservoirs of mobilizable energy found in adipocytes (9 kcal). They are involved in the transport of fat-soluble vitamins (A, D, E and K), they are the precursors of steroids and prostaglandins (Alamu et al. 2013). They are important structural constituents of the brain, the nervous system and the cell membrane (Marieb and Hoehn 2014). Potential deficiencies in omega-3 (linolenic acid) and omega-6 (α-linoleic acid) fatty acids (Cuvelier et al. 2004) were recently found in developing countries with limited access to the sea. Locusts could play a vital role in these countries, by supplementing diets with essential fatty acids (Van Huis et al. 2014). The lipid levels in migratory locusts are comparable to those of beef, fish, lamb, chicken, milk, and eggs (De Foliart 1991; Mohamed et al. 2010).

12.3.3 Carbohydrates

Carbohydrates are the main source of energy in the human body and can reduce protein consumption and promote detoxification (Alamu et al. 2013). Acridians generally contain a low carbohydrate content of about 1.1 g/100 g (Insect Food 2014). Carbohydrate levels vary depending on the species: 24.90 g/100 g in Z. *variegatus* (Ekop et al. 2010); 2.30 ± 1.79 g/100 g in A. *ruficornis*; 1.43 ± 1.04 g/100 g in K. *angulifera* (Badanaro et al. 2015); 7.13 g/100 g in *Anacridium melanorhodon* (Nafisa Mohamed 2002); 4.05–5.51 g/100 g with an average of 4.78 g/100 g dry matter in *Locusta migratoria* (Mohamed 2015). The carbohydrate content is influenced by the stage of postembryonic development. In *Zonocerus variegatus*, the first larval stage has a carbohydrate content of 0.42 g/100 g, which decreases in the second stage (0.39 g/100 g) and increases in the third (0.87 g/100 g), fourth (9.65 g/100 g), and fifth (9.77 g/100 g) stages of development. This content decreases to 8.81 g/100 g in the sixth instar and increases to 10.02% in adults (Ademolu et al. 2010; Ladeji et al. 2003).

Acridians consumption does not appear to be a good source of carbohydrate for humans, as they require about 400–500 g of carbohydrate (Alamu et al. 2013).

12.3.4 Vitamins

Vitamins cannot be synthesized in the human body and must be provided continuously by food (Alamu et al. 2013). Locust consumption provides vitamins (Ali et al. 2010), including vitamins B1 (thiamine), B2 (riboflavin), B3 (niacin), B6 (pyridoxine), C, D, E, K and carotene (Weiping et al. 2017). Vitamins A, B, C, and B2 are present in Z. *variegatus* (Banjo et al. 2006; Alamu et al. 2013) with proportions of 6.82 mg/100 g for vitamin A; 0.07 mg/100 g for Vitamin B2 and 8.64 mg/100 g for Vitamin C (Length 2006). The contents of these vitamins are lower in *Cyrtacanthacris aeruginosus unicolor* [Vitamin A (1.00 mg/100 g), Vitamin C (1.00 mg/100 g)] but with higher levels of Vitamin B2 (0.08 mg/100 g) in *C. aeruginosus unicolor* than in Z. *variegatus* (Length 2006). Vitamin levels depend on stages of development of Acridians. In Z. *variegatus*, the vitamin C content increases from stage 1 ($1.6 ± 0.02 × 10^{-3}$ mg/100 g) to the adult stage ($11.7 ± 0.20 × 10^{-3}$ mg/100 g dry matter). There is no significant difference in the level of vitamin B between the different stages of development ($0.025 ± 0.1$ to $0.095 ± 0.00 × 10^{-3}$ mg/100 g dry matter). For vitamin A, its content is $111.79 ± 1.0$ at stage I, $293.22 ± 2.0$ at stage II, $366.13 ± 0.0$ at stage III, $4441.18 ± 0.1$ at stage IV, $576.93 ± 0.0$ at stage V, $627.83 ± 0.1$ at stage VI, and $814.49 ± 3.0 × 10^{-3}$ mg/100 g dry matter at the adult stage (Ademolu et al. 2010).

Edible Acridians can therefore be used as qualitative and quantitative sources of vitamins for humans. They can be used as a source of vitamins in food supplements for malnourished humans (Stadlmayr et al. 2012). High intake of vitamin D and calcium, or high consumption of dairy products, is associated with reduced risk of type 2 diabetes (Bertiere 2009).

Vitamins are a group of organic compounds that are necessary for the metabolism of the human body (Alamu et al. 2013). Vitamin C maintains the flexibility of the blood vessels and improves circulation in the arteries. Vitamins A and C act as antioxidants, scavengers of oxygen-free radicals. B vitamins act as coenzymes in several enzymatic systems of the body (Stadlmayr et al. 2012; Badanaro et al. 2015). Vitamins are essential to stimulate the metabolic processes and strengthen the function of the immune system. They are present in most edible Acridians, as is the case of thiamine, with levels from 0.11 to 8.9 mg/100 g of dry matter, similar to that of whole bread which contains 0.16 mg and 0.19 mg/100 g of vitamins B1 and B2 respectively (Bukkens 2005).

12.3.5   Fibers

Dietary fiber is present in edible Acridians: the content is 8.63 ± 0.30 g/100 g in *A. ruficornis* and 5.60 ± 0.40 g/100 g in *K. angulifera* (Badanaro et al. 2015). In *Z. variegatus* its levels vary according to the post-embryonic stages of development: 0.85 ± 0.1 in stage I larvae, 0.92 ± 0.0 in stage II, 1.45 ± 0.05 g/100 g in stage III, 0.89 ± 0.01 g/100 g in stage IV, 0.90 ± 0.00 g/100 g in stage V larvae, 0.98 ± 0.01 g/100 g in stage VI larvae and 1.23 ± 0.02 g/100 g in adults (Ademolu et al. 2010; Idowu et al. 2004; Caddan 1988). Dietary fiber is a polysaccharide component that is not hydrolyzed by digestive enzymes. This resistance to digestion, combined with an ability to swell with water (retention of water for intestinal contents) has beneficial effects on intestinal transit (Badanaro et al. 2015). The presence of crude fiber in Acridians is of benefit for human nutrition and is probably due to chitin, a substance common to all insects and recognized as a dietary fibre that plays a key role in digestion (Majeti and Kumar 2000; Badanaro et al. 2015).

12.3.6   Energy Values

Energy intake varies according to the organic matter (1 g of carbohydrate yields 4 kcal, 1 g of lipid = 9 kcal, 1 g of proteins = 4 kcal in the bomb calorimetric) (Bourdel 2012). Acridians consumption provides 559 kcal of energy/100 g (Insect Food 2014). The gross energy of the migratory locust (*Locusta migratoria*, raw adult) is between 143 and 195 kcal/100 g of fresh weight with an average of 0.4908 kcal/100 g which can vary according to the food consumed by the latter (Oonincx et al. 2011; Mohamed 2015; Oonincx and Van der Poel 2011). The specie *K. angulifera* contains 385.35 ± 3.82 kcal/100 g whereas *A. ruficornis* (Badanaro et al. 2015) and *L. migratoria* contain 342.79 ± 6.19 kcal/100 g and

0.4803–0.5003 kcal/100 g respectively (Elagba 2015). Acridians are a better source of energy compared to conventional African foods such as vegetables (spinach, eggplant), fruits (orange, banana), tubers (cassava, yam), cereals (rice) and meats (chicken, fish, pork, beef) (Stadlmayr et al. 2012) (Table 12.2).

12.4   Mineral Composition of Edible Acridians

Acridians consumed in Africa contain 15 mineral salts (Hercberg 1988) with levels varying according to the specie, developmental stage and the habitat. These are aluminum, barium, boron, cobalt, calcium, chromium, copper, potassium, iron, manganese, magnesium, phosphorus, sodium, zinc, and lead.

12.4.1   Aluminum

Aluminum, a trace element essential for metabolism, can be found in *Locusta migratoria* with a content of 0.0443 ± 0.001 mg/100 g (Elagba 2015). Until 2008, the tolerable weekly intake of aluminum was 1 mg/kg of body weight per week. It is a mineral that is toxic at very high doses. Its accumulation in the body can lead to osteomalacia (Alfrey et al. 1976) and probably to Alzheimer's disease, amyotrophic lateral sclerosis, and Parkinson's disease.

12.4.2   Barium

It is not necessary for the proper functioning of the human organism. Absorption of a large quantity of Barium can cause paralysis, and in some cases death (INRS 2012). Barium is present in *L. migratoria* at a low level, 0.2192 ± 0.036 mg/100 g (Elagba 2015).

12.4.3   Boron

Boron is a trace mineral that participates in the metabolism of calcium, copper, magnesium, amino acids, glucose, triglycerides, and estrogens. It has a positive impact on calcification and stabilization of bone mass. It does not appear to be bioaccumulated (Moseman 1994). Locusts like *L. migratoria* are a source, with a content of 0.0298 ± 0.006 mg/100 g (Elagba 2015).

Table 12.2 Nutritive value of some edible Acridians in Africa

Acridian	Proteins (g/100 g)	Lipids (g/100 g D.W.)	Carbohydrates (g/100 g D.W.)	Fibres (g/100 g)	Energy (kcal/100 g)	Vitamins (g/100 g)	Authors
Z. variegatus (L.)	26.28–58.0	3.80 15.5–24.90	24.90	6.82 (A, B, C, B2)			Alamu et al. (2013), Bonjo et al. 2006), Length (2006), Badanaro et al. (2015)
Kraussaria angulifera (Krauss, 1877)	(64.69 ± 0.08)	11.71 ± 0.27	1.43 ± 1.04	5.60 ± 0.4	385.35 ± 3.82		Badanaro et al. (2015)
Acanthacris ruficornis citrina (Serville, 1838)	58.32 ± 0.47	9.00 ± 0.22	2.30 ± 1.79	8.63 ± 0.3	342.79 ± 6.19		Badanaro et al. (2015)
Anacridium melanorhodon (Walker, 1870)	67.75(fried) and 66.24 (boiled)	7.47	7.13 (fried) and 5.47 (boiled)				Nafisa Mohamed (2002)
Locusta migratoria (Linnaeus, 1758)	50.42 ± 2	19.6	4.78	15.65 ± 1.7	0.4803–0.5003		Elagba (2015), Mohamed (2015), Oonincx and van der Poel (2011), Zakari et al. (2015)
Cyrtacanthacris aeruginosus unicolor	12.10	3.50			0.08–1 (A, B, C, B2)		Womeni et al. (2009), Length (2006)
Schistocerca gregaria (Forskål, 1775)	54.5 (male) and 86.6 (Female)	9.80					Zakari et al. (2015)

NB: all the values have been converted in g/100 g and in kcal

12.4.4 Cobalt

The daily human requirement for cobalt does not exceed 0.6–0.15 mg. *L. migratoria* is a source because it contains a content of 0.006 ± 0.001 mg/100 g (Elagba 2015). Cobalt is an essential trace element that enters into the structure of vitamin B12 and indirectly to the production of red blood cells and regulation of the functioning of various enzymes. At low doses, it is beneficial for the human body. The absence or the great scarcity of this element causes a decrease in appetite and milk production. On the other hand, cobalt (II) deficiency promotes the toxic action of selenium that creeps into the body (Peters and Elliot 1983).

12.4.5 Calcium

The calcium content of *Acanthacris ruficornis* (256.93 ± 2.82 mg/100 g) is higher than that obtained in *K. angulifera* (137.5 ± 0.28 mg/100 g) (Badanaro et al. 2015), *Cyrtacanthacris aeruginosus* (4.40 mg/100 g) (Womeni et al. 2009) and *Zonocerus variegatus* where the level varies according to the larval stage (Fig. 12.1): the sixth larval stage has a higher content than the other larval stages (Banjo et al. 2006; Ademolu et al. 2010). Calcium plays a role in the solidification of bones and teeth, but also in membrane permeability, transmission of nerve impulses and muscle contraction (Alamu et al. 2013).

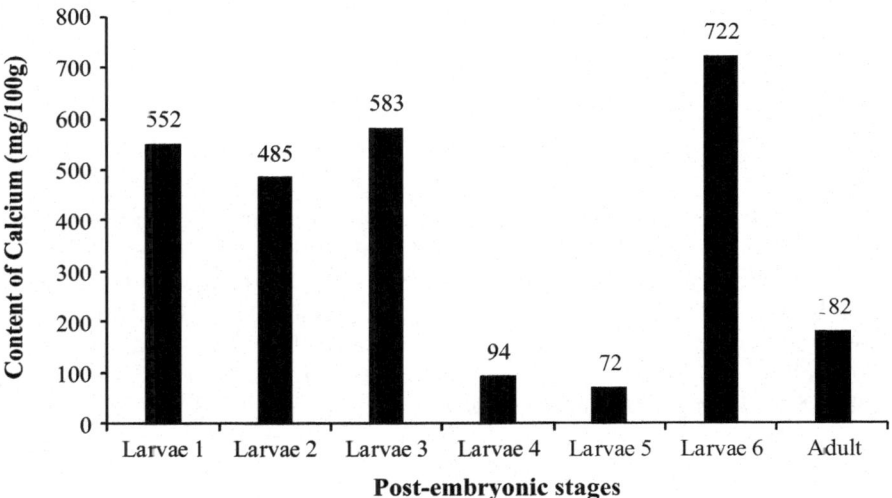

Fig. 12.1 Content variation of calcium in *Z. variegatus* larvae and adults

12.4.6   Chromium

Chromium is a trace element present in *L. migratoria* (0.006 ± 0.001 mg/100 g) (Elagba 2015) and *Z. variegatus* (0.057–0.083 mg/100 g). The content depends on the locality where the Acridian lives (Idowu et al. 2007). An adult needs 50 to 70 × 10^{-3} mg of chromium a day; which seems a little low, considering the absence of toxicity (Martin 2018). Chromium is a trace element that facilitates the action of insulin. In case of hypoglycemia, if our stocks of chromium are insufficient, the mobilization of insulin cannot be done properly and sugars stored in the body are not mobilizable. A diet rich in chromium could protect against atherosclerosis because deficiency leads to an increase in triglycerides and a decrease in "good cholesterol." Chromium is also involved in lowering body fat (Grant et al. 1997).

12.4.7   Copper

In the human body, it is essential for the action of many enzymes. It is involved in the metabolism of several nutrients: Carbohydrates, lipids and iron, contribute to the formation of red blood cells, immune defense, bone mineralization, regulation of neurotransmitters and the production of melanin. Copper also has an antioxidant role. It is a cofactor of superoxide dismutase. It is present in small quantities in Acridians: 3.73 ± 0.12 mg/100 g in *A. ruficornis*; 3.55 ± 0.34 mg/100 g in *K. angulifera* (Badanaro et al. 2015) and 0.9 ± 0.3 mg/100 g in *Zonocerus variegatus* (Olaofe et al. 1998; Idowu et al. 2007). Its content varies with the habitat of the Acridian (Idowu et al. 2007).

12.4.8   Potassium

It is involved in the maintenance of osmotic pressure, transmission of nerve impulses and gluconeogenesis in humans (Saris and Morrison 2010). The potassium level must be balanced to ensure the proper functioning of the heart, muscles and kidneys. Potassium can be obtained in Acridians where the content varies according to the specie and stage of development: 7.61 to 20.3 × 10^2 mg/100 g dry weight in *Zonocerus variegatus*. This content is higher than in *K. angulifera* (886.15 ± 2.37 mg/100 g) and *A. ruficornis* (103.42 ± 0.79 mg/100 g) (Badanaro et al. 2015). The early stages of *Z. variegatus* have higher levels [20.03 × 10^2 ± 0.00 × 10^2 in stage I, 24.2 × 10^2 ± 0.2 × 10^2 in stage II and 23.0 × 10^2 ± 2.0 × 10^2 in stage III] than the solitary stages [8.66 × 10^2 ± 0.1 × 10^2 in stage IV, 7.67 × 10^2 ± 0.2 × 10^2 in stage V, 15.0 × 10^2 ± 0.1 × 10^2 in stage VI and 7.61 × 10^2 ± 0.00 × 10^2 mg/100 g in adults] (Ademolu et al. 2010).

12.4.9 Iron

Iron strengthens the immune system as a cofactor of enzymes and antioxidants (Talwar et al. 1989). The human body needs it in small quantities and it can be obtained by consuming Acridians such as *Zonocerus variegatus* (1.19×10^2 mg/100 g to 9.23×10^2 mg/100 g) (Ademolu et al. 2010); *Cyrtacanthacris aeruginosus* (0.35×10^2 mg/100 g) (Womeni et al. 2009); *Locusta migratoria* (8–20 mg/100 g dry weight) (Oonincx et al. 2010); *Acanthacris ruficornis* ($10.65\ 10^2 \pm 0.06 \times 10^2$ mg/100 g) and *K. angulifera* ($23.05 \times 10^2 \pm 0.07 \times 10^2$ mg/100 g) (Badanaro et al. 2015). In *Z. variegatus*, the level is higher in stage 1 (9.1×10^2 mg/100 g), stage 2 (8.29×10^2 mg/100 g), stage 3 (9.23×10^2 mg/100 g) and stage 6 (7.71×10^2 mg/100 g) larvae); than in stage 4 (1.26×10^2 mg/100 g), stage 5 (1.19×10^2 mg/100 g) and adult (1.84×10^2 mg/100 g) larvae (Ademolu et al. 2010). Acridians can be of great help because iron deficiency is the most common and widespread nutritional disorder in the World; especially in the diet of pregnant women in developing countries, and particularly in Africa (Orr 1986).

12.4.10 Manganese

It activates many enzymatic systems by limiting the action of free radicals and thus prevents cell aging. It also participates in the proper functioning of fat metabolism, synthesis of connective tissue and regulation of glucose in the blood, regulates allergies by strengthening immune defenses, contributes to the production of sex hormones, as well as secretion of milk after childbirth and promotes the secretion of acetylcholine (http://www.linfo.re/magazine/sante-beaute/le-role-du-manganese-in-our-organism). Acridians manganese quantities are about 2.94 ± 0.10 mg/100 g in *A. ruficornis*, 0.13 ± 0.10 mg/100 g in *K. angulifera* (Badanaro et al. 2015) and 0.0040 ± 0.001 mg/100 g in *L. migratoria* (Elagba 2015).

12.4.11 Magnesium

Magnesium is needed for more than 300 biochemical reactions in the body; it prevents cardiomyopathy, muscle degeneration, spermatogenic deterioration, immune dysfunction, and coagulation disorder (Chaturvedi et al. 2004; Kekeunou and Tamesse 2016). It is involved in the reduction of blood pressure (Hercberg 1988), bone formation and skeletal integrity (Saris et al. 2000; Kemi et al. 2006); It can be obtained by consuming *Zonocerus variegatus* (0.96×10^2 to 0.39×10^2 mg/100 g), *Cyrtacanthacris aeruginosus* (0.09×10^2 mg/100 g)

(Womeni et al. 2009), *Acanthacris ruficornis* (126.37 ± 0.35 mg/100 g) and *K. angulifera* (64.94 ± 0.73 mg/100 g) (Badanaro et al. 2015). In *Z. variegatus*, its content varies with the developmental stage: $0.96 \times 10^2 \pm 0.03 \times 10^2$ in. stage I, $1.09 \times 10^2 \pm 0.01 \times 10^2$ in. stage II, $0.34 \times 10^2 \pm 0.02 \times 10^2$ in. stage III, $0.34 \times 10^2 \pm 0.02 \times 10^2$ in. stage IV, $0.39 \times 10^2 \pm 0.01 \times 10^2$ in. stage V, $0.31 \times 10^2 \pm 0.01 \times 10^2$ in. stage VI and $0.39 \times 10^2 \pm 0.1 \times 10^2$ mg/100 g in. dry weight in adults (Ademolu et al. 2010).

12.4.12 Phosphorus

Phosphorus is a constituent of nucleic acids, proteins and phospholipids. It intervenes in the transfer of energy through muscle activity. Phosphorus ensures the strength of bones and teeth, combined with calcium. It has a high content ranging from 45 to 218 mg/100 g of dry weight analyzed in *Zonocerus* (Alamu et al. 2013), 100.2 mg in *Cyrtacanthacris aeruginosus*, from 26.5 to 32.6×10^{-1} mg/100 g in *Locusta migratoria*, 225.43 mg/100 g in *Acanthacris ruficornis* and 146.47 mg/100 g in *K. angulifera* (Badanaro et al. 2015). In *Z. variegatus* this content is higher in adults ($218 \times 10^2 \pm 1.0 \times 10^2$) and is $45 \times 10^2 \pm 5 \times 10^2$, $90 \times 10^2 \pm 2 \times 10^2$, $115 \times 10^2 \pm 2 \times 10^2$, $122 \times 10^2 \pm 1 \times 10^2$, $163 \times 10^2 \pm 3 \times 10^2$, $190 \times 10^2 \pm 3 \times 10^2$ mg/100 g body weight. Dry in the first six stages of development respectively (Ademolu et al. 2010). Acridians can therefore be an important source of phosphorus.

12.4.13 Sodium

The human body needs Sodium for maintenance of fluid balance, neuromuscular function and transport of micronutrients across membranes (Saris and Morrison 2010). Acridians are a significant source of sodium with contents varying with the specie. *Z. variegatus* (13.75×10^2 to 2.97×10^2 mg/100 g dry weight) (Ademolu et al. 2010) is richer than *Acanthacris ruficornis* (43.95 ± 1.36 mg/100 g) and *Kraussaria angulifera* (59.21 ± 0.40 mg/100 g) (Badanaro et al. 2015). In *Z. variegatus* its content varies according to the larval stage with a higher concentration in the young larvae [Stage 1 larva: $13.50 \times 10^2 \pm 0.00 \times 10^2$ mg/100 g, Stage 2 larvae: $12.55 \times 10^2 \pm 0.05 \times 10^2$ mg/100 g, Stage 3 larvae: $13.75 \times 10^2 \pm 0.05 \times 10^2$ mg/10 0 g] than in solitary stages [Stage 4 larvae: $2.97 \times 10^2 \pm 0.01 \times 10^2$ mg/100 g, Stage 5 larvae: $3.05 \times 10^2 \pm 0.05 \times 10^2$ mg/100 g; Stage 6 larvae: $6.94 \times 10^2 \pm 0.04 \times 10^2$ mg/100 g and adults: $3.06 \times 10^2 \pm 0.00 \times 10^2$ mg/100 g dry weight] (Banjo et al. 2006; Ademolu et al. 2010).

12.4.14   Zinc

Zinc is a trace element that strengthens the immune system, and antioxidants (Talwar et al. 1989). It prevents cardiomyopathy, muscle degeneration, stunting, spermatogenesis and coagulation disorders (Chaturvedi et al. 2004; Kekeunou and Tamesse 2016). It also provides bone formation and skeletal integrity (Kemi et al. 2006), which is also essential for the prevention of osteoporosis (Saris et al. 2000; Heaney and Nordin 2002). Acridian consumption is a source of Zinc supply. It can be found in *Z. variegatus* (0.093–0.207 mg/100 g), *A. ruficornis* (19.65 ± 0.16 mg/100 g), and *K. angulifera* (13.09 ± 0.07 mg/100 g) (Badanaro et al. 2015) and *L. migratoria* (0.0879 ± 0.009 mg/100 g) (Elagba 2015). Its content depends on the origin of the Acridian (Idowu et al. 2007) and the stage of development: $0.29 \times 10^2 \pm 0.04 \times 10^2$ (Larva 1), $0.32 \times 10^2 \pm 0.02 \times 10^2$ (Larva 2), $0.15 \times 10^2 \pm 0.05 \times 10^2$ (Larva 3), $0.15 \times 10^2 \pm 0.00 \times 10^2$ (Larva 4), $0.18 \times 10^2 \pm 0.02 \times 10^2$ (Larva 5), $0.19 \times 10^2 \pm 0.02 \times 10^2$ (Larva 6) and $0.17 \times 10^2 \pm 0.02 \times 10^2$ (Adult) in mg/100 g dry weight of *Z. variegatus* (Ademolu et al. 2010).

12.4.15   Lead

It is a toxic substance that is particularly harmful for young children, in whom it is at the origin of intellectual disabilities (WHO). There is no threshold under which exposure to lead would be safe for humans. Respiratory and blood uptake is increased by iron and calcium deficiency. Acridians contain lead, including *L. migratoria* (0.0213 ± 0.009 mg/100 g) (Elagba 2015) and *Z. variegatus* in which its content of 0.068–0.300 mg/100 g depends on the locality (Idowu et al. 2007).

12.5   Effect of Cooking on Nutrient Availability

Baking of Acridians affects the availability of nutrients. In boiling *Anacridium melanorhodon*, protein (66.24 g/100 g) and ash (5.53 g/100 g) is low, but high in crude fibre (8.38 g/100 g) and moisture (7.47 g/100 g) contents. Meanwhile frying process revealed a high protein (67.75 g/100 g), ash (6.02%) and a low fibre (7.32 g/100 g) and moisture contents (5,47 g/100 g). Boiling also reduces the tannin content to 5.8 mg/100 g compared to frying (9.0 mg/100 g) and has no significant effect on the phytic acid content. The boiling of locust flour increased protein digestibility in vitro (49.89 g/100 g) compared to frying (41.13 g/100 g) and also revealed a higher proportion of Na, Mn, and Cu and lower proportion of Ca, K, P, Fe, and Zn

Table 12.3 Mineral content of some edible Acridians in Africa

Mineral	Zonocerus variegatus (L.) (mg/100 g)	Locusta migratoria (mg/100 g)	Cyrtacanthacris aeruginosus (mg/100 g)	Kraussaria angulifera (mg/100 g)	Acanthacris ruficornis (mg/100 g)
Iron (Fe) (mg/100 g)	1.19×10^2 to 9.23×10^2	8–20	0.35×10^2	23.05×10^2	10.65×10^2
Zinc (Zn)	0.093–0.207	0.0879 ± 0.009		13.09 ± 0.07	19.65 ± 0.16
Potassium (K)	7.61 to 20.3×10^2			886.15 ± 2.37	103.42 ± 0.79
Magnesium (Mg)	0.96×10^2 to 0.39×10^2		0.09×10^2	64.94 ± 0.73	126.37 ± 0.35
Calcium (Ca)	72–722		4.40	137.5 ± 0.28	256.93 ± 2.82
Sodium (Na)	13.75×10^2 to 2.97×10^2			59.21 ± 0.40	43.95 ± 1.36
Copper (Cu)	0.9 ± 0.3			3.55 ± 0.34	3.73 ± 0.12
Phosphorus (P)	45–21	26.5 to 32.6×10^{-1}	100.2	225.43	146.47
Aluminum (Al)		0.0443 ± 0.001			
Manganese (Mn)		0.0040 ± 0.001		0.13 ± 0.10	2.94 ± 0.10
Boron (B)		0.0298 ± 0.006			
Barium		0.2192 ± 0.036			
Chrome (Cr)	0.057 to 0.083	0.006 ± 0.001			
Lead (Pb)	0.068 to 0.300	0.0213 ± 0.009			
Cobalt		0.006 ± 0.001			
References	Ademolu et al. (2010), Banjo et al. (2006), Alamu et al. (2013), Womeni et al. (2009), Olaofe et al. (1998)	Elagba (2015)	Womeni et al. (2009)	Badanaro et al. (2015)	Badanaro et al. (2015)

NB: all the values have been converted in mg/100 g

than the fried sample. The moisture absorption capacity of the baked and fried samples was rated excellent (Nafisa Mohamed 2002). Cooking modifies the nutrient content of Acridians. *Z. variegatus* increases the content of linoleic acid from 16.0 to 17.5 g/100 g dry weight (Ekop et al. 2010).

12.6 General Discussion and Conclusion (Continue)

The low water levels reported in Acridians compared to conventional fresh meat and fish (65–75%) is favorable to a better conservation of Acridians (Stadlmayr et al. 2012; Womeni et al. 2009; Desrosier 2014; Badanaro et al. 2015). Qualitatively, edible Acridian are a source of carbohydrates, fats, proteins, vitamins, water, and minerals. Acridians, usually known for their apocalyptic actions in agriculture (ancestral "plague"), could become an asset in the fight against nutritional diseases in Africa. Their direct or indirect incorporation (as nutritional supplements) into human nutrition can help in the fight against malnutrition in Africa, especially with regard to rickets that affects most malnourished children and severely compromises their growth (Véronique 1997). The preparation of nutritional supplements from Acridians can help in the resolution of nutritional deficiencies in pregnant women and African children. However, the presence of harmful heavy metals or tannins detected in some Acridians shows that insects can also become dangerous to human health. They can accumulate these metals and become indicators of pollution of the natural environment because most of these metals found in food can come from environmental pollution related to industrial development (Idowu et al. 2007). This raises the need for further health studies on these foods before recommending their general consumption in African societies.

Conflicts of Interest The authors declare no conflict of interest.

Author contributions: K.S.; S.N.A.; M.N.M.: conception of review, design, analysis, and interpretation of data, drafting of the manuscript and critical revision. M.B.A.L.; W.A.C.; T.J.L.: critical revision.

References

Ademolu KO, Idowu AB, Olatunde GO (2010) Nutritional value assessment of variegated grasshopper, *Zonocerus variegatus* (L.) (Acridoidea: Pygomorphidae), during post-embryonic development. Afr Entomol 18(2):360–364

Adeoye OT (2014) Eco-diversity of edible insects of Nigeria and its impact on food security. J Biol Life Sci 5(2):175–187

Adeyeye EI (2005) Amino acid composition of variegated grasshopper, *Zonocerus variegatus*. Trop Sci 45:141–143

Adeyeye EI (2011) Fatty acid composition of *Zonocerus variegatus*, *Macrotermes bellicosus* and *Anacardium occidentalis* Kernel. Int J Pharma BioSci 2:135–144

Alamu OT, Amao AO, Nwokedi CI, Oke OA, Lawa IO (2013) Diversity and nutritional status of edible insects in Nigeria: a review. Int J Biodiv Conserv 5(4):215–222

Ali A, Adji Mouhamadou B, Saibou C, Aoudou Y, Tchiegang C (2010) Physio-chemical properties and safety of grasshoppers, important contributors to food security in far north region of Cameroon. Res J Animal Sci 4(5):108–111

Alfrey AC, Le Gendre GR, Kaehny WD (1976) The dialysis encephalopathy syndrome. Possible aluminum intoxication. N Engl J Med 294:184–188

Aman P, Michel F, Roel U, Séverin H, Priyanka M, Simon L, Malik H et al (2016) Grasshoppers as a food source? A review. Biotechnol Agron Soc Environ 20(S1):337–352

Badanaro F, Bilabina I, Awaga Kwami L, Sanbena Bassan B, Amevoin K, Amouzou K(2015) Identification et composition nutritionnelle de quelques espèces d'orthoptères consommées au Togo. Vol.03 num.01. ISSN 2424–7235

Bang H, Dyerberg J, Nieslsen A (1971) Plasma lipid and lipoprotein pattern in greenlandic West-coast Eskimos. Lancet, 1143p

Bani G (1995) Some aspects of entomophagy in the Congo. Food Insect Newsletter 8:4–5

Banjo A, Lawal O, Songonuga E (2006) The nutritional value of fourteen species of edible insects in southwestern Nigeria. Afr J Biotechnol 5(3):298–301

Barreteau D (1999) L'homme et l'animal dans le bassin du lac Tchad. Actes du colloque du rëseua Mega-Tchad, Orléans 15–17. Collection Colloques et Séminaires, no. 00/354. Université Nanterre, Paris, pp 133–169

Bertiere S (2009) Matière grasse laitière et maladies cardiovasculaires : synthèse d'un symposium scientifique international. Chole-doc, Janvier-Février

Bourdel MI (2012) Troubles nutritionnels chez le sujet âgé. La revue du praticien 62. 8p

Bukkens SGF (1997) The nutritional value of edible insects. Ecol Food Nutr 36:287–319

Bukkens SGF (2005) Insects in the human diet: nutritional aspects. In: Paoletti MG (ed) Ecological implications of minilivestock: role of rodents, frogs, snails, and insects for sustainable development. Science Publishers, Enfield, CT, pp 545–577

Caddan AM (1988) Moisture sorption characteristics of several food fibres. J Food Sci 53(4):1150–1155

Chakravorty J, Ghosh S, Jung C, Meyer-Rochow VB (2014) Nutritional composition of Chondacris rosea and Brachytrupes orientalis. J Asia Pac Entomol 17(3):407–415

Chaturvedi VC, Shrivastava R, Upreti RK (2004) Viral infections and trace elements: a complex interaction. Curr Sci 87:1536–1554

Chavanduka DM (1976) Insects as a source of protein to the African. Rhod Sci News 9:217–220

Chiffaud J, Mestre J (1990) Le Criquet Puant Zonocerus Variegatus (Linne, 1758): Essai De Synthèse Bibliographique. CIRAD-PRIFAS, Montpellier, France. 140 pp

Courade G (1996) ' Entre libéralisation et ajustement structurel : la sécurité alimentaire dans un étau', Cahiers Agricultures, vol. 5, n° 4, p. 221–227. http://revues.cirad.fr/index.php/cahiers-agricultures/article/download/29950/29710

Cuvelier C, Cabaraux JF, Dufrasne I, Hornick JL, Istasse L (2004) Acides gras: nomenclature et siurces alimentaires. Annal Med Veter 148:133–140

De Foliart GR (1991) Insect fatty acids: similar to those of poultry and fish in their degree of unsaturation, but higher in the polyunsaturables. Food Insects Newsletter 4:1–4

De Visscher MN (1990) Résultat de l'enquête SAS 1989 sur Zonocerus variegatus (Linne, 1758), Acridien ravageur communément appelé "Criquet puant" ou "Criquet bariole". PRIFAS, Montpellier, p 25

Demaison L, Moreau D (2002) Dietary n-3 polyunsaturated fatty acids and coronary heart disease-related mortality: a possible mechanism of action. Cell Mol Life Sci 5:463–477

Desrosier NW (2014) The technology of food preservation. http://www.britannica.com/EBchecked/topic/212684/foodpreservation/50564/Asepticprocessing#toc50568. Accessed 07 Aug 2014

Ekop EA, Udoh AL, Akpan PE (2010) Proximate and anti-nutrient composition of four edible insects in Akwa Ibom State, Nigeria. World J Appl Sci Technol 2(2):224–231

Elagba HAM (2015) Determination of nutritive value of the edible migratory locust Locusta migratoria, Linnaeus, 1758 (Orthoptera: Acrididae). Int J Adv Pharm Biol Chem 4(1):144–148

FAO/WHO (1973) Energy and protein requirements. Nutrition meetings report series, no. 52. WHO technical report series, no. 522

Fasoranti JO, Ajiboye DO (1993) Some edible insects of Kwara State Nigeria. Am Entomol 39:113–116

Finke MD, DeFoliart GR, Benevenga NJ (1989) Use of a four-parameter logistic model to evaluate the quality of proteins from three insect species when fed to rats. J Nutr 119:864–871

Fortin J (2005) L'encyclopédie visuelle des aliments. Editions du Chariot d'Or:688

Gelfand M (1971) Insects. In: Gelfand M (ed) Diet and tradition in African culture. E&S Livingstone, Edinburgh, pp 163–171

Gillingham LG, Harris-Janz S, Jones PJ (2011) Dietary monounsaturated fatty acids are protective against metabolic syndrome and cardiovascular disease risk factors. J Lipids 46(3):209–228

Grant CM, MacIver FH, Dawes IW (1997) Glutathion synthetase is dispensable for growth under both normal and oxidative stress conditions in the yeast Saccharomyces cerevisiae due to an accumulation of the dipeptide gamma-glutamylcysteine. Mol Bio Cell 8(9):1699–1707

Habou ZA, Tougiani A, Seydou R, Toudou A (2015) Une évaluation de Criquets comestibles au Niger : Ornithacris turbida cavroisi (Finot, 1907) Anacridium melanorhodon (Walker, 1870) et Accanthacris ruficornis citrina (Serville, 1838). J Appl Biosci 90:8348–8354

Harris WV (1940) Some notes on insects as food. Tanganyika Notes Rec 9:45–48

Ha YL, Grimm NK, Pariza MW (1987) Anticarcinogens from fried ground beef: heat-altered derivatives of linoleic acid. Carcinogenesis 8:1881–1887

Hassan KA, Hassan AB, Eltayeb MM, Osman GA, Hassan NM, Babiker EE (2007) Solubility and functional properties of boiled and fried Sudanese tree locust flour as function of NaCl concentration. J Food Technol 5:201–214

Hercberg S (1988) La carence en fer et nutrition humaine. EMI, Lavoisier. 203p

Heaney PR, Nordin BEC (2002) Calcium effects on phosphorus absorption: implications for the prevention and co-therapy of osteoporosis. J Am Coll Nutr 21(3):239–244

Hujon-Song (2010) Grasshopper systematics: past, present and future. J Orthop Res 19:57–68. https://doi.org/10.1665/034.019.0112

Idowu AB, Modder WWD (1996) Possible control of the grasshopper stinking, Zonocerus variegatus (L.) (Orthoptera: Pyrgomorphidae), in Ondo State through human consumption. Nigerian Field 61:7–14

Idowu MA, Idowu AB, Faseki MF (2004) Nutrient composition and processing of variegated grasshopper Zonocerus variegates (L) for human consumption. Nigerian J Entomol 21:65–70

Idowu TA, Wewe N, Amusan AAS (2007) Heavy metal content of the variegated grasshopper Zonocerus variegatus (L) (Orthoptera: Pyrgomorphidae) collected from various locations in Abeokuta, Ogun State. Nigerian J Entomol 24:35–41

INRS (2012) Baryum et composés [archive], Fiche no 125

Insect Food (2014) The nature cricket box. http://www.insectes-food.com/insectes-edibles-europe-/6-locust.html. Accessed 20 July 2014

Kekeunou S, Tamesse JL (2016) Consumption of the variegated grasshopper in Africa: importance and threats. J Insects Food Feed 2(3):213–222

Kekeunou S, Weise S, Messi J, Tamo M (2006) Farmers' perception on the importance of variegated grasshopper (Zonocerus variegatus (L.)) in the agricultural production systems of the humid forest zone of Southern Cameroon. J Ethnobiol Ethnomed 2:17. https://doi.org/10.1186/1746-4269-2-17

Kelemu S, Niassy S, Torto B, Fiaboe K, Affognon H, Tonnang H, Maniania NK, Ekesi S (2015) African edible insects for food and feed: inventory, diversity, commonalities and contribution to food security. J Insects Food Feed 1(2):103–119

Kemi VE, Kärkkäinen MUM, Lamberg-Allardt CJE (2006) High phosphorus intake acutely and negatively affects calcium and bone metabolism in a dose-dependentmanner in healthy young females. Br J Nutr 96:545–552

Kritchevsky D, Tepper SA, Wright S, Tso P, Czarnecki SK (2000) Influence of conjugated linoleic acid (CLA) on establishment and progression of atherosclerosis in rabbits. J Am Coll Nutr 19:472–477

Ladeji O, Solomon M, Hugh M (2003) Proximate chemical analysis of zonocerus variegatus pdf. Nigerian. J Biotechnol 14(1):42–45

Lévy-Luxereau A (1980) Note sur quelques criquets de la région de Maradi (Niger) et leur noms Hausa. J Agric Trad Bot appl 37:263–272.

Length F (2006) The nutritional value of fourteen species of insects in southwestern Nigeria. Afr J Biotechnol 5(3):298–301

Malaisse F (1997) Se Nourrir en Forêt Claire Africaine. Approche Ecologique et Nutritionnelle. Gembloux, Belgique : Presses agronomiques de Gembloux; Wageningen, CTA, Pays Bas, 384 p

Malaisse F (2004) Ressources alimentaires non conventionnelles. Tropicultura, spe, 30–36 Edible insects in Eastern and Southern Africa: challenges and opportunities Authors; Muniirah Mbabaziac, Yusuf B Byaruhangaa PhD and Prof. Thomas Omara Alwalab

Marieb E, Hoehn K (2014) Anatomie et physiologie humaines : plateforme numérique MonLab-Licence étudiant 60 mois. Pearson Education France, Harlow

Martin A (2018) Apports nutritionnels conseillés pour la population française, Edition TEC & Doc 3e éd., 610p

Majeti NV, Kumar R (2000) A review of chitin and chitosan applications. React Fun Polymers 46:1–27

Mbata KJ (1995) Traditional uses of arthropods in Zambia. Food Insect Newsletter 8:1–7

Michaelsen KF, Hoppe C, Roos N, Kaestel P, Stougaard M, Lauritzen L, Mølgaard C, Girma T, Friis H (2009) Choice of foods and ingredients for malnourished children 6 months to 5 years of age. Food Nutr Bull 30(3):343–404

Mohamed EHA (2015) Determination of nutritive value of the edible migratory locust Locusta migratoria, Linnaeus, 1758 (Orthoptera : Acrididae). Int J Adv 4(1):144–148

Mohamed EHA, Al-Maqbaly R, Mansour HM (2010) Proximate composition, amino acid and mineral contents of five commercial Nile fishes in Sudan. Afr J Food Sci 4(10):650–654

Moseman RF (1994) Chemical disposition of boron in animals and humans. Environ Health Perspect 102:113–117

Nafisa Mohamed ARAE (2002) Nutritional evaluation and functional properties of flour of boiled and fried sari el lail locust (anacridium melanorhodon. Master of Science in Food Science and Technology, University of Khartoum, 75 pp

Nkouka E (1987) Les insectes comestibles dans les sociétés d'Afrique Centrale. Revue Scientifique et Culturelle du CICIBA, Muntu 6:171–178. Nutr. 119; 864-871

Nonaka K (1996) Ethnoentomology of the Central Kalahari San. African Study Monographs, 22:29–46

Olaofe O, Adeyemi F, Adediran G (1994) Amino acid and mineral composition and functional properties of some oil seeds. J Agric Food Chem 42:878–884

Olaofe O, Arogundade LA, Adeyeye EI, Falusi OM (1998) Composition and food properties of the variegated grasshoppers, Zonocerus variegatus. AGRIS 38(4):233–237

Oonincx DG, van Itterbeeck J, Heetkamp MJ, van den Brand H, van Loon JJ, van Huis A (2010) An exploration on greenhouse gas and ammonia production by insect species suitable for animal or human consumption. PLoS One 5(12):e14445. https://doi.org/10.1371/journal.pone.0014445

Oonincx DGAB, van der Poel AFB (2011) Effects of diet on the chemical composition of migratory locusts (Locusta migratoria). Zoo Biol 30:9–16. USDA database (www.ars.usda.gov) Release 23

Orr B (1986) Improvement of women's health linked to reducing widespread anemia. Int Health News 7(3)

Peters JP, Elliot JM (1983) Effet of vitamin B12 status on performance of the lactating ewe and gluconeogenesis from propionate. J Dairy Sci 66:1917–1925

Quin PJ (1959) Food and feeding habits of the pedi with special reference to identification, classification, preparation and nutritive value of the respective foods. Witwatersrand University Press, Johannesburg, p 278

Ramos-Elorduy JB, Pino Moreno JM, Martínez Camacho VH (2012) Could grasshoppers be a nutritive meal? Food Nutr Sci 3:164–175

Rumpold BA, Schlüter OK (2013) Potential and challenges of insects as an innovative source for food and feed production. Innov Food Sci Emerg Technol 17:1–11

Ros E (2003) Dietary cis-monounsaturated fatty acids and metabolic control in type 2 diabetes. Am J Clin Nutr 78:617–625

Saris NE, Mervaala E, Karppanen H, Khawaja JA, Lewenstam A (2000) Magnesium: an update on physiological, clinical, and analytical aspects. Clinica Chimica Acta 294:1–26

Saris A, Morrison J (2010) Food security in Africa: market and trade policy for staple foods in eastern and southern Africa. FAO and Edward Elgar, Northampton, MA, 434 pp.

Solomon M, Ladeji O, Umoru H (2008) Nutritional evaluation of the giant grasshopper (Zonocerus variegatus) protein and the possible effects of its high dietary fibre on amino acids and mineral bioavailability. Afr J Food Agric Nutr Dev 8(2):238–248

Stadlmayr B, Charrondiere UR, Enujiugha VN, Bayili RG, Fagbohoun EG, Samb B, Baddy P, Barikmo I, Ouattara F, Oshaug A, Akinyele I, Annor GA, Bomfeh K, Ene-Obong H, Smith IF, Thiam I, Burlingame B (2012) West African food composition table. FAO, Rome. 171 p

Talwar GP, Srivastava LM, Mudgil KD (1989) Textbook of biochemistry and human biology. Prentice Hall of India Private Limited, New Delhi

Tamesse JL, Kekeunou S, Tchatchouang LJ, Ndegue OLM, Aissatou LM, Tomboouck D, Youssa B (2016) Insects as Food, traditional medecin and cultural rites in the west and south regions of Cameroon. J Insect Food feed 2(3):153–160

Tchibozo S, Malaisse F, Mergen P (2016) Insectes consommés par l'Homme en Afrique occidentale francophone Edible insects by Human in Western French Africa. Geo-Eco-Trop 40-2:105–114

Teffo LS, Toms RB, Eloff JN (2007) Preliminary data on the nutritional composition of the edible stink-bug, Encosternum delegorguei Spinola, consumed in Limpopo province, South Africa. South Afr J Sci 103:434–436

Tome D (2009) Besoins en protéines et en acides aminés & qualité des protéines alimentaires. Chole-doc 111 Janvier-Fevrier

United Nations (2011) World Population Prospects: The 2010 Revision. New York. Department of Economic and Social Affairs, Population Division

Van Huis A, Van Itterbeeck J, Klunder H, Mertens E, Halloran A, Muir G and Vantomme P (2014) Edible insects: future prospects for food and feed security. FAO forestry paper no. 171. FAO: Rome. 230–235

Van Huis A (2003) Insects as food in sub-saharan Africa. Insect Sci Appl 23(3):163–185

Véronique B (1997) Les insectes, une ressource alimentaire d'avenir. Insectes 10 N° 106–1997

Verkerk MC, Tramper J, Van Trijp JCM, Martens DE (2007) Insect cells for human food. Biotechnol Adv 25:198–202

Weiping Y, Junna L, Huaqing L, Biyu L (2017) Nutritional Value, Food Ingredients, Chemical and Species Composition of Edible Insects in China. Chapter from the book Future Foods 29p

Womeni HM, Michel L, Tiencheu B, Tchouanguep MF, Pierre V, Jacques F, Michel P (2009) Oils of insects and larvae consumed in Africa: potential sources of polyunsaturated fatty acids. OCL 16(4):230–235

Zakari AH, Abasse T, Ramatou S, Adam T (2015) Une évaluation de Criquets comestibles au Niger: Ornithacris turbida cavroisi (Finot, 1907), Anacridium melanorhodon (Walker, 1870) et Accanthacris ruficornis citrina (Serville, 1838). J Appl Biosci 90:8348–8354. ISSN 1997-5902

Chapter 13
Nutrient Composition of Black Soldier Fly (*Hermetia illucens*)

Matan Shelomi

Abstract The highly polyphagous nature of black soldier fly (BSF), *Hermetia illucens,* means it can efficiently convert a large amount of nearly any organic biomass, including manures and lignocellulosic wastes from pre- and post-consumer food waste, into biomass rich in protein and fat. This waste management aspect of BSF is immense, but as the biomass can be returned to the food supply as feed for poultry and fish, the use of BSF to reduce the costs of meat production is gaining significant international attention. We review the extant literature on the subject with a focus on the nutritional value of BSF or BSF meal, which can be used as a whole or partial replacement of soybean meal and fish meal in animal feeds. BSF is high in protein of high quality and sufficient digestibility, with amino acid profiles sufficient for most livestock and animals. BSF is high in fat, predominantly saturated fats followed by monounsaturated fats, normally with high omerga-6 to omega-3 polyunsaturated fat ratios. Defatted larvae or meal can produce a higher protein product. BSF are acceptable sources of vitamin E and certain minerals. The exact macronutrient and micronutrient composition of BSF can be altered by changing the composition of the substrate on which they are fed, the extent of which depends on whether the primary purpose of the bioconversion facility is waste management or feed production. Tests with animals generally show none of the positive effects of replacing standard diets with BSF on the health of the animal and features of the meat. BSF do not bioaccumulate pesticides, drugs, or mycotoxins, but do accumulate cadmium. The fate of prions in BSF is unknown and demands study.

Keywords Animal feed · Bioconversion · Biorefinery · Black soldier fly · Edible insects · Fisheries · *Hermetia illucens* · Nutrition · Poultry feed · Swine feed · Waste management

M. Shelomi (✉)
Department of Entomology, National Taiwan University, Taipei, Taiwan
e-mail: mshelomi@ntu.edu.tw

© Springer Nature Switzerland AG 2020   195
A. Adam Mariod (ed.), *African Edible Insects As Alternative Source of Food, Oil, Protein and Bioactive Components*, https://doi.org/10.1007/978-3-030-32952-5_13

Abbreviations

AA	Amino acid
BSF	Black soldier fly
BSFL	Black soldier fly larvae
BSFM	Black soldier fly meal
MUFA	Monounsaturated fatty acid
PUFA	Polyunsaturated fatty acid
SFA	Saturated fatty acid

13.1 The Black Soldier Fly as Food and Feed

The black soldier fly (BSF), *Hermetia illucens* (L. 1758), is a highly polyphagous scavenger, whose larvae are capable of eating almost any organic waste that is not too hard or wet (Nguyen et al. 2015), including manure, corpses, lignocellulosic matter, and pre- and post-consumer food waste. High activities of a wide assortment of digestive enzymes in their gut have enabled them to become arguably the most efficient digester of organic material in the Diptera (Kim et al. 2011). The adults (Fig. 13.1) do not feed, meaning they are less likely to spread diseases relative to house flies or corpse flies, nor do they sting or irritate or otherwise annoy. These characteristics make BSF useful in waste removal: at the simplest, waste such as manure or offal is spread outside to be colonized by wild flies, and the larvae remove the bulk of the waste, saving disposal costs and eliminating harborage of disease-vectoring filth flies (Bradley and Sheppard 1984; Sheppard et al. 1994). This benefit alone could address developing world health problems associated with inadequate waste management, including agricultural waste and human feces (Banks et al. 2014;

Fig. 13.1 Black soldier fly adult and larvae. (Source: https://commons. wikimedia.org)

Diener et al. 2011). The larvae also feed on food waste and vegetable matter, reducing the nitrites and volatiles and diverting the nutrients into their own biomass, reducing the financial and environmental costs associated with composting (Green and Popa 2012; Popa and Green 2012).

Expanding on these principles, BSF is highly desired in bioconversion systems, in which waste is not only removed, but also valorized into desirable products. The frass (insect feces) and unprocessed waste that the black soldier fly larvae (BSFL) do not consume can be ground into a fertilizer, for example, or used for biodiesel or biogas production (Li et al. 2015).The main product is the larvae themselves, which can be fed directly to fish, poultry, swine, and other livestock. Even shrimp (Cummins et al. 2017) and alligators (Bodri and Cole 2007) can be safely fed BSFL. By using the manures, corpses, and offal of these livestock to feed the larvae that are then fed to back the livestock, one produces a semi-closed nutrient loop that greatly saves on costs while reducing waste (Sheppard 1983). BSFL can also be ground into pellets or meal (BSFM) for better transportation and/or later processing, and oil can be extracted from the larvae in presses. BSFM and BSF oil are comparable to soy or fishmeal and soy or fish oil respectively.Fishmeal in particular is widely used as animal feed, but its costs continue to climb as overfishing drives down fish populations (Tacon et al. 2006). The larvae can also be converted to biodiesel, and are a remarkably effective way to convert otherwise recalcitrant lignocellulosic matter into fuel (Li et al. 2015; Zheng et al. 2012a). The unsustainability of fish meal and the rising calls for sustainable sources of protein, including a rapidly growing edible insect movement, mean more and more attention is being given to BSF around the world. BSF may even surpass the silkworm or cricket in terms of valuable, commercially farmed insect.

The structure of a BSF bioconversion facility or "biorefinery" depends on the needs of the user. At the smallest end of the scale range are kitchen countertop-sized machines for urbanites interested in adding insects to their diet. Small scale, single-household or single-farm biorefineries exist to allow one to convert waste into larvae, and possibly to use them for feeding of livestock such as poultry or fish. These refineries range from specially fabricated containers to homemade constructions built of cheap materials, instructions for which can be obtained from various online sources. Large scale, warehouse-sized fly factories exist as well: an example is a pilot-scale facility in Sidoarjo, Indonesia, associated with a wholesale market and funded by the Swiss Federal Institute of Aquatic Science and Technology. It takes fruit and vegetable waste from the market and converts it to fly larvae at the rate of several tons of biowaste per day, using ventilated, vertically stacked trays of larvae on waste. The final-stage larvae are either given alive to local fish and poultry farmers as feed, or pelletized as feed with a longer shelf life (Dortmans et al. 2017). Conceivably a city or other such municipal-level government could establish a large-scale biowaste recycling facility for the purpose of waste management and the ancillary bonus of feed production, comparable and arguably superior to extant recycling systems in which biowaste is fed to pigs (who produce prodigious amounts of waste themselves), composted into fertilizer, or used to make biogas; and certainly better than disposing of it in a landfill, burning it, or open dumping.

Importantly, a BSF facility is low tech, low skill, and low cost, making it easy to establish in developing nations (Dortmans et al. 2017; Nyakeri et al. 2017).

The biology of BSF and the engineering of a bioreactor will not be elaborated on here, though the references of this chapter include some handy guides (Dortmans et al. 2017; Sheppard et al. 2002; Nakamura et al. 2016; Nyakeri et al. 2017; Caruso et al. 2014; Veldkamp et al. 2012). Briefly, larvae are deposited on shredded or milled biowaste with moisture content of 70–80% and allowed to feed and grow. The sources of the larvae vary. BSF is widespread around the world (Marshall et al. 2015), so in someareas simply having the compostable material out and exposed will attract the flies to lay eggs. For others areas, BSF eggs or larvae can be ordered online and used to start a breeding colony. In larger facilities, dedicated breeding chambers for the adults are established, as they mate in flight and have minimum light and space requirements for mating and egg-laying (Nakamura et al. 2016; Zhang et al. 2010). Hatchings are kept on a substrate for five days, then either diverted to waste management or to produce new adults. The larvae on the biowaste grow for approximately 12 days, with a waste conversion ratio on a wet weight basis of up to 25% (higher if water weight is excluded). In their final stage, the larvae "self-harvest," wandering away from the substrate to find a place to pupate. Small-scale bioreactors take advantage of this behavior by simply placing a ramp that leads from the substrate to a collection bag for the larvae to crawl on. Larger facilities will harvest the pre-pupae in other ways. The larvae can be used as is, sun-dried or roasted or flash frozen for transport as whole larvae, or shredded and milled or pelleted for production of BSFM or other products. An oil press, centrifuge, or solvent extraction can be used to defatten the larvae for better storage or to extract and valorize the oil, which can have feed or culinary uses and is highly suitable for biodiesel (Li et al. 2011a, b). One study fed BSF the degreased solids from restaurant waste, whose grease was used for biodiesel production, and then used the BSF for biodiesel production, effectively doubling the yield (Zheng et al. 2012b). Thus biowaste of all kinds, from manure to food waste, can be recycled into the feed, fertilizer, and energy streams.

While relatively few obstacles exist to establish BSF as waste management systems, several questions remain before BSF can enter the food chain worldwide: as feed and as human food. Are BSFL or BSFM nutritious foods for livestock? Can BSF pass diseases or toxins from their feeding substrate into the organism that feeds on them? Does the replacement of fish meal or poultry feed with BSF products affect the nutritional quality of the meat downstream? These nutrition-related questions are the subject of this chapter.

13.2 Macronutrients: Protein and Fat

The macronutrients are carbohydrates, protein, and fat. Insects are not notable sources of carbohydrates and we need say no more about them here, other than to note that the exoskeletons of insects, as the shells of crabs and lobsters and the cell walls of fungi, are made of chitin. Chitin is a long-chain glucose polysaccharide with amino acids attached, that is most likely indigestible in humans: nutritionally it functions similar to insoluble fiber at best, with the associated benefits to regular bowel movements, to indigestible keratin (the protein of hair and fingernails) at worse. Normal consumption of insects should not cause problems, and black soldier fly larvae, like other maggots, have soft exoskeletons with relatively little chitin compared to more robust insects such as crickets. Chitin in BSFL has been estimated at 21 g/kg wet body weight (Finke 2013). However, BSF pupae are more robust than BSFL as well as other maggots and caterpillars. Chitin in BSF would be considered a fiber or an inedible portion that can negatively affect protein digestibility (Kroeckel et al. 2012) and certainly would affect texture and flavor of the product. We also note here the results of one study that measured energy content: BSFL are 199.4 kcal/100 g serving (Finke 2013).

The use of insects as food, both for humans and as animal feed, centers on insects being a sustainable source of protein relative to vertebrates. Even soy fares poorly in comparison to insects due to its high land and water requirements. Insects are complete proteins, with all the essential amino acids (AA) for humans, including the conditional amino acids such as arginine essential for infants, poultry, swine, and some pets (Spranghers et al. 2017), though with insufficient taurine for cats (Finke 2013). The predominant amino acids according to one study are lysine, valine, and arginine (Spranghers et al. 2017); aspartic acid and glutamic acid in two others (Liland et al. 2017; Finke 2013); aspartic acid, glutamic acid, and proline in a fourth (De Marco et al. 2015). Insect proteins are generally highly digestible, with protein digestibility corrected amino acid scores as high as 1 (the maximum score) for silkworm pupae. The exact score for BSF is unknown, but an experiment with pigs found calculated the apparent digestibility of nitrogen in BSF at 76.0, compared to 77.2 for soybean meal (Newton et al. 1977). A study comparing several different insects' suitability as dog and cat food found that housfly pupae (*Musca domestica*) followed by BSF pupae had the highest AA scores. BSF pupae had the lowest in vitro organic matter digestibility and low nitrogen digestibility, while BSFL were comparable to other insects (Bosch et al. 2014).

The critical nutritional value for insects as food or feed is the ratio of protein to fat. For BSF, this value varies considerably: the larvae can be anywhere from 17.5 to 63% protein and 12.8–49% fat (Diener et al. 2009; Spranghers et al. 2017; St-Hilaire et al. 2007a, b; Mutafela 2015; Sheppard et al. 1994; Barragan-Fonseca et al. 2017; Nyakeri et al. 2017; Tschirner and Simon 2015; Arango Gutiérrez et al. 2004; Bosch et al. 2014; Finke 2013; Oonincx et al. 2015), though protein values are usually between 30 and 50%. Note that the nitrogenous amino acids on chitin can sometimes be miscalculated as crude protein, so chitin corrections are sometimes applied (Spranghers et al. 2017). The high fat values mean defatting BSFL or BSFM before producing animal feed may be necessary for industrial-scale feed production. The high oil content would cause the product to go rancid, and can negatively affect the flavor of the product. Defatted BSFL would have a much higher protein to fat ratio and thus be much more suitable as a fishmeal substitute. A study on BSFM improved the 20% crude fat and 36% crude protein content to 8.8% crude fat and 47% crude protein by solvent extraction (Bussler et al. 2016), while another estimate improvements reaching over 60% crude protein (Spranghers et al. 2017).

The fats in a food or feed are comprised of saturated fatty acids (SFA), in which the carbon-carbon bonds of the fatty acids are all single bonds, and unsaturated fatty acids, in which one (monounsaturated fatty acids, MUFA) or more (polyunsaturated fatty acids, PUFA) of the bonds are double bonds. These double bonds in turn can be cis- or trans-configured, though transfats are not common in nature and are not an issue in BSF products. For human health, the question of which fats in which ratios are best for health, particularly cardiovascular health, seems to have a different and conflicting answer every few decades. Transfats were once considered to be the healthful alternative to saturated fats: they are now considered nearly toxic in terms of how far they raise one's risk of coronary disease (Schleifer 2012). Dietary recommendations to reduce the number of saturated fats one consumes have been criticised for causing the transfat debacle and for reducing consumption of PUFAs, which are considered to be protective of the heart and reduce the risk of cardiovascular disease... for now. Studies with PUFA supplements, including the very popular and lucrative omega-3 fish oil capsules, have failed to show any benefits (Abdelhamid et al. 2018). The effects of a diet high in fish, which is known to reduce the risk of heart disease, may be less due to increased PUFA and more due to reduced red meats and fats in general. To ensure that this chapter does not become outdated by the time it is printed, we will not go too far into the question of which fats are "good" and which "bad," other than to note that most humans can get all of the necessary fats they need from the vegetables in their diet, and that supplementation for already overweight Westerners is not necessary or effective in improve health. For animals, however, the nutritional requirements may differ, and so fatted BSFL, defatted BSFL, and BSF oil can be used in various ratios to create the ideal feed for livestock.

The fatty acid profiles of BSF vary considerably, but the general rule is that SFAs dominate considerably, and MUFA levels are usually higher than PUFA levels (Liland et al. 2017; Mohd-Noor et al. 2017). The most common acids are the SFAs lauric acid [C12:0] (28.1–38.43%) and palmitic acid [C16:0] (22.0%) (Moula et al. 2018; Ushakova et al. 2016; Oonincx et al. 2015). One study found BSF fed fish offal were 30% lipid, of which 3% was omega-3 fatty acids likely incorporated from this diet (St-Hilaire et al. 2007a, b). A study comparing BSF prepupae reared on chicken feed, biogas digestate, vegetable waste, restaurant waste found that the insect lipids contained high lauric acid levels and were comprised predominantly of SFAs, with saturated/unsaturated fatty acid ratios ranging from 2.5 to 5 (Spranghers et al. 2017). The substrates, being mostly plant-derived, had more unsaturated fats than saturated, but the BSF changed this ratio, as would any animal feeding on plants. The ratio of MUFA/PUFA ranged from 0.8 to 1.98, and the ratio of omega-6 to omega-3 PUFAs, which modern nutritionists recommend be lower, ranged from 1.98 to 15.1 (Spranghers et al. 2017; Oonincx et al. 2015). Nutrient utilization by black soldier flies fed with chicken, pig, or cow manure.

Why is there such variation? Different methods of measuring these macronutrients certainly play a role, as well as whether values reported are for dry or moist weight, are chitin corrected, etc. Also, some studies measured the protein/fat and PUFA/MUFA ratios of penultimate stage larvae, others of pre-pupae that have started to wander, others of the pupae themselves. All stages are edible, but they are not alike, particularly the pupae (Bosch et al. 2014). Different populations of BSF from around the world also differ in their fitness, development rates, and waste conversion abilities (Zhou et al. 2013). The main cause of the variation, however, comes from what the BSFL fed on. Simply put, the higher the protein content of the substrate and the more is available to the larvae, the higher the protein content of the larvae (Diener et al. 2009; Spranghers et al. 2017; Mutafela 2015; Oonincx et al. 2015). The higher or lower the PUFA/MUFA or omega-6/omega-3 ratios of the substrate, the higher or lower these ratios are in the larvae (St-Hilaire et al. 2007a, b; Spranghers et al. 2017). This causative relationship means one can titrate the nutritional values of BSF to an extent by controlling the feed they are given, in order to produce a highest quality feed product (Mohd-Noor et al. 2017). For example, adding up to 50% brown algae (*Ascophyllum nodosum*) to the BSF substrate increases their omega-g levels (Liland et al. 2017). However, this modulation defeats the purpose of using BSF as waste management, especially if one needs to add commercial feed to the BSF feedstock. If one is raising BSF for the express purpose of using them as feed, then knowing or estimating the protein/fat and MUFA/PUFA ratios of the feed is important. A silver lining is that crude protein content, while variable, is always high and the AA profile of high quality in BSFL regardless of what they are fed: if they are fed a relatively protein-poor diet of primarily plants, the BSF protein levels will be higher to that of initial substrate, and the AA profiles more consistent even than those of commercial soybean meals (Spranghers et al. 2017), although AA profiles can still be improved with substrate supplementation (Liland et al. 2017).

To summarize, BSF is an excellent source of protein, but may need defatting for certain target consumers. No study concluded that BSF is unsuitable for livestock, and nearly all recommended its use as a partial replacement of fishmeal or soybean meal.

13.3 Micronutrients: Vitamins and Minerals

We found few studies looking at vitamins in BSF. Levels of vitamin A, C, and D are low. One study looked at wild BSF from the Bondo area of Kenya, attracted to an open system of composting with BSF for use by poultry and fish farmers. Wild BSFL here have 0.24 mg/100 g thiamine, 2.2 mg/100 g riboflavin, and 1.3 mg/100 g vitamin E (Nyakeri et al. 2017). Analysis of commercially produced BSFL found 0.77 mg/100 g thiamine, 1.62 mg/100 g riboflavin, 3.85 mg/100 g pantothenic acid, 7.1 mg/100 g niacin, 0.6 mg/100 g pyridoxine, 0.27 mg/100 g folic acid, 0.035 mg/100 g biotin, 5.58 mg/100 g vitamin B_{12}, 110 mg/100 g choline, 8.38 mg/100 g carnitine, and 0.62 mg/100 g vitamin E (Finke 2013). A controlled-diet study found that total vitamin E was predominantly alpha-tocopherol, and that vitamin E levels could be increased with seaweed supplementation (Liland et al. 2017). Vitamin E appears to be the only vitamin for which BSFL meet 50–100% of the NRC requirements of laboratory rats for growth (Finke 2013).

The mineral content of BSF varies substantially with diet, but does not always match the mineral content of the diet itself: calcium levels among the four substrates used in Spraghers et al. (2017) varied along with the calcium levels of the BSF, which ranged from 1.23–66.15 g/kg dry weight, but did not correlate. Iron too varied among the larvae (0.11–0.43 g/kg) but did not correlate with the iron content of the substrate. Other minerals did not vary greatly among the larvae fed on different diets. The average values in g/kg dry weight with standard deviation are as follows: Ca = 31.2 ± 26.7, Cu = 0.01, Fe = 0.25 ± 0.16, K = 6.21 ± 0.37, Mg = 2.59 ± 0.42, Mn = 0.22 ± 0.15, Na = 0.71 ± 0.13, P = 4.39 ± 0.44, S = 0.2 ± 0.08, Zn = 0.088 ± 0.049 (Spranghers et al. 2017). Another study using seaweed supplementation of a plant-based diet found that most mineral concentrations in the larvae correlated linearly with the mineral concentration in the diet. That study's values for BSF fed a control diet are as follows: Ca = 8.4 ± 0.4, Cu = 7.9 ± 0.2, Fe = 0.12 ± 0.02, K = 10.2 ± 0.2, Mg = 2.1 ± 0.1, Mn = 0.19 ± 0.01, Na = 1.0 ± 0.1, P = 6.8 ± 0.1, Se = 0.1, Zn = 68 ± 1.6 (Liland et al. 2017). Note that one study measure sulfur, the other selenium. The Kenya study found that wild BSFL have 0.57% Fe, 2.27% K, 0.56% Mn, and 3.07% Na (Nyakeri et al. 2017). An analysis of commercial "phoenix worms," which is the nickname given to BSFL used as pet reptile feed, had the following results: Ca = 9.3, Cl = 1.1, Cu = 0.004, Fe = 0.066, K = 4.53, Mg = 1.74, Mn = 0.062, Na = 0.89, P = 3.56, Se = 0.032, and Z = 0.056 (Finke 2013). As for trace elements, adding iodine-rich seaweed to the substrate increases the normally niliodine levels of the BSFL, as expected (Liland et al. 2017). Lutein (0.059 mg/100 g) and zeaxanthin (0.128 mg/100 g) are present in BSFL, but not beta carotene (Finke 2013). Lastly, sterols in BSFL hovered around 0.2% dry matter by weight, predominantly β-sitosterol and stigmasterol, though these can vary with diet (Liland et al. 2017).

13.4   Effects on Livestock

The above values are meaningless without practical tests of the effect of BSF supplementation of the diet on the animals eating it. The good news is that this has been done for many animals, especially fish and poultry. The even better news is that the results are positive: BSF supplementation or partial replacement of fishmeal or soymeal in a diet has little to no negative effects on the health of the animals that eat them, nor does it significantly affect the flavor or nutritional profile of the products of these animals, although not all testes recommend 100% diet replacement.On occasion one even sees positive effects! One would have no reason to suspect any direct or indirect nutritional costs of BSF supplementation on humans relative to other insect proteins. I leave it to the marketing executives to determine whether or not BSF should be designated a "superfood," but in terms of animal feed, the results are nearly unanimous: BSF supplementation is safe and effective, though the exact percentage of diet replaceable with BSF depends on the livestock and the response being measured. The results of these tests are presented grouped by the clade of livestock, and include tests on the effect of BSF on livestock growth, health, and subsequent nutrient quality such as fatty acid profiles.

A study with 10–15% BSF diets replacing equivalent amounts of soybean meal in the diet of broiler quails (*Coturnix coturnix japonica*) found no significant differences in the health and performance of the quails, nor did it negatively affect the breast weight or yield, or meat toughness and cooking loss (Cullere et al. 2016). A subsequent study found no changes to quail breast meat cholesterol or protein content, but increased SFA and MUFA and decreased PUFA levels (Cullere et al. 2017). Another study found significant but nonlinear effects on feed consumption, egg weight, and egg production per day for quail fed a diet with 0 to 100% replacement of fishmeal with BSFM, with 25–50% replacement ideal (Widjastuti et al. 2014). A study with chickens found a positive effect of partial BSFL replacement of the commercial diet on the weekly weights of the chickens and no effect on the protein content, fatty acid profiles, or omega-6/3 ratios of the meat (Moula et al. 2018). Another study found increasing SFA, reduced PUFA, and no change to MUFA in broiler chicken breast meat with increased replacement of soybean oil with BSFL up to 100%, but found no other differences, including in crude protein content, and concluded that adding BSFL "guaranteed satisfactory productive performances, carcass traits and overall meat quality" (Schiavone et al. 2017a). Use of defatted BSFL over unprocessed or partially defatted BSFL was recommended (Schiavone et al. 2017b). Another study focusing on layer hens found that up to 100% of soybean cake could be replaced with BSFM with no effects on hen performance or health (Maurer et al. 2016). Studies from Nigeria on broiler hens found no effects on the performance or health of the hens even at 100% replacement of fishmeal with BSFM, going so far as to recommend "massive" commercial production (Awoniyi 2007; Oluokun 2000).

Tests with European seabass (*Dicentrarchus labrax*) found no negative effects on growth performance or feed utilization and digestibility of fishmeal replacement with BSF up to 19.5% BSFM replacing 45% FM (Magalhães et al. 2017). Tests in turbot fish (*Psetta maxima*) found reduced growth performance with increasing levels of BSFM in a diet replacing fishmeal despite maintaining the same protein/fat percentages. This effect was blamed on chitin, which reduced the digestibility of the feed (Kroeckel et al. 2012). Tests on carp (*Cyprinus carpio* var. Jian) with soybean oil replaced from 0 up to 100% with BSF oil found no significant differences on growth performance or fish meat fatty acid composition across the board (Li et al. 2016). Tests with rainbow trout (*Oncorhynchus mykiss*) replacing fishmeal with BSF prepupae up to 50% found significantly reduced growth on fish fed BSF, but these effects vanished if the larvae had been grown with fish offal (Sealey et al. 2011). These "enriched" BSF match what would be found in BSF bioreactors associated with fisheries, in which the fishery waste such as offal and dead fish is recycled via BSF and fed back to the fish. Another study with rainbow trout found no negative effects of a 15% BSF protein diet on feed conversion rates, and calculated a 38% reduction in the necessary fish oil, but did note that the fish meats had reduced omerga-3 fatty acids. They concluded that BSF is a suitable trout feedstuff (St-Hilaire et al. 2007a, b). Yet another study on rainbow trout found no effects on fish growth performance or health or other characteristics, nor any effects on meat quality other than reduced PUFA levels, for BSFL replacement of fishmeal up to 50% (Renna et al. 2017). Tests in shrimp (*Litopenaeus vannamei*) using diets of 35% protein feed with varying percentages of BSFM to fishmeal found that >95% growth rates and protein content could be achieved so long as the percentage of BSFM in the diet was limited to approximately 25% (Cummins et al. 2017).

Some sensory tests have been done on the organoleptic properties of the livestock meats to humans. A quail study found that sensory characteristic of breast meat did not differ between BSF-fed and control diet quails, nor were any off-flavors perceived. The rainbow trout study found human taste testers cannot tell the difference between fish meat fed control diets or BSF diets (Sealey et al. 2011). Taste tests from boiled and fried cuts of broiler hens with 0–100% fishmeal replacement with BSFM found no effect of BSFM on the appearance and flavor of male chickens. For female chickens, significant differences were found, but not in an linear relationship: 50–75% BSFM to fishmeal produced the best meat (Awoniyi 2007).

None of these findings matter if the livestock will not eat the diet. Quails had no preference between control diets and 15% BSF diets (Cullere et al. 2016), and in another study they preferred diets at 25 or 50% BSFM (Widjastuti et al. 2014). Pigs showed no preference between BSFM diets and diets with of soybean meal with added grease, but significantly preferred BSFM to soymeal without grease (Newton et al. 1977). Turbot on the other hand strongly preferred diets with more fishmeal and less BSFM (Kroeckel et al. 2012). Evidence suggests livestock will usually eat diets supplemented with BSFL products to a point, and of course live larvae are a natural food source for fish, poultry, and reptiles or amphibians in the wild.

13.5   Safety and Legal Issues

The main benefit of BSF is their ability to valorize nearly all forms of biowaste into edible matter. This benefit, however, is also the main obstacle to their widespread adoption: many countries have legal limits on what kind of substrate can be fed to any organism that will enter the food supply, be it as human food or as animal feed (Van Raamsdonk et al. 2017). Edible and nutritious BSF can be reared on manure (human and animal), on cadavers, on blood and abattoir waste, on catering waste, and conceivably even on medical biowaste, but none of these diets are legally acceptable for feed under European or American law (Kupferschmidt 2015; Laurenza and Carreño 2015). Compare this to crickets, which are mostly reared on commercial chicken feed or even commercial cricket feed, supplemented occasionally with fruits and vegetables. In general, any animal used as feed or food must itself have been fed a good-quality diet: manure and corpses are not acceptable.

The policies are sensible, given that contaminants in feed can enter the food and eventually sicken the final consumer, meaning humans. However, BSF, more than other organisms, can produce a safe product even if fed unsafe feed. The laws are an obstacle to adoption of BSF but are not insurmountable in the face of evidence. BSFL is already the only insect product legalized for use as animal feed in the USA, albeit only for salmon. Edible insect lobbying groups exist, and they are working to introduce legislation that will facilitate BSF and other insects' entries into the food and feed supply. What legislators want to see are scientific studies confirming that BSF is safe to eat, even if fed on a substrate that alone would not be. The results of these studies to date are summarized below.

Studies on BSF and microbes include seeing if they reduce the microbial load of the waste substrate, with ecological and public health benefits, and whether they contain these microbes as edible pupae. A study using labeled *Escherichia coli* O157:H7 and *Salmonella enterica* serovar Enteritidis found that BSFL feeding drastically reduced the populations of these microbes in chicken manure, but not in cow or swine manure, though the latter was blamed on manure pH affecting the larvae (Erickson et al. 2004). A later study successfully demonstrated *E. coli* reduction in BSFL-treated dairy cow manure (Liu et al. 2008). A study on human fecal sludge found reductions in *Salmonella* spp. but not in *Enterococcus spp.* or *Ascaris suum* (Lalander et al. 2013), and another found reductions in several strains of *Salmonella*, including an antibiotic multiresistant strain, from feces treated with BSF (Gabler and Vinnerås 2014). Viruses were similarly reduced (Lalander et al. 2015). A study examining strains of *Bacillus subtilis* inoculated into poultry manure fed to BSFL found that the bacteria were present in the larval gut and negatively affected larval development. A study examining microbes in BSFL and in the guts of chickens fed differing amounts of BSFL found several microbes in both the larvae and the chickens, but these did not differ from the microbes found in hens not fed BSF; rather, they were ubiquitous environmental microbes that hens and humans alike are exposed to daily and toward which they have sufficient immunity to prevent illness (Awoniyi 2007). As an aside, one study identified several antibacterial and antifungal compounds from lyophilized, septically challenged BSFL active on a very broad range of microbes, including gram-positive and gram-negative bacteria and even against methicillin resistant *Staphylococcus aureus* (Park et al. 2014).

The biggest unknowns are prions: the pathogenic, misfolded proteins responsible for diseases such as bovine spongiform encephalitis (also known as "mad cow disease"), scrapie, and Creutzfeld–Jakob disease. These diseases are incurable and fatal, meaning prevention is taken extremely seriously. Prevention has focused on eliminating the practice of feeding potentially infected nervous tissues of livestock to other animals, both humans and pets and livestock. The very reason that fishmeal and soybean meal have become so essential in animal feed is as a replacement of the meat-and-bone meal though to have been responsible for the mad cow disease outbreaks of 1980s and 1990s. Under current laws, these tissues could not be fed to BSF either, especially if these larvae would then be fed to livestock, due to the potential that prions will persist in the larvae (Vos 2000). Insects are not known to develop prion diseases, but at this point in time whether BSF can bioaccumulate prions is unknown.

We should note here that microbial issues can be dealt with through processing of BSF products or their subsequent cooking (Swinscoe et al. 2018). As in essentially all foods, BSF's microbial load is reduced if cooked, and higher if raw and/or improperly stored. Processing BSFL into a final product such as pellets or meal can solve some of the above safety issues, so long as proper hygiene and good practices are observed during packaging, storage, and handling of the final product (Swinscoe et al. 2018). It is also worth noting that cases of enteric myiasis with BSF have been reported, discouraging consumption of BSF eggs or live larvae (Lee et al. 1995; Meleney and Harwood 1935; Yang 2014; Werner 1956). Prions, however, survive heat, so demonstrating that BSF does not accumulate prions is essential to clearing the final hurdle to allowing post-consumer waste into the BSF feedstock options. If it turns out that cow prions can remain in BSF, then household and catering waste as well as abattoir waste will remain unacceptable for BSF, but the abundance of plant-based waste as well as manure would still be suitable even for BSF destined as a human food.

None of the studied pesticides (azoxystrobinchlorpyrifos, chlorpyrifos-methyl, pirimiphos-methyl, propiconazole), mycotoxins (aflatoxin B1/B2/G2, deoxynivalenol, ochratoxin A, zearalenone), or pharmaceutical drugs (carbamazepine, roxithromycin, trimethoprim) can accumulate in BSFL (Purschke et al. 2017; Lalander et al. 2016; Bosch et al. 2017). This is both good for the consumer of the BSFL, and for the environment, as these compounds are being destroyed by the larvae and are thus eliminated from the compost. Obviously certain insecticides such as pyrethroids will kill BSF, interestingly at higher rates than they kill house flies (Tomberlin et al. 2002). Regarding heavy metals, cadmium in particular and, in some but not all studies, lead appeared to bioaccumulate in the larvae and pupae, while arsenic, chromium, mercury, nickel, and zinc were not incorporated (Gao et al. 2017; Purschke et al. 2017; Charlton et al. 2015; Diener et al. 2015). Thus cadmium levels in a substrate may need to be monitored for the health of the larvae and the final consumers.

Regarding allergies: certain people do have food allergies to insects, such as silkworm pupae available for consumption in East Asia (Ji et al. 2008). Chitin is one of the potential allergens in insects, and people with dust-mite allergies may also be allergic to eating insects, though this has not yet been demonstrated. Certain panallergens in insects, crustaceans, and mollusks such as arginine kinase or tropomyosin indicate that those with shellfish allergies may also be allergic to insects (Ribeiro et al. 2018). Still other people are solely allergic to insects. Note that there have not been any recorded allergies to any Diptera maggots; on the contrary, some studies found boosted immunity in consumers of housefly larva powder (Nkegbe et al. 2018). Whether BSF fed on other allergy inducing foods such as peanuts become themselves allergenic is unknown and unlikely, but not impossible in the event of those foods being the last meal of a larva before it is processed and consumed by a human of extreme sensitivity. Direct allergies to insects are more likely, and so those with preexisting shellfish allergies may wish to avoid or be cautious with edible insects of any kind, including BSF. For livestock that have been fed BSF or other insects, the risk of allergens persisting into the meat is zero. Given that early exposure to allergens can prevent development of severe allergies later in life, perhaps ensuring or at least not preventing infants from swallowing insects periodically can help bring insect-related allergy incidences down (Du Toit et al. 2015).

13.6 Conclusion

From a legal and nutritional aspect alike, we see a tension between using BSF as a waste management tool, where they are reared on whatever waste is available, and using BSF as a feed source, where certain feeding substrates are more highly desired on a nutritional standpoint while others are effectively banned for safety reasons. The choice of substrate depends on the local needs in both these regards, as does the scale of the bioreactor setup from single-home to community sized. Ultimately, the potential of BSF is immense and its increasing establishment as a staple microlivestock animal is inevitable (Wang and Shelomi 2017). With each passing month, more and more organizations are establishing BSF facilities around the world, in every continent, from Germany to Guam to Ghana. With new start-ups, new charitable development projects, new processing technologies, new book chapters, and so on, the rising tide of BSF is clear.

References

Abdelhamid AS, Brown TJ, Brainard JS, Biswas P, Thorpe GC, Moore HJ et al (2018) Omega-3 fatty acids for the primary and secondary prevention of cardiovascular disease. Cochrane Database Syst Rev (7). https://doi.org/10.1002/14651858.CD003177.pub3

Arango Gutiérrez GP, Vergara Ruiz RA, Mejía Vélez H (2004) Compositional, microbiological and protein digestibility analysis of the larva meal of Hermetia illuscens L.(Diptera: Stratiomyiidae) at Angelópolis-Antioquia, Colombia. Rev Fac Nac Agron Medellin 57:2491–2500

Awoniyi TAM (2007) Health, nutrional and consumers' acceptability assurance of maggot meal inclusion in livestock diet: a review. Int J Trop Med 2(2):52–56

Banks IJ, Gibson WT, Cameron MM (2014) Growth rates of black soldier fly larvae fed on fresh human faeces and their implication for improving sanitation. Trop Med Int Health 19(1):14–22. https://doi.org/10.1111/tmi.12228.

Barragan-Fonseca KB, Dicke M, van Loon JJA (2017) Nutritional value of the black soldier fly (Hermetia illucens L.) and its suitability as animal feed—a review. J Insects Food Feed 3(2):105–120

Bodri MS, Cole ER (2007) Black soldier fly (Hermetia illucens Linnaeus) as feed for the American Alligator (Alligator mississippiensis Daudin). Georgia J Sci 65(2):82

Bosch G, Zhang S, Oonincx DGAB, Hendriks WH (2014) Protein quality of insects as potential ingredients for dog and cat foods. J Nutr Sci 3:e29. https://doi.org/10.1017/jns.2014.23.

Bosch G, van der Fels-Klerx HJ, Rijk TCD, Oonincx DGAB (2017) Aflatoxin B1 tolerance and accumulation in black soldier fly larvae (Hermetia illucens) and yellow mealworms (Tenebrio molitor). Toxins 9(6):185

Bradley SW, Sheppard DC (1984) House fly oviposition inhibition by larvae of *Hermetia illucens*, the black soldier fly. J Chem Ecol 10(6):853–859. https://doi.org/10.1007/Bf00987968

Bussler S, Rumpold BA, Jander E, Rawel HM, Schluter OK (2016) Recovery and techno-functionality of flours and proteins from two edible insect species: meal worm (Tenebrio molitor) and black soldier fly (Hermetia illucens) larvae. Heliyon 2(12):e00218. https://doi.org/10.1016/j.heliyon.2016.e00218.

Caruso D, Devic E, Subamia I, Talamond P, Baras E (2014) Technical handbook of domestication and production of Diptera black soldier fly (BSF), *Hermetia illucens*, Stratiomyidae. IRD Press, Bogor, p 141

Charlton AJ, Dickinson M, Wakefield ME, Fitches E, Kenis M, Han R et al (2015) Exploring the chemical safety of fly larvae as a source of protein for animal feed. J Insects Food Feed 1(1):7–16

Cullere M, Tasoniero G, Giaccone V, Miotti-Scapin R, Claeys E, De Smet S et al (2016) Black soldier fly as dietary protein source for broiler quails: apparent digestibility, excreta microbial load, feed choice, performance, carcass and meat traits. Animal 10(12):1923–1930. https://doi.org/10.1017/S1751731116001270

Cullere M, Tasoniero G, Giaccone V, Acuti G, Marangon A, Dalle Zotte A (2017) Black soldier fly as dietary protein source for broiler quails: meat proximate composition, fatty acid and amino acid profile, oxidative status and sensory traits. Animal 12:1–8. https://doi.org/10.1017/S1751731117001860.

Cummins VC, Rawles SD, Thompson KR, Velasquez A, Kobayashi Y, Hager J et al (2017) Evaluation of black soldier fly (Hermetia illucens) larvae meal as partial or total replacement of marine fish meal in practical diets for Pacific white shrimp (Litopenaeus vannamei). Aquaculture 473:337–344

De Marco M, Martínez S, Hernandez F, Madrid J, Gai F, Rotolo L et al (2015) Nutritional value of two insect larval meals (Tenebrio molitor and Hermetia illucens) for broiler chickens: apparent nutrient digestibility, apparent ileal amino acid digestibility and apparent metabolizable energy. Anim Feed Sci Technol 209:211–218. https://doi.org/10.1016/j.anifeedsci.2015.08.006.

Diener S, Zurbrugg C, Tockner K (2009) Conversion of organic material by black soldier fly larvae: establishing optimal feeding rates. Waste Manag Res 27(6):603–610. https://doi.org/10.1177/0734242X09103838.

Diener S, Solano NMS, Gutierrez FR, Zurbrugg C, Tockner K (2011) Biological treatment of municipal organic waste using black soldier fly larvae. Waste Biomass Valoriz 2(4):357–363. https://doi.org/10.1007/s12649-011-9079-1.

Diener S, Zurbrügg C, Tockner K (2015) Bioaccumulation of heavy metals in the black soldier fly, Hermetia illucens and effects on its life cycle. J Insects Food Feed 1(4):261–270

Dortmans B, Diener S, Verstappen B, Zurbrügg C (2017) Black soldier fly biowaste processing— a step-by-step guide. Eawag-Swiss Federal Institute of Aquatic Science and Technology, Dübendorf, p 87

Du Toit G, Roberts G, Sayre PH, Bahnson HT, Radulovic S, Santos AF et al (2015) Randomized trial of peanut consumption in infants at risk for peanut allergy. N Engl J Med 372(9):803–813

Erickson MC, Islam M, Sheppard C, Liao J, Doyle MP (2004) Reduction of Escherichia coli O157:H7 and Salmonella enterica serovar Enteritidis in chicken manure by larvae of the black soldier fly. J Food Prot 67(4):685–690

Finke MD (2013) Complete nutrient content of four species of feeder insects. Zoo Biol 32(1):27–36. https://doi.org/10.1002/zoo.21012

Gabler F, Vinnerås B (2014) Using Black Soldier Fly for waste recycling and effective Salmonella spp. reduction. Swedish University of Agricultural Sciences, Uppsala

Gao Q, Wang XY, Wang WQ, Lei CL, Zhu F (2017) Influences of chromium and cadmium on the development of black soldier fly larvae. Environ Sci Pollut Res 24(9):8637–8644. https://doi.org/10.1007/s11356-017-8550-3.

Green TR, Popa R (2012) Enhanced ammonia content in compost leachate processed by black soldier fly larvae. Appl Biochem Biotechnol 166(6):1381–1387. https://doi.org/10.1007/s12010-011-9530-6.

Ji K-M, Zhan Z-K, Chen J-J, Liu Z-G (2008) Anaphylactic shock caused by silkworm pupa consumption in China. Allergy 63(10):1407–1408. https://doi.org/10.1111/j.1398-9995.2008.01838.x.

Kim W, Bae S, Park K, Lee S, Choi Y, Han S et al (2011) Biochemical characterization of digestive enzymes in the black soldier fly, Hermetia illucens (Diptera: Stratiomyidae). J Asia Pac Entomol 14(1):11–14. https://doi.org/10.1016/j.aspen.2010.11.003.

Kroeckel S, Harjes AGE, Roth I, Katz H, Wuertz S, Susenbeth A et al (2012) When a turbot catches a fly: Evaluation of a pre-pupae meal of the Black Soldier Fly (Hermetia illucens) as fish meal substitute—growth performance and chitin degradation in juvenile turbot (Psetta maxima). Aquaculture 364:345–352. https://doi.org/10.1016/j.aquaculture.2012.08.041.

Kupferschmidt K (2015) Buzz food. Science 350(6258):267–269. https://doi.org/10.1126/science.350.6258.267.

Lalander CH, Diener S, Magri ME, Zurbrugg C, Lindstrom A, Vinneras B (2013) Faecal sludge management with the larvae of the black soldier fly (Hermetia illucens)—from a hygiene aspect. Sci Total Environ 458-460:312–318. https://doi.org/10.1016/j.scitotenv.2013.04.033

Lalander CH, Fidjeland J, Diener S, Eriksson S, Vinneras B (2015) High waste-to-biomass conversion and efficient Salmonella spp. reduction using black soldier fly for waste recycling. Agron Sustain Dev 35(1):261–271. https://doi.org/10.1007/s13593-014-0235-4.

Lalander C, Senecal J, Gros Calvo M, Ahrens L, Josefsson S, Wiberg K et al (2016) Fate of pharmaceuticals and pesticides in fly larvae composting. Sci Total Environ 565:279–286. https://doi.org/10.1016/j.scitotenv.2016.04.147.

Laurenza EC, Carreño I (2015) Edible insects and insect-based products in the EU: safety assessments, legal loopholes and business opportunities. Eur J Risk Regul 6(2):288–292

Lee HL, Chandrawathani P, Wong WY, Tharam S, Lim WY (1995) A case of human enteric myiasis due to larvae of Hermetia illucens (Family: Stratiomyiadae): first report in Malaysia. Malays J Pathol 17(2):109–111

Li Q, Zheng LY, Cai H, Garza E, Yu ZN, Zhou SD (2011a) From organic waste to biodiesel: black soldier fly, Hermetia illucens, makes it feasible. Fuel 90(4):1545–1548. https://doi.org/10.1016/j.fuel.2010.11.016.

Li Q, Zheng LY, Qiu N, Cai H, Tomberlin JK, Yu ZN (2011b) Bioconversion of dairy manure by black soldier fly (Diptera: Stratiomyidae) for biodiesel and sugar production. Waste Manag 31(6):1316–1320. https://doi.org/10.1016/j.wasman.2011.01.005.

Li W, Li Q, Zheng LY, Wang YY, Zhang JB, Yu ZN et al (2015) Potential biodiesel and biogas production from corncob by anaerobic fermentation and black soldier fly. Bioresour Technol 194:276–282. https://doi.org/10.1016/j.biortech.2015.06.112.

Li SL, Ji H, Zhang BX, Tian JJ, Zhou JS, Yu HB (2016) Influence of black soldier fly (Hermetia illucens) larvae oil on growth performance, body composition, tissue fatty acid composition and lipid deposition in juvenile Jian carp (Cyprinus carpio var Jian). Aquaculture 465:43–52. https://doi.org/10.1016/j.aquaculture.2016.08.020.

Liland NS, Biancarosa I, Araujo P, Biemans D, Bruckner CG, Waagbø R et al (2017) Modulation of nutrient composition of black soldier fly (Hermetia illucens) larvae by feeding seaweed-enriched media. PLoS One 12(8):e0183188. https://doi.org/10.1371/journal.pone.0183188.

Liu QL, Tomberlin JK, Brady JA, Sanford MR, Yu ZN (2008) Black soldier fly (Diptera: Stratiomyidae) larvae reduce Escherichia coli in dairy manure. Environ Entomol 37(6):1525–1530. https://doi.org/10.1603/0046-225x-37.6.1525

Magalhães R, Sánchez-López A, Leal RS, Martínez-Llorens S, Oliva-Teles A, Peres H (2017) Black soldier fly (Hermetia illucens) pre-pupae meal as a fish meal replacement in diets for European seabass (Dicentrarchus labrax). Aquaculture 476:79–85

Marshall S, Woodley N, Hauser MJTJ o t ES o O (2015) The historical spread of the Black Soldier Fly, Hermetia illucens (L.) (Diptera, , Stratiomyidae, Hermetiinae), and its establishment in Canada. J Entomol Soc Ontario 146:51–54

Maurer V, Holinger M, Amsler Z, Früh B, Wohlfahrt J, Stamer A et al (2016) Replacement of soybean cake by *Hermetia illucens* meal in diets for layers. J Insects Food Feed 2(2):83–90

Meleney HE, Harwood PD (1935) Human intestinal myiasis due to the larvae of the soldier fly, Hermetia illucens Linne (Diptera, Stratiomyidae). Am J Trop Med Hygiene 1(1):45–49

Mohd-Noor S-N, Wong C-Y, Lim J-W, Uemura Y, Lam M-K, Ramli A et al (2017) Optimization of self-fermented period of waste coconut endosperm destined to feed black soldier fly larvae in enhancing the lipid and protein yields. Renew Energy 111:646–654

Moula N, Scippo M-L, Douny C, Degand G, Dawans E, Cabaraux J-F et al (2018) Performances of local poultry breed fed black soldier fly larvae reared on horse manure. Animal Nutr 4(1):73–78. https://doi.org/10.1016/j.aninu.2017.10.002.

Mutafela RN (2015) High value organic waste treatment via black soldier fly bioconversion: onsite pilot study. Master of Science, KTH Royal Institute of Technology, Stockholm

Nakamura S, Ichiki RT, Shimoda M, Morioka S (2016) Small-scale rearing of the black soldier fly, Hermetia illucens (Diptera: Stratiomyidae), in the laboratory: low-cost and year-round rearing. Appl Entomol Zool 51(1):161–166. https://doi.org/10.1007/s13355-015-0376-1.

Newton GL, Booram CV, Barker RW, Hale OM (1977) Dried Hermetia Illucens larvae meal as a supplement for swine. J Anim Sci 44(3):395–400. https://doi.org/10.2527/jas1977.443395x.

Nguyen TT, Tomberlin JK, Vanlaerhoven S (2015) Ability of black soldier fly (Diptera: Stratiomyidae) larvae to recycle food waste. Environ Entomol 44(2):406–410. https://doi.org/10.1093/ee/nvv002.

Nkegbe EK, Adu-Aboagye G, Affedzie OS, Nacambo S, Boafo AB, Kenis M et al (2018) Potential health and safety issues in the small-scale production of fly larvae for animal feed—a review. Ghan J Animal Sci 9(1):10

Nyakeri EM, Ogola HJ, Ayieko MA, Amimo FA (2017) An open system for farming black soldier fly larvae as a source of proteins for smallscale poultry and fish production. J Insects Food Feed 3(1):51–56

Oluokun JA (2000) Upgrading the nutritive value of full-fat soyabeans meal for broiler production with either fishmeal or black soldier fly larvae meal (Hermetia illucens). Nigerian J Animal Sci 3(2)

Oonincx DGAB, van Broekhoven S, van Huis A, van Loon JJA (2015) Feed conversion, survival and development, and composition of four insect species on diets composed of food by-products. PLoS One 10(12):e0144601. https://doi.org/10.1371/journal.pone.0144601.

Park SI, Chang BS, Yoe SM (2014) Detection of antimicrobial substances from larvae of the black soldier fly, Hermetia illucens (Diptera: Stratiomyidae). Entomol Res 44(2):58–64. https://doi.org/10.1111/1748-5967.12050.

Popa R, Green TR (2012) Using black soldier fly larvae for processing organic leachates. J Econ Entomol 105(2):374–378. https://doi.org/10.1603/EC11192.

Purschke B, Scheibelberger R, Axmann S, Adler A, Jager H (2017) Impact of substrate contamination with mycotoxins, heavy metals and pesticides on growth performance and composition of black soldier fly larvae (Hermetia illucens) for use in the feed and food value chain. Food Addit Contam Part A Chem Anal Control Expo Risk Assess 34(8):1410–1420. https://doi.org/10.1080/19440049.2017.1299946

Renna M, Schiavone A, Gai F, Dabbou S, Lussiana C, Malfatto V et al (2017) Evaluation of the suitability of a partially defatted black soldier fly (Hermetia illucens L.) larvae meal as ingredient for rainbow trout (Oncorhynchus mykiss Walbaum) diets. J Animal Sci Biotechnol 8(1):57

Ribeiro JC, Cunha LM, Sousa-Pinto B, Fonseca J (2018) Allergic risks of consuming edible insects: a systematic review. Mol Nutr Food Res 62(1):1700030. https://doi.org/10.1002/mnfr.201700030

Schiavone A, Cullere M, De Marco M, Meneguz M, Biasato I, Bergagna S et al (2017a) Partial or total replacement of soybean oil by black soldier fly larvae (Hermetia illucens L.) fat in broiler diets: effect on growth performances, feed-choice, blood traits, carcass characteristics and meat quality. Ital J Anim Sci 16(1):93–100. https://doi.org/10.1080/1828051X.2016.1249968.

Schiavone A, De Marco M, Martínez S, Dabbou S, Renna M, Madrid J et al (2017b) Nutritional value of a partially defatted and a highly defatted black soldier fly larvae (Hermetia illucens L.) meal for broiler chickens: apparent nutrient digestibility, apparent metabolizable energy and apparent ileal amino acid digestibility. J Animal Sci Biotechnol 8(1):51

Schleifer D (2012) The perfect solution: how trans fats became the healthy replacement for saturated fats. J Technol Cult 53(1):94–119

Sealey WM, Gaylord TG, Barrows FT, Tomberlin JK, McGuire MA, Ross C et al (2011) Sensory analysis of rainbow trout, Oncorhynchus mykiss, fed enriched black soldier fly prepupae, Hermetia illucens. J World Aquacult Soc 42(1):34–45. https://doi.org/10.1111/J.1749-7345.2010.00441.X

Sheppard C (1983) Housefly and lesser fly control utilizing the black soldier fly in manure management-systems for caged laying hens. Environ Entomol 12(5):1439–1442

Sheppard DC, Newton GL, Thompson SA, Savage S (1994) A value-added manure management-system using the black soldier fly. Bioresour Technol 50(3):275–279

Sheppard DC, Tomberlin JK, Joyce JA, Kiser BC, Sumner SM (2002) Rearing methods for the black soldier fly (Diptera: Stratiomyidae). J Med Entomol 39(4):695–698

Spranghers T, Ottoboni M, Klootwijk C, Ovyn A, Deboosere S, De Meulenaer B et al (2017) Nutritional composition of black soldier fly (Hermetia illucens) prepupae reared on different organic waste substrates. J Sci Food Agric 97(8):2594–2600. https://doi.org/10.1002/jsfa.8081.

St-Hilaire S, Cranfill K, McGuire MA, Mosley EE, Tomberlin JK, Newton L et al (2007a) Fish offal recycling by the black soldier fly produces a foodstuff high in omega-3 fatty acids. J World Aquacult Soc 38(2):309–313. https://doi.org/10.1111/J.1749-7345.2007.00101.X

St-Hilaire S, Sheppard C, Tomberlin JK, Irving S, Newton L, McGuire MA et al (2007b) Fly prepupae as a feedstuff for rainbow trout, Oncorhynchus mykiss. J World Aquacult Soc 38(1):59–67. https://doi.org/10.1111/j.1749-7345.2006.00073.x

Swinscoe I, Oliver DM, Gilburn AS, Lunestad B, Lock E-J, Ørnsrud R et al (2018) Seaweed-fed black soldier fly (*Hermetia illucens*) larvae as feed for salmon aquaculture: assessing the risks of pathogen transfer. J Insects Food Feed 5:1–14. https://doi.org/10.3920/JIFF2017.0067

Tacon AGJ, Hasan MR, Subasinghe RP (2006) Use of fishery resources as feed inputs for aquaculture development: trends and policy implications. FAO Fisheries Circular no. 1018. FAO, Rome

Tomberlin JK, Sheppard DC, Joyce JA (2002) Susceptibility of black soldier fly (Diptera : Stratiomyidae) larvae and adults to four insecticides. J Econ Entomol 95(3):598–602. https://doi.org/10.1603/0022-0493-95.3.598

Tschirner M, Simon A (2015) Influence of different growing substrates and processing on the nutrient composition of black soldier fly larvae destined for animal feed. J Insects Food Feed 1(4):249–259

Ushakova NA, Brodskii ES, Kovalenko AA, Bastrakov AI, Kozlova AA, Pavlov DS (2016) Characteristics of lipid fractions of larvae of the black soldier fly Hermetia illucens. Dokl Biochem Biophys 468(1):209–212. https://doi.org/10.1134/S1607672916030145.

Van Raamsdonk LWD, Van der Fels-Klerx HJ, De Jong J (2017) New feed ingredients: The insect opportunity. Food Addit Contamin 34(8):1384–1397

Veldkamp T, Van Duinkerken G, Van Huis A, Lakemond CMM, Ottevanger E, Bosch G et al (2012) Insects as a sustainable feed ingredient in pig and poultry diets: a feasibility study. Wageningen UR Livestock Research, Lelystad, p 48

Vos E (2000) EU food safety regulation in the aftermath of the BSE crisis. J Consum Policy 23(3):227–255

Wang Y-S, Shelomi M (2017) Review of black soldier fly (Hermetia illucens) as animal feed and human food. Foods 6(10):91

Werner FG (1956) Two cases of intestinal myiasis in man produced by Hermetia (Diptera: Stratiomyiidae). Psyche 63(3):112–112

Widjastuti T, Wiradimadja R, Rusmana D (2014) The effect of substitution of fish meal by black soldier fly (*Hermetia illucens*) maggot meal in the diet on production performance of quail (*Coturnix coturnix japonica*). Anim Sci 57:125–129

Yang P (2014) Two records of intestinal myiasis caused by Ornidia obesa and Hermetia illucens in Hawaii. Proc Hawaiian Entomol Soc 46:29

Zhang JB, Huang L, He J, Tomberlin JK, Li JH, Lei CL et al (2010) An artificial light source influences mating and oviposition of black soldier flies, Hermetia illucens. J Insect Sci 10:202

Zheng LY, Hou YF, Li W, Yang S, Li Q, Yu ZN (2012a) Biodiesel production from rice straw and restaurant waste employing black soldier fly assisted by microbes. Energy 47(1):225–229. https://doi.org/10.1016/j.energy.2012.09.006

Zheng LY, Li Q, Zhang JB, Yu ZN (2012b) Double the biodiesel yield: Rearing black soldier fly larvae, Hermetia illucens, on solid residual fraction of restaurant waste after grease extraction for biodiesel production. Renew Energy 41:75–79. https://doi.org/10.1016/j.renene.2011.10.004.

Zhou F, Tomberlin JK, Zheng L, Yu Z, Zhang J (2013) Developmental and waste reduction plasticity of three black soldier fly strains (Diptera: Stratiomyidae) raised on different livestock manures. J Med Entomol 50(6):1224–1230

Chapter 14
Production, Nutrient Composition, and Bioactive Components of Crickets (Gryllidae) for Human Nutrition

Monica A. Ayieko and Mary A. Orinda

Abstract The use of insects as alternative protein food is gaining popularity, as insects are able to convert the ingested organic matter from plants into high-quality animal protein. There are several farming activities that have almost perfected cricket production and are produced at larger scales, others are still producing small quantities in cages, basins, and buckets at the household levels, enough to feed the families and the surrounding communities. Cricket protein has not shown any sign of antinutrients commonly observed with plant products. Crickets need to be fed on balanced diet for optimum growth. A balanced cricket diet should contain protein, lipids and carbohydrates and vitamins and minerals.

Keywords Nutrient · Bioactive components · Crickets · Nutrition · Harvesting

14.1 Introduction

One of the current world major concerns of the twenty-first century is production of adequate animal protein to feed the increasing population. Production of large animals for human consumption is proving to be challenging due to the limiting production resources such as land and water. In African the level of household poverty is relative to the amount of animal protein the family consumes. Insects for food and feed are therefore taking the center stage as an alternative source of animal protein due to numerous positive advantages. Cricket production for human consumption is probably the most undertaken insects rearing initiative in the world. Different species of crickets are being reared under different types of structures (Ayieko et al. 2016). The use of insects as alternative protein food is gaining popularity in the twenty-first century because insects are able to convert the ingested organic matter from plants into high quality animal protein and other necessary bioactive nutrients more efficiently than other larger livestock.

M. A. Ayieko (✉) · M. A. Orinda
School of Agricultural and Food Sciences, Jaramogi Oginga Odinga University of Science and Technology, Bondo, Kenya
e-mail: mayieko@jooust.ac.ke

© Springer Nature Switzerland AG 2020   213
A. Adam Mariod (ed.), *African Edible Insects As Alternative Source of Food, Oil, Protein and Bioactive Components*, https://doi.org/10.1007/978-3-030-32952-5_14

Indeed, in assets have been recorded environmentally friendly (Zielińska et al. 2018). There are several farming activities that have almost perfected cricket production and are produced at larger scales, others are still producing small quantities in cages, basins and buckets at the household levels, enough to feed the families and the surrounding communities. Of the insects being reared at the household levels, crickets are probably the easiest to domesticate under a wide range of climatic conditions. The high feed conversion rate of insects is particularly advantageous for animal protein production. The insects survive well on a range of feed such as chicken mash, cereals, fruits, and vegetables ordinarily consumed by human beings (Miech 2018). This chapter therefore presents a summary on basic production of crickets for human consumption, the nutritional values of the crickets and bioactive components. The chapter discusses cricket production for human nutrition under the following topics:

1. The biology of crickets.
2. Nutrient and bioactive components of crickets.
3. Hygienic, safety and safe housing for crickets.
4. Safe feeding of crickets.
5. Harvesting and post-harvest management.
6. Processing and storage.
7. Consumption (processed vs. whole).
8. Economic sense and cents of production.
9. Challenges of cricket production.

14.2 The Biology of Crickets

Crickets are cold blooded nocturnal insects of the order Orthoptera and fall in the family of Gryllidae (true crickets). There are about 1900 species of crickets. They live in moist or damp places under logs, shallow burrows or rocks in the wild. They are omnivorous scavengers that feed on both animal and plant matter. Occasionally they exhibit or display predatory behavior upon the crippled or weak crickets or when the food source is irregular. Crickets reproduce sexually and lay eggs in a moist medium (Otte 2007).

14.3 Cricket Life Cycle

There are many different species of cricket, but each undergoes the same three key life cycle stages: egg, nymph (pinheads) and adult. The time to complete each stage varies slightly according to species. Crickets go through incomplete metamorphosis, meaning they do not enter into a pupal stage, but hatch from the egg looking like adult crickets except for the wings. There are seven to ten moults (instars) depending with the species. After each moult, the crickets almost double in size (https://www.hunker.com) (Fig. 14.1).

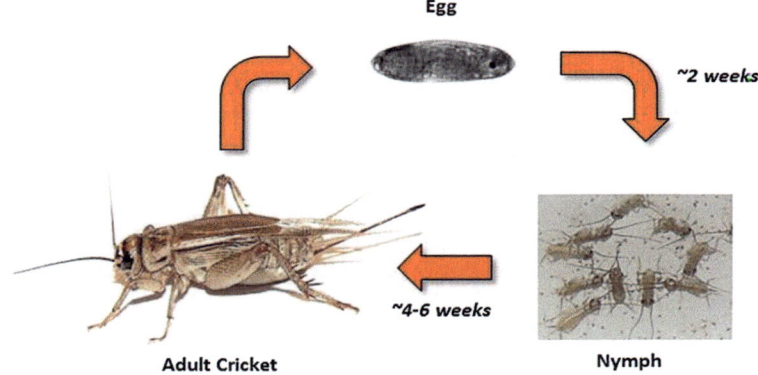

Fig. 14.1 Cricket life cycle

14.3.1 Eggs

The female cricket deposits eggs into a moist medium, usually about half an inch deep using a special organ known as an "ovipositor" on its rear. A female cricket can lay up to 200 eggs at a time. A cricket egg is pale yellow and cylindrical in shape measuring about 2–3 mm in length. The female cricket stores sperms from different males in her body and chooses the one to fertilize each egg as it passes through the seminal receptacle. The eggs hatch in about 14–17 days, but this will depend with environmental factors such as temperature and humidity. At temperatures of between 28–35 °C with relative humidity (RH) of 65%, the eggs hatch after 2 weeks. An egg that is about to hatch has a black dot on one end (Gershman 2010).

14.3.2 Nymphs

Nymph crickets are very small when they first emerge from their eggs, measuring only about 1/8 in. in length. The nymphs are also known as pinheads because they measure approximately the size of the head of the pin. Cricket nymphs commonly become pray for larger crickets. Though the cricket nymphs appear like adult crickets in many ways, they are not as developed. The female nymphs do not have an ovipositor for egg-laying, and neither the male nor the female nymphs have a developed set of wings. The nymphs (pinheads) start nibbling at 14 days (2 weeks) after hatching. Their metabolic system starts to function at this age, however, prior to this they can depend majorly on the nutrients from the egg. The nymphs go through between seven to ten instars before reaching adulthood depending on species. Crickets start to develop wings at about 1 month of age (Malinoshevskij 2018).

14.3.3   Adults

Adult crickets are identified by distinctive fully developed wings and usually measure about 23 cm in length. Their bodies have three distinct segments: the head, the thorax and the abdomen. They also have three pairs of legs and two antennae. In most species of cricket, the adults have wings. Adult crickets can jump up to 60 cm above the ground. Male crickets tend to emerge beforehand in order to facilitate mating (protandry). Female crickets are usually bigger, approximately 15–25% bigger than males because they accumulate more food for egg production in addition to having a longer development time and, in many cases, extra instar (Sakaluk 1980).

14.4   Digestion and Feeding Behavior

The digestive system of a cricket is divided into foregut, midgut, and hindgut. Crickets respond to food through biting using mandibles. Their mouths are made of the labrum, labium, maxillae, and mandibles. The labrum, or upper lip, covers and protects the subsequent mouth parts. The rest of the mouth parts support the mandibles in their purpose of chewing. Chemical sensors on their mouth parts help with tasting before biting. When food is swallowed, it enters into the cricket's monogastric digestive system. Food will go through the pharynx, down the oesophagus, into the crop where the food is stored and hydrolysis occurs. Digestion in crickets occurs in the crop but most enzymes, with the exception of salivary enzyme are secreted from the midgut which are present in fluid that is regurgitated from the midgut into the crop. As the food moves, the crop becomes the proventricular whose interior walls contain tiny cuticular teeth which, function as a grinding organ and in combination with the secreted enzymes breaks down food into smaller particles (Woodring et al. 2006). The food moves into the intestines, colon, and finally the rectum. Most of the absorption of the nutrients occurs at the anterior end of the midgut as the digested food enters from the crop. Once food reaches the rectum, it is dehydrated and fecal pellets are formed. The rectum can retain 90% of the water in the fecal matter, an adaptation that allows the cricket to live even in dry areas (Teo and Woodring 1994).

14.5   Reproduction of Crickets

Crickets are polyandrous in nature and reproduces sexually. Polyandry is important because of the material and genetic benefits. Material benefits take the form of sperm replenishment and resource acquisition in the form of nuptial gifts while genetic benefits deal with control of inbreeding.

Mating usually starts within 24–72 h after maturation which is then followed by laying. Peak laying usually occurs 4–8 days after copulation. Mate selection depends on the male size and chirping type. It is of importance to note that female crickets prefer bigger males because they produce larger spermatophore which increases fecundity. The process follows a courtship process which starts with the male calling out for females (stridulation) into their territories. The courtship involves a series of chirping, body vibrations and production of pheromones which ends into copulation. Copulation occurs when the male reaches back with his abdomen and the female bends her abdomen downward to link the genitalia. Copulation lasts only a few seconds, during which the male threads the tube of a small spermatophore into the female's genitalia (Worthington et al. 2015).

14.6 Nutrient and Bioactive Components of Crickets

The practice of eating insects as well as their incorporation in other foods is increasing worldwide. From different parts of Africa, insects are highly seasonal and are eaten at different morphological stages. While edible insects were formerly consumed as a cultural delicacy mostly in developing countries, they are gaining recognition as important sources of nutrients, edible insects are highly nutritious (Kelemu et al. 2015). Other than their high seasonality due to the challenges of domestication and inadequate technology to support that, there are several hindrances in the use of insects as food. The key to this is the attitudinal barrier caused by cultural factors brought about by fear and curiosity. Consumers' acceptance of edible insects can be enhanced by processing and blending with familiar food products for instance pastry products such as biscuits, this is however expected to result in changes perceptions and sensory attributes. However, many factors influence the nutritive value of processed insects. Processing edible insects changes their digestibility, bioavailability and total nutritional value due to formation of disulfide linkages within the protein matrix during processing. The introduction of new or modified products in the market may also pose several challenges, especially if some of the ingredients used are new to the consumer. Cricket protein has not shown any sign of antinutrients commonly observed with plant products. However, the substrates used for rearing the insects is critical. The most important activity to minimize this in crickets is to starve the insects for 24 h before harvesting. An alternative method is to feed the crickets on fresh clean vegetables or fruits such as melons and pumpkins, cabbages, kales or spinach 24 h to clean out the guts of unwanted microbiological. Use of melons, celery leaves, or dhania (coriander leaves) enhances the flavor of crickets. Careful selection of desired plants products to mix with cricket powder is paramount. Whether the characteristics of the new product will be judged positively by the potential consumer will depend on the skills and ingenuity of preparation. Consumer preference studies are therefore important to determine how the product is perceived and the actual nutritional composition after processing.

As alluded, processing insects introduces several changes. However, different processing methods have varying effects on decontamination as well. Microbiological quality of processed insects using different cooking methods including degutting tend to considerably reduce the levels of *coliforms*, *Staphylococcus*, and *Escherichia coli*. This is a, particularly, pointer to the danger of traditional way of slightly roasting insects such as locust and crickets which may contain some levels of Enterobacteriaceae. Simply boiling the crickets for about 5 min is adequate to eliminate microbes (Vandeweyer et al. 2018).

Laboratory experiments indicate that it is important to guard against extreme processing of edible insects to minimize losing the functional characteristics of insects that are highly coveted. Other effects of functional properties of crickets include water absorption, water holding and oil absorption capacity which is critical in product development and storage quality. These are critical in improving the palatability and increased acceptance by consumers. During product formation with the insect powder, dispensability and solubility of the protein is desirable for enhances emulsification. At the lab levels, it was observed that processed products made from coarsely ground cricket flour were less acceptable than those made from fine cricket powdered. Therefore, the functional size of the insect powder is also key to product acceptance (https://www.cricketflours.com).

An increasing number of consumers are becoming aware of the value of eating edible insects and insect-based foods, hence the demand to know the shelf stability and their nutritional value. Shelf life is an important property of any food and is of interest to everyone in the food chain from the producer to the consumer, biscuits are highly hygroscopic, fragile and characterized by low moisture content, acceptability tests in the form of appropriate sensory analysis, are an important part of the shelf life evaluation as well as moisture analysis. Moisture acts as a key influencing factor for shelf life. Increased moisture of food due to water sorption from the atmosphere or by mass transport from neighboring components results in soggy, soft textured products. Physicochemical changes during storage can also cause loss of shelf-life resulting deterioration of its quality (Tan 2017). Water is a constituent of food which affects food safety, stability, quality and physical properties. Off-flavors, off-odors, and loss of crispiness in packaged food are the major causes of consumer rejection. These might be caused by oxidative rancidity of packaged food, permeation, migration and reactions between packaging components with the food components. Most food products rely on their particular packaging to achieve their expected shelf life. The shelf-life of biscuits depends upon inherent characteristics of the product, barrier and other functional properties of the packaging material, packaging operations adopted, distribution and storage patterns followed, and lastly economic considerations.

The safety of food with regard to edible insects is also a major concern for consumers as well as the producers; the fact that they are not novel food and that they are reared on substrates that have been used as animal feed over time makes them generally safe to consume. However, since processing involves a lot of handling, it is important to ascertain safety. To ensure safety of food products, microbial tests

Table 14.1 Nutritional analysis of crickets fed on kales and chicken mash

S/no	Parameters	Dry weight	Wet weight
1	Retinol (Vitamin A) (μg/g)	0.35 ± 0.05	0.22 ± 0.01
2	Vitamin B2 (Riboflavin) (μg/g)	6.3 ± 0.9	5.9 ± 0.3
3	Vitamin B1 (Thiamine) (μg/g)	15.2 ± 1.4	11.6 ± 0.5
4	Vitamin E (μg/g)	331.3 ± 7.6	129 ± 0.2 2.3
5	Protein content (%)	47.1 ± 0.6	14.9 ± 0.7
6	Ash content (%)	3.8 ± 0.0	1.3 ± 0.1
7	Fat content (%)	25.8 ± 0.4	4.1 ± 0.1
8	Fiber content (%)	5.5 ± 0.1	2.5 ± 0.4
9	Moisture content (%)	8.1 ± 0.2	67.7 ± 2.2
10	Carbohydrates (%)	9.7	9.5
11	Manganese (μg/g)	58.7 ± 1.8	25.5 ± 1.7
12	Sodium (μg/g)	8502.3 ± 131.1	3745.4 ± 29.7
13	Copper (μg/g)	29.4 ± 0.6	7.9 ± 0.1
14	Calcium (μg/g)	3147.7 ± 2.7	1345.9 ± 56 6
15	Potassium (μg/g)	9797.5 ± 82.0	2473.8 ± 65.1
16	Iron (μg/g)	51.8 ± 2.1	40.1 ± 0.7
17	Phosphorus (μg/g)	331.3 ± 7.6	129.2 ± 2.3
18	Zinc (μg/g)	21.79[a]	–

[a]Please note that wet weight was not analyzed at the same time due to some technological challenges that could not be verified. As observed, nutritional outcome of the crickets tends to vary with the substrates. The table shows crickets were fed on green vegetables supplemented with chicken feed

are necessary; this way the risk of contamination during processing can be prevented (Belluco et al. 2015).

As mentioned earlier, crickets are very rich in several food nutrients that have attracted human interest. Several studies exist with slightly different nutritional values reported but the fact remain that the insects are richer nutritionally than most protein foods available (Table 14.1).

14.7 Hygienic, Safety and Safe Housing for Crickets

Housing is one of the most important components of sustainable cricket production. A suitable house should address critical abiotic issues like favorable temperature, humidity and light regime to ensure reliability and cost-effectiveness of production in terms of low mortality, optimal development and production. Optimum housing is also important to ensure that the insects do not come into contact with external microbial contamination or allow predators' interference with the colony. The development of ideal housing system for crickets is still under trial but the following factors should guide on the choice of housing system for a healthy production of crickets for human consumption (Orinda et al. 2017).

- Affordability: Affordability of housing is key to successful rearing of crickets for human food. Several materials ranging from wooden housing to plastic cages have been tried. The cost of materials will depend mainly on the availability of the choice material in the area. Several farmers have successfully reared crickets in concrete pens or earthen pens made of twigs and mud.

  Space availability which influences the choice of the caging system or reactor to use. A good house should have enough space to enable the farmer move freely while working and also house the rearing containers.

- Maintenance of favorable conditions in terms of temperature & relative humidity (RH%). A good house should be able to maintain temperature range of between 28 and 35 °C and relative humidity of 60–65%.
- Safety and security: The house should be away from insecticide sprayed fields, free from predators and away from flood prone areas. Security concern should also deal with environmental impacts in case of escape.
- Convenience in terms of accessibility by the farmer and proximity to source of feed such as vegetable fields.

14.8 Safe Feeding for Crickets

Crickets need to be fed on balanced diet for optimum growth. A balanced cricket diet should contain both macro nutrients (protein, lipids and carbohydrates) and micronutrients (vitamins and minerals). Physical qualities of feed such as hardness, shape, smell, taste, and anti-nutritional components influence the crickets' capacity to consume and digest the feed. Crickets are omnivorous meaning they eat both animal and plant matter indicating a wider nutritional spectrum. This makes them easy to rear in large scale since they can be fed on variety of feed types. They respond to feed through biting and as such texture of the feed is important because these are biting insects. Pinheads can be given softer feed but the texture can be made coarser as they approach adults otherwise the adult crickets will start nibbling the hideouts. The feed can be made into smaller pellets. Balanced diet is paramount because dietary composition has effects on the nutrient composition of the crickets.

A suitable feed substrate should ensure:

- High survivability and longer lifespan, that is, less mortality due to quality of the diet. Low quality diet increases crickets' vulnerability to diseases and parasites.
- High fecundity (fertility) to cater for the subsequent generations for continual production.
- Shorter development period, that is, the time it takes for the crickets to reach maturity. Crickets fed on low-quality feed take longer to reach maturity.
- Crickets of high nutritional value and larger biomass.
- Less or reduced cannibalism. Crickets fed on low quality diet will cannibalize on others to compensate for the missing nutrients.

14.9   Harvesting and Post-harvest Management

This should be done at the right time before natural attrition sets in. Harvest to dispose old stock after laying for 2 weeks. The harvested crickets can be consumed or sold off to processors. A farmer can harvest some crickets when they are about to emerge into adults when the chitin is still soft to cut down on cost of labor and feed (efficiency of feed conversion starts to decline after this stage), however, he must maintain a small population called mating or breeding colony to sustain the colony. In this case, the harvesting should be done when approximately 85% of the whole colony is winged. The mating stock should be maintained at approximately 30–35% of the total population because of the possibility of mortality and natural variances due to environmental factors and fecundity (Hanboonsong et al. 2013).

14.9.1   Harvesting Process

1. Harvesting should be done in the morning when the crickets are less agile. Crickets are cold blooded which means their body temperatures depends on the temperature of the surrounding. In the morning when it is still chilly, they would prefer hiding in the hideouts. This makes collection easier.
2. To easily collect the crickets, gently lift the hideouts from the rearing containers and then shake them gently in another clean bucket with a lid.
3. After collecting all the crickets, transfer them into a clean plastic or zip-lock paper.
4. Put the plastic paper containing crickets in a deep freezer for 30 min or until the crickets become immobile or frozen. This is a humane and food-safe way of immobilizing the crickets and does not interfere with their physical and nutritional composition. Since the house crickets are ectotherm, sub-zero temperature leads to torpor, a naturally occurring state of low-activity until the temperature rises. Blanch the crickets. Blanching is the process whereby the crickets are put in a boiled water for 3 min and then removed and immediately plunged into cold water for 1 min to halt the cooking process.
5. The blanched crickets can further be frozen, dried or roasted.

14.10   Processing and Storage

Processing is the act of obtaining raw materials from the whole insect that can now be used in food preparation. A lot of issues should be considered such as processing whole or parts of the insect, for example, removal of the legs or wings before processing and usability of these removed parts. The other concern should be the method and duration preservation to prevent the potential growth of microbial fauna. Fortunately, right methods such as boiling and refrigeration can help prevent this.

14.11   Consumption (Processed vs. Whole)

Despite the fact that there has been general consensus on the benefits of consuming crickets as source of protein, majority of the world population still view it with a lot of skepticism. The irk and yuck factor is still a hindrance to the appreciation of the noble idea mainly because of general perceptions or concerns of toxicity, allergens, anti-nutrients, accumulation of traces of pesticides, and zoonotic diseases.

14.12   Challenges of Cricket Production

The most challenging issues about cricket production is feed. Most farmers in Kenya are using chicken feed and several people just add extra ingredients to enrich. Many farmers add fish meal or soya meal to improve the protein percentage. Some farmers are successful with this method while others are not reaping the benefits of enriching the chicken meal comparable to what they add. Several other methods including adding dried Moringa leaves ground into flour and added to the chicken meal has worked well. This also adds medicinal value to maintain good health of the colony and controlling to some extent, fungal growth in the colony.

1. Behavioral and attitude change and mass production:
   As much as cricket rearing is picking up well with several households, the larger market has not fully accepted cricket production as a viable enterprise. The good thing is that many households have embraced cricket production in Kenya and are trying out in small scales. Currently cricket production is the one-most undertaken because it is simple and easy to get starting flock around the firms. One does not need special seeds (eggs) imported from somewhere else to start cricket production. It is better to pick the local crickets in the area and develop your own brooding stock than picking from some other places. However, farmers from the same are (geographical and environment area) can exchange eggs, a management strategy that often used to minimize inbreeding after years of production from the same place (Rumpold and Schlüter 2013).

2. Strain or species to be used for large scale production: There are several species of crickets with different strains or genetic variations. They also have different adaptation potentials to domestication of captive rearing. A prospective farmer must first make sure that the chosen strain is readily available within the locality for ease of acclimatization. If a foreign species is used or introduced, then exhaustive safety assessment should be done to assess the likely environmental impact that might arise even if acclimatization may not be a problem.

   To ensure optimum and efficient mass production, a farmer must take into account the following characteristics of the cricket species.

   (a) Ease of mass rearing in terms of adaptability to the local climatic conditions.
   (b) Availability and ease of colony formation. This is very important especially in a case where a farmer wants to neutralize inbreeding.

(c) Produces faster, that is, shorter life cycles ensuring high production turn over/cycles. This also ensures quick or faster return on cricket investment.
(d) High fecundity, that is, the chosen species should be able to lay many eggs, thus forming big colonies within a short duration.
(e) Gains larger biomass faster with high nutritional value.
(f) Marketability, that is, readily acceptable to consumers or palatable.
(g) Resilience: Less susceptible to pests and diseases.

3. Welfare of the crickets: House crickets are animals and therefore welfare is required. They should be free from hunger, thirst, discomfort, pain, injury, diseases, fear, distress and should have the possibility to express normal behavior.
4. Food safety, quality, and legislation (policy):

(a) What are the food safety management system?—design and layout of the housing that ensures cleanliness, cleaning procedures and pest control in the rearing facility.
(b) HACCP. What are the control measures that eliminate or reduce food safety hazard? This should specify specific temperature and time for processing to reduce bacterial contamination level.
(c) Lack of certification processes for insect-based food or products.

References

Ayieko MA, Ogola HJ, Ayieko IA (2016) Introducing rearing crickets (gryllids) at household levels: adoption, processing and nutritional values. J Insects Food Feed 2(3):203–211
Belluco S, Losasso C, Maggioletti M, Alonzi C, Ricci A, Paoletti MG (2015) Edible insects: a food security solution or a food safety concern? Animal Front 5(2):25–30
Gershman SN (2010) Numbers of matings give female field crickets a direct benefit but not a genetic benefit. J Insect Behav 23(1):59–68. https://doi.org/10.1007/s10905-009-9195-y
Hanboonsong Y, Jamjanya T, Durst PB (2013) Six-legged livestock: edible insect farming, collection and marketing in Thailand. Food and Agriculture Organization of the United Nations, Regional Office for Asia and the Pacific, Bangkok
https://www.cricketflours.com/how-to-make-cricket-flour/
https://www.hunker.com/12334517/life-cycle-of-a-cricket
Kelemu S, Niassy S, Torto B, Ekesi S (2015) African edible insects for food and feed: inventory, diversity, commonalities and contribution to food security. J Insects Food Feed 1(2):103–119
Malinoshevskij V (2018) How long do crickets live? The life cycle of a cricket. IN. https://pests-guide.com/crickets/how-long-do-crickets-live/. Accessed 19 Aug 2019
Miech P (2018) Cricket farming. An alternative for producing food and feed in Cambodia. PhD thesis, Swedish University of Agricultural Sciences, Uppsala
Orinda M, Mosi RO, Ayieko M, Amimo FA (2017) Effects of housing on growth performance of common house cricket and field cricket. J Entom Zool Stud 5(5):1138–1142
Otte D (2007) Australian crickets (Orthoptera: Gryllidae). Acad Nat Sci 17–24. ISBN 978-1-4223-1928-4
Rumpold BA, Schlüter OK (2013) Potential and challenges of insects as an innovative source for food and feed production. Innov Food Sci Emerg Technol 17:1–11
Sakaluk SK (1980) Sexual behavior and factors affecting female reproduction in house and field crickets. MSc thesis, Brock University St. Catharines, ON, Canada

Tan HSG (2017) Eating insects: consumer acceptance of a culturally inappropriate food. PhD thesis, Wageningen University, Wageningen, NL

Teo H, Woodring JP (1994) Comparative total activities of digestive enzymes in different gut regions of the house cricket, *Acheta domesticus* L. (Orthoptera: Gryllidae). Ann Entomol Soc Am 87(6):886–890

Vandeweyer D, Wynants E, Crauwels S, Verreth C, Viaene N, Claes J, Lievens B, Campenhout LV (2018) Microbial dynamics during industrial rearing, processing, and storage of tropical house crickets (*Gryllodes sigillatus*) for human consumption. Appl Enviro Microb 84(12):1–13

Woodring J, Hoffmann KH, Lorenz MW (2006) Activity, release and flow of digestive enzymes in the cricket, *Gryllus bimaculatus*. Physiol Entomol 32(1):56–63. https://doi.org/10.1111/j.1365-3032.2006.00541.x

Worthington AM, Jurenka RA, Kelly CD (2015) Mating for male-derived prostaglandin: a functional explanation for the increased fecundity of mated female crickets? J Exp Bio 218:2720–2727. https://doi.org/10.1242/jeb.121327

Zielińska E, Karaś M, Jakubczyk A, Zieliński D, Baraniak B (2018) Edible insects as source of proteins. In: Mérillon JM, Ramawat K (eds) Bioactive molecules in food. Reference series in phytochemistry. Springer, Cham

Chapter 15
Nutrient Composition and Bioactive Components of Ants (*Oecophylla smaragdina* Fabricius)

Abdalbasit Adam Mariod

Abstract Weaver ants are fit to be eaten by people and are rich in protein and fatty acids. The whole lipid content varies from 6.3% to 15.2%, and the predominant lipid components were triacylglycerol 37.4–79.4%, followed by phospholipids 6.1–21.5%. Sterols constitute 0.5–0.8%, and oleic acid is the primary fatty acid. The ants contain high level of acidic amino acids and few glycine residues. Ant eggs include three lectins. All the lectins have been observed to be composed of four sub-devices of unequal sizes, and are glycoprotein.

Keywords *Oecophylla smaragdina* · Nutritional value · Uses

15.1 Introduction

The weaver ants belonging to the genus Oecophylla consist of two extant species: *O. smaragdina* which is distributed at some stage in tropical Asia, Australasia, and some Pacific islands and O. *longinoda* distributed at some stage in tropical Africa (Sribandit et al. 2008). *O. longinoda* is sent in the Afro-tropics and *O. smaragdina* from India and Sri Lanka in South Asia, via southeastern Asia to northern Australia and Melanesia (Crozier et al. 2010). In Australia, *Oecophylla smaragdina* is located inside the tropical coastal regions up to some distance south as Broome in Western Australia and throughout the coastal tropics of the Northern Territory down to Yeppoon in Queensland. The species share similar organic and ecological traits. They may be both polydomous cover ants that build leaf nests on their host bushes. Nests are built via drawing collectively leaves and fixing them with silk constructed from their larvae (Offenberg et al. 2006).

A. A. Mariod (✉)
Indigenous Knowledge and Heritage Centre, Ghibaish College of Science and Technology, Ghibaish, Sudan

© Springer Nature Switzerland AG 2020

A. Adam Mariod (ed.), *African Edible Insects As Alternative Source of Food, Oil, Protein and Bioactive Components*, https://doi.org/10.1007/978-3-030-32952-5_15

15.2 Nutrient Composition and Bioactive Components of Ants

Sihamala et al. (2010) reported total lipid contents, lipid classes, and fatty acid compositions of hot-air-dried fit-to-be-eaten black ants (*Polyrhachis vicina*). These authors noticed that the full lipid content range from 6.3% to 15.2%, and the foremost lipid components were triacylglycerol ranging from 37.4% to 79.4%, followed by phospholipids ranging from 6.1% to 21.5%; diacylglycerol was in the range of 6.1–18.1%, and cholesterol ester was in the range of 4.9–13.5%. They reported that oleic acid as the most predominant fatty acid, ranging from 52.1% to 63.0%, followed by palmitic acid ranging from 16.5% to 20.8%; linoleic acid ranged from 2.1% to 7.0% (Sihamala et al. 2010). Four fibroin proteins of weaver ants (WAF1–four) were identified, which incorporate 391, 400, 395, and 443 amino acid residues, respectively (Sutherland et al. 2007). The amino acid compositions of ants are pretty unique from silkworm fibroin, wherein it consists of high level of acidic amino acids and few glycine residues, leading to a one-of-a-kind conformation and properties as compared to silkworm fibroin. Variations of weaver ant fibroins in relation to different fibroins, in particular to silkworm fibroins, brought about the question of their viable packages, particularly for biomedical packages just like those of silkworm fibroins (Siri and Maensiri 2010). Ant eggs contain a glycoprotein which changed into extracted with phenol-saline and purified on jacalin-Sepharose 4B. The glycoprotein contained D-mannose, *o*-galactose, D-glucose, and 2-acetamido-2-deoxy-D-glucose, and gave precipitin bands with the cY-*o*-galactosyl-particular lectias. The presence of non-reducing α-D-galactosyl cease groups become corroborated by methylation evaluation and enzymic degradation. The formation of the jacalin-glycoprotein complex was dependent on time, pH, and the ionic strength of the medium (Ray and Chati-Erjee 1989).

Three lectins were extracted and isolated from ant egg and purified by gel filtration on Sephadex G-75 accompanied by way of ion-exchange chromatography on DEAE-cellulose to obtain the 100% ammonium sulfate fraction from the crude extract, DEAE-cellulose (Hassan et al. 1995). Fatty acid analysis showed that white ant fat was predominantly unsaturated with 60% and 57% unsaturated fats respectively. White ants had 13.4% and 6.7% linoleic acid respectively, and 44% and 48% of oleic acid respectively (*http://www.fao.org*).

15.3 Ants as Food and Medicine

Weaver ants are one of the most valued insects eaten by human beings (entomophagy). Weaver ants can be applied without delay as a protein and a meals supply since the ants (in particular the ant larvae) are fit for human consumption for humans and high in protein and fatty acids (Raksakantong et al. 2010). It has moreover been proven that the harvest of weaver ants can be maintained whilst on the same time using the ants for biocontrol of pest insects in tropical plantations, for the reason that queen larvae and pupae which might be the primary goal of harvest, are not critical for colony survival (Offenberg and Wiwatwitaya 2010). Van Mele, and Cuc (2000) use *O. smaragdina* in citrus, as a fruit high-quality improver and a biological control agent, they said that the use of ants gives advantages in terms of a better environment, fewer health risks for farmers and customers, without affecting farmers' income.

A growing interest in the eating of ants has brought about higher call in Thailand with increasing expenses as an end result. For that reason, the gathering of ants is turning into extra profitable and the harvest strain on local *O. smaragdina* populations may increase, potentially leading to an unsustainable over exploitation of these ants in herbal habitats (Sribandit et al. 2008). Because of their predatory nature Oecophylla ants are identified as biological control agents in tropical tree vegetation as they may be able to promote spread of plants against many unique insect pests (Van Mele 2008). In this manner they are used in a roundabout way as an opportunity to chemical insecticides. It is less well known that the ants can be utilized, as a commercial product. There exist a minimum of three unique markets for using these units in Southeast Asia: (1) in Chinese and Indian conventional drugs, (2) as a valued feed for songbirds in Indonesia (Césard 2004), and (3) as a prized human delicacy in Thailand and other Asian international locations. In Chhattisgarh, India, conventional healers believe that regular consumption of *O. smaragdina* will prevent rheumatism—a view shared by practitioners of conventional Chinese medicine (Oudhia 2002). The Indian healers also put together oils in which they dip accumulated ants. After 40 days oils are used externally to remedy rheumatism, gout, ringworm or other skin sicknesses, or else as an aphrodisiac (Oudhia 2002).

The lifestyle of consumers together with Oecophylla ants in meals and/ or conventional remedy has been pronounced from diverse cultures in Thailand, India, Myanmar, Borneo, Philippines, Papua New Guinea, Australia, and Congo (De Foliart 2008). In Thailand *O. smaragdina* is considered a delicacy and has been eaten by human beings for hundreds of years. Imagos in addition to brood is used in a spread of Thai dishes and are without difficulty acquired from many neighborhood markets throughout the country during the ant harvest season. Larvae and pupae are desired over imagos and the queen caste preferred over the worker castes and adult males. The season in which *O. smaragdina* produce new queens therefore defines the ant harvest season (Sribandit et al. 2008). The ants are used as ingredients in soups, salads, and fried dishes and on occasion eaten uncooked together with spices as a snack. The subculture of eating ants is well known among the Isaan human beings of Northeast Thailand and the people in Northern Thailand, however, has unfold to other components of the country with the migration of human beings from those cultures.

The harvest of Oecophylla ants in Northeast Thailand is significant and not most effective for local subsistence; however, it is a good source of income. There is an economically urged growing interest in harvesting ants and consequently an increasing pressure on natural ant populations. Ant farming can be a method to preserve sustainability and on the identical time decorate profitability (Sribandit et al. 2008). Siri and Maensiri (2010) produced herbal fibers of weaver ants and they tested its utility as a natural matrix to assist in vitro cell adhesion. They determined that the morphological shape of natural ant fibrous mat made it served as a great matrix to assist cell adhesion and proliferation. Apparently, old and freshly made fibers differed in their water wettability, which is critical for its applications.

15.4 Ants Other Uses

Weaver ant silk is produced by final-instar larva, not adult, for nest production rather than for cocoon spinning. Silk is produced within the labial or larval salivary glands of the instar larva. The formation of silk in salivary glands begins and finishes within the center and the end of the fifth instar. The secretory cells in salivary glands launch a homogeneous substance, that is polymerized in the lumen of to shape compact birefringent tactoids. After water is absorbed, tactoids are aggregated to shape spiral-form filament with zigzag sample. The common structure of fibroins incorporates a sign peptide and a coiled coil area containing about 30 heptad repeats flanked by variable areas. The heptad repeat generally consists of hydrophobic residues inside the α and d positions and hydrophilic residues within the remaining positions (Woolfson et al. 2005). Weaver ant fibroins are small proteins (about 400 amino acid residues), do not include primary series repeats and occasional protein identification to different fibroins. Four fibroin proteins of weaver ants (WAF1–4) were identified, which contain 391, 400, 395, and 443 amino acid residues, respectively (Sutherland et al. 2007).

References

Césard N (2004) Harvesting and commercialization of Kroto (*Oecophylla smaragdina*) in the Malingping area, West Java, Indonesia. In: Kusters K, Belcher B (eds) Forest products, livelihoods and conservation. Case studies of nontimber product systems. Center for International Forestry Research, Bogor, pp 61–77

Crozier RH, Newey PS, Schlüns EA, Robson SKA (2010) A masterpiece of evolution—*Oecophylla* weaver ants (Hymenoptera: Formicidae). Myrmecological News 13:57–71

De Foliart GR (2008) The human use of insects as a food resource: a bibliographic account in progress. Downloaded from http://www.foodinsects.com. Accessed 22 April 2016

Hassan et al. (1995). http://www.fao.org/ag/humannutrition/abd150869408f6a5.pdf

Offenberg J, Wiwatwitaya D (2010) Sustainable weaver ant (*Oecophylla smaragdina*) farming: harvest yields and its effects on worker ant densities. Asian Myrmecol 3:55–62

Offenberg J, Nielsen MG, Macintosh DJ, Aksornkoae S, Havanon S (2006) Weaver ants increase premature loss of leaves used for nest construction in rhizophora trees. Biotropica 38:782–785

Oudhia P (2002) Traditional medicinal knowledge about red ant Oecophylla smaragdina (Fab.) [Hymenoptera; Formicidae] in Chhattisgarh, India. Insect Environment 8(3):114–115

Raksakantong P, Meeso N, Kubola J, Siriamornpun S (2010) Fatty acids and proximate composition of eight Thai edible terricolous insects. Food Res Int 43(1):350–355

Ray and Chati-Erjee (1989) Purification of Ant-Egg Glycoprotein and Its Interaction with Jacalin. Carbohydr Res 191:305–314

Sihamala O, Bhulaidok S, Li-rong S, Duo L (2010) Lipids and Fatty Acid Composition of Dried Edible Red and Black Ants. Agri Sci China 9(7):1072–1077

Siri S, Maensiri S (2010) Alternative biomaterials: Natural, non-woven, fibroin-based silk nanofibers of weaver ants (Oecophylla smaragdina). Int J Biol Macromol 46:529–534

Sribandit W, Wiwatwitaya D, Suksard S, Offenberg J (2008) The importance of weaver ant (Oecophylla smaragdina Fabricius) harvest to a local community in Northeastern Thailand. Asian Myrmecology 2:129–138

Sutherland TD, Weisman S, Trueman HE, Sriskantha A, Trueman JWH, Haritos VS (2007) Conservation of essential design features in coiled coil silks. Mol Biol Evol 24:2424–2432

Van Mele, Cuc (2000) Evolution and status of Oecophylla smaragdina (Fabricius) as a pest control agent in citrus in the Mekong Delta, Vietnam. International Journal of Pest Management 46(4):295–301

Van Mele, P (2008) A historical review of research on the weaver ant Oecophylla in biological control. Agric For Entomol 10:13–22

Woolfson DN (2005) The design of coiled coil structures and assemblies. In: Parry DAD, Squire JM, editors. Fibrous proteins: coiled-coils, collagen and elastomers. San Diego, CA: Elsevier. p. 79–112

Chapter 16
Nutrient Composition and Bioactive Components of the Migratory Locust (*Locusta migratoria*)

Suzy Munir Salama

Abstract The present chapter concerns the nutrient content and the biologically active compounds of the migratory locust (*Locusta migratoria*) as one of the mostly consumable edible insect in many countries. Studies showed that the body of *Locusta migratoria* contains appreciable concentration of proteins, monosaturated and polysaturated fatty acids, fibre, vitamins and minerals, and can provide humans with the calories required in comparison with the traditional foodstuff as beef, chicken and pork. Additionally, the body of the migratory locust contains biologically active compounds such as chitin, retinol, vitamin D, vitamin B_{12}, variety of carotenoids and antioxidant peptides that can protect against many ailments such as chronic kidney and neurodegenerative disorders, cardiovascular diseases, diabetes and skin problems. Moreover, it was found that the dry matter of *Locusta migratoria* has antioxidant activity against oxidative stress via the ability to chelate metal ions recording the highest chelating capacity to copper ions. Researchers proved that the nutrient composition and the bioactive compound constituents of the migratory insect's body vary according to the diet as well as the developmental stages of the insect.

Keywords *Locusta migratoria* · Nutrients · Bioactive ingredients

16.1 Introduction

As evidenced by fossil analysis, humans used edible insects as a source of food before farming and hunting tools had been invented (Mitsuhashi 2008). More than 2000 edible insect species are traditionally used worldwide in more than 110

S. M. Salama (✉)
Department of Biomedical Science, Faculty of Medicine, University of Malaya, Kuala Lumpur, Malaysia
e-mail: suzymunir@um.edu.my

© Springer Nature Switzerland AG 2020

A. Adam Mariod (ed.), *African Edible Insects As Alternative Source of Food, Oil, Protein and Bioactive Components*, https://doi.org/10.1007/978-3-030-32952-5_16

countries in the present years (Kouřimská and Adámková 2016). In comparison with the traditional sources of food used globally such as poultry, pork and beef, insects are considered as secondary source of food. Currently edible insects are promoted to be included among other food sources for humans in many countries (Feng et al. 2018).

Edible insects contain all the nutrients required for the human as well as the pet's body. Therefore, practising entomophagy (eating insects) became traditional compared to meat eating as source of proteins, lipids and energy (Dreon and Paoletti 2009). In addition, studies on edible insects exposed that they contain a variety of biologically active compounds that possess pharmacological importance in preventing many diseases as cancers and autoimmune diseases (Holick 2004). Many edible insects have been farmed to be used as source of food in Europe such as crickets, locusts and beetles (Kouřimská and Adámková 2016).

The migratory locust (*Locusta migratoria*) belongs to order Orthoptera, that is mostly available on trees, bushes and shrubs. The percentage of human consumption to the edible insects of Orthoptera represents 13% compared to other consumable insects from other orders worldwide (Av et al. 2013). Basically locusts are one of the most consumable edible insects due to their high content of proteins, fats, minerals, vitamins and fibre (Finke 2002). In this review, the nutrient composition and the bioactive compounds of *Locusta migratoria* will be shown based on the reported studies of nutritionists and pharmacologists.

16.2 Nutrient Composition of *Locusta migratoria*

16.2.1 Protein Content

Studies showed that the protein content of the insect's body differs with its developmental stage where the adult contains the highest protein content followed by the pupa down to the larva (Yin et al. 2017). Based on the analysis of researchers to the body of the African migratory insect *Locusta migratoria*, 100 g of the dry matter contains 50–62% protein (Kouřimská and Adámková 2016; Mohamed 2015a). In comparison with the protein content of egg and beef (95% and 98% respectively), the protein content of *Locusta migratoria* represents more than half. In addition, the protein content of *Locusta migratoria* is higher compared to that of many plant species (Kouřimská and Adámková 2016).

Proteins as biological macromolecules are composed mainly of amino acids. Edible insects' bodies contain amino acids up to 70%, while their content of essential amino acids forms up to 30% approaching the proper ratio reported by the World Health Organization (WHO) and Food and Agriculture Organization (FAO) (Yin et al. 2017). Research analysis from about 100 edible insect species reported a range of essential amino acids 50–95% of the total number of amino acids (Xiaoming et al. 2010). *Locusta migratoria* was found to contain more than 15% of phenylalanine and tyrosine amino acids (Ancsin and Wyatt 1996).

16.2.2   Fat Content

The fat content of the insect's body differs from other animals and varies from one insect species to another. Additionally, the season and the developmental stage of the insect affect their fat constituent (Yin et al. 2017). Mohamed (2015a) reported that the migratory insect *Locusta migratoria* represents a good source of fat constituting the second main component of the insect's body. Basically, triglycerides compose 80% of the total fat present in the body of the insect, while phospholipids form less than 20% (Kouřimská and Adámková 2016; Ekpo et al. 2009). The fat content of the dry matter of *Locusta migratoria* constitutes 13–20% (Kouřimská and Adámková 2016; Mohamed 2015a). Parallel study revealed that *Locusta migratoria* contains 25 identified fatty acids. Saturated fatty acids constitute approximately 84 mg/g and unsaturated fatty acids 121 mg/g, 12 mg/g for omega-6 and 25 mg/g for omega-3 (Mohamed 2015b).

16.2.3   Carbohydrate Content

Carbohydrates constitute a low percentage in edible insects (Yin et al. 2017) in comparison with other nutrients. Studies proved that the percentage of carbohydrates in *Locusta migratoria* ranges between 4% and 6%. Chitin as the main carbohydrate component of insect's body constitutes an average of 17% in the adult insect and this ratio decreases in the pupa and larva (Yin et al. 2017).

16.2.4   Fibre Content

The exoskeleton of insects contains chitin which is considered as the main fibre content of the insect's body (Av et al. 2013). Chitin is an insoluble fibre that when taken in food can be digested through chitinase enzyme which is actively functioning in the gastric juice of native tropical people who traditionally use insects as their main source of food (Muzzarelli et al. 2001). Additionally, chitin is called animal fibre due to its similar action to cellulose inside the human body (Kouřimská and Adámková 2016). Studies reported that the fibre content of *Locusta migratoria* is approximately 16% (Mohamed 2015a).

16.2.5   Minerals Content

Based on the studies, the body of edible insects contain many macro-minerals and micro-minerals that are required by the human body especially iron, copper, zinc and magnesium. Locusts in particular contain 27 mineral elements (Yin et al. 2017).

Mohamed (2015a) published that 100 g of the dry matter of *Locusta migratoria* recorded high phosphorus content (27–33 ppm) compared to other minerals (Ba, Fe, Zn, Al, B, Cr, Pb, Co and Mn) which measured 0.04–2 ppm approximately (Mohamed 2015a). Basically, the mineral content of 100 g of dry matter of *Locusta migratoria* ranges between 8 and 20 mg (Av et al. 2013). Recently, it was reported that *Locusta migratoria* contains equal amount of zinc and higher ratio of iron measured in mg/100 g dry matter and compared to poultry, beef and pork (Mwangi et al. 2018).

16.2.6   Vitamins Content

Vitamins play crucial role in maintaining the physiological activities of the human body (Yin et al. 2017). Edible insects contain array of vitamins that differ with season and the quality of the insect's feed. Studies on edible insects revealed many vitamins such as thiamine, riboflavin and vitamin B_{12}, retinol, β-carotene, pantothenic acid, biotin, tocopherols and vitamin C (Kouřimská and Adámková 2016). Further, researches published that 100 g of caterpillars can significantly provide the human body with the daily vitamins required (Mohamed 2015a; van Huis 1996). Previous studies on *Locusta migratoria* published that carotenoids are significantly found in the fatty tissues and blood of adults with particular reference to β-carotene (Goodwin and Srisukh 1949). Recent studies showed that adult migratory locusts contain double their content of vitamin D_3 than that of nymphs, while this content increases when the insect is exposed to ultraviolet irradiance (Oonincx et al. 2018). Another recent study revealed that 100 g of *Locusta migratoria* insect contains 0.84 µg of vitamin B_{12} (Schmidt et al. 2019). As per past studies, the concentration of retinol in invertebrates does not exceed 1 mg/kg of dry matter (Pennino et al. 1991). This result was confirmed by Oonincx and Van der Poel who published that the retinol content of *Locusta migratoria* ranges between 0.1 and 0.2 mg/kg dry matter approximately (Oonincx and Van der Poel 2011).

16.2.7   Energy Content

Edible insects have shown good source of energy to the human body via their high fat content and the amount of supplied energy depending on the stage of the insect consumed. Pupae followed by larvae provide the body with more energy than adult insects depending on the fat content as mentioned earlier in this review (Yin et al. 2017). Studies revealed that consumption of 100 g of dry matter of adult *Locusta migratoria* provides average amount of energy equal to 491 calories (Mohamed 2015a), while 100 g of fresh weight provides energy more than 800 kJ based on the type of diet (Av et al. 2013; Oonincx and Van der Poel 2011). The estimated nutrient composition of *Locusta migratoria* is illustrated on Fig. 16.1.

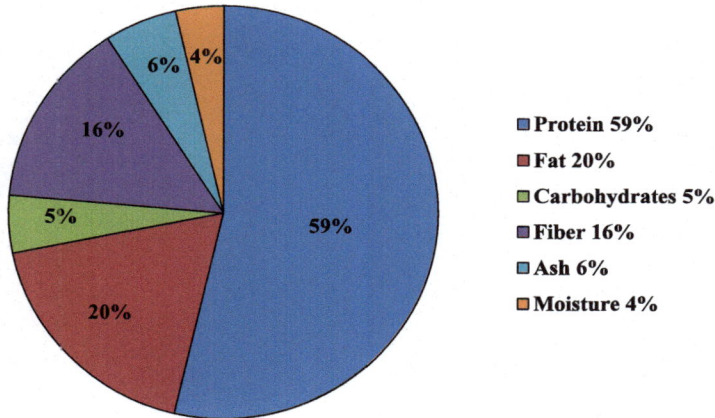

Fig. 16.1 Nutritive composition of 100 g of dry matter of *Locusta migratoria*

16.2.8 Effect of Diet Quality on the Nutritive Constituents of Locusta migratoria

Basically, the nutrient composition of edible insects differs according to their diet. For example the protein content of the dry matter can vary 8 folds depending on the quality of diet (Bukkens and Paoletti 2005). Oonincx and Van der Poel detailed that supplementing the diet of *Locusta migratoria* insect with carrots or wheat bran improved fat content of the insect's body, while wheat bran decreased protein content. Additionally, wheat bran decreased α- and β-carotene concentration, while carrot did not impart changes to the concentration of carotene composition. Addition of wheat bran and carrot in the diet of *Locusta migratoria* increased vitamin A content of the insect's body. Further, some minerals were remarkably affected by diet (Oonincx and Van der Poel 2011).

16.3 Bioactive Compounds of Locusta migratoria

The bioactive constituents of *Locusta migratoria* insect are illustrated on Fig. 16.2. Based on the studies, the migratory insect contains the following biologically active compounds:

16.3.1 Chitin

Chitin is a polysaccharide called *N*-acetyl-D-glucosamine and has the ability to absorb toxins. Besides its high nutritive value, improves peristalsis motion of the intestine, regulates the bacterial flora of the intestine, promotes the immune function

Fig. 16.2 The bioactive
contents of the body of
Locusta migratoria

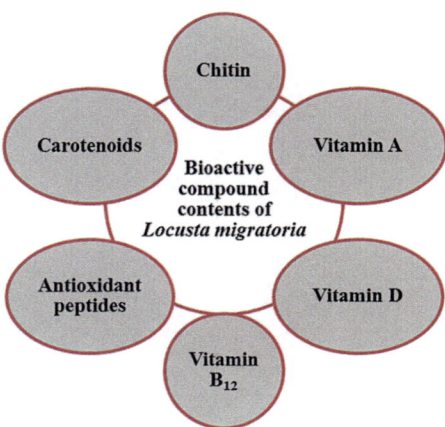

of the body and prevents the increase in blood pressure (Yin et al. 2017). Additionally, it is good anticoagulant, enhances healing of wounds and reduces cholesterol and triglycerides of the blood serum via reduction of lipid absorption from the intestine (Koide 1998; Pandey and Poonia 2018). Moreover, the role of polysaccharides in preventing cancer and oxidative stress, and regulating immune responses have been investigated (Feng et al. 2016).

16.3.2   Vitamin A

Although the retinol content of the dry matter of *Locusta migratoria* measured low concentration as mentioned above and reported by Oonincx and Van der Poel (Oonincx and Van der Poel 2011), but researchers stated that mammals including humans are able to convert some carotenoids into vitamin A via enzymatic action (Harrison and Kopec 2018). Studies proved that supplementing in vitro tumour cell lines with vitamin A has induced apoptosis proving the anti-tumour activity of vitamin A (Klamt et al. 2005). Also a recent pilot study on the anti-aging activity of retinol on the human skin showed that tropical treatment of the skin with serum containing retinol concentration of 0.3% for 2 months has improved the health of the skin (Zasada and Budzisz 2019).

16.3.3   Vitamin B_{12}

As mentioned earlier in the present review, the dry matter of *Locusta migratoria* contains appreciable concentration of vitamin B_{12}. Recent research reports showed that deficiency of vitamin B_{12} results in many health problems such as neurological and heart disorders, and anaemia (Smith et al. 2018). In addition, vitamin B_{12} deficiency associated with aging is autistic and schizophrenic (Zhang et al. 2016). Furthermore, Sugihara et al. (2017) reported that vitamin B_{12} plays key regulator in immunomodulation and health of cell immunity.

16.3.4 Vitamin D

Locusta migratoria obtains vitamin D either from diet or de novo synthesis (Oonincx et al. 2018). Vitamin D represents a group of steroids that play an important role in maintaining the absorption of Ca, Mg and phosphates through the intestine (DeLuca and Schnoes 1983). Besides, it maintains the health of skin and bones, and protects against autoimmune heart and cancer diseases (Holick 2004). Moreover, it plays a vital role in apoptosis, cellular proliferation and immunity (Oonincx et al. 2018).

16.3.5 Carotenoids

The body of *Locusta migratoria* contains 7–8 µg of β-carotene/100 g dry matter and varies according to diet (Mohamed 2015a; Oonincx and Van der Poel 2011). Further, the dry matter of *Locusta migratoria* contains small concentration of other carotenoids depending on the insect's feed such as trans-β-carotene (3–5 mg/kg), β-cryptoxanthin (0.12–0.3 mg/kg), α-carotene (0.02–0.08 mg/kg), lutein (1–2 mg/kg), zeaxanthin (0.1–0.15 mg/kg) and lycopene (0.01–0.02 mg/kg). Early studies reported that carotenoids are biologically active compounds against many cancers and increase the immune responses of the body (Bendich and Olson 1989). Later studies showed that carotenoids play a role in protecting against cancers and chronic heart diseases (Wang 2012).

16.3.6 Antioxidant Peptides

Based on recent studies, the percentage of peptides in the dry matter of locusts may reach 40%. In addition, researchers found that peptides isolated from *Locusta migratoria* acquire significant antiradical activity and chelating power to metal ions recording the highest copper ion chelating capacity more than 80% (Zielińska et al. 2017). This antioxidant power of the insects' peptides indicates that *Locusta migratoria* can be natural therapy against diseases linked to oxidative stress such as cardiovascular, neurodegenerative and chronic kidney diseases, diabetes and cancer (Liguori et al. 2018).

16.4   Conclusion

The present review shows that the migratory locust, *Locusta migratoria* contains appreciable concentrations of nutritive compounds including proteins, polysaccharides, fatty acids, fibre and energy. In addition, it contains important bioactive compounds that play pivotal role in maintaining the health of the human body. The nutritional and chemical composition of *Locusta migratoria* body indicates that it can be a good source of food for human and promising therapy in protecting the body against many ailments as well.

References

Ancsin JB, Wyatt GR (1996) Purification and characterization of two storage proteins from *Locusta migratoria* showing distinct developmental and hormonal regulation. Insect Biochem Mol Biol 26(5):501–510

Bendich A, Olson JA (1989) Biological actions of carotenoids. FASEB J 3(8):1927–1932

Bukkens SG, Paoletti M (2005) Insects in the human diet: nutritional aspects. In: Ecological implications of minilivestock. Science publishers, Inc., Enfield, NH, USA, pp 545–577

DeLuca HF, Schnoes HK (1983) Vitamin D: recent advances. Annu Rev Biochem 52(1):411–439

Dreon A, Paoletti M (2009) The wild food (plants and insects) in Western Friuli local knowledge (Friuli-Venezia Giulia, North Eastern Italy). Contrib Nat Hist 12(12):461–488

Ekpo K, Onigbinde A, Asia I (2009) Pharmaceutical potentials of the oils of some popular insects consumed in southern Nigeria. Afr J Pharm Pharmacol 3(2):051–057

Feng Y, Chen X, Zhao M (2016) Edible insects of China. Science Press, Beijing

Feng Y, Chen XM, Zhao M, He Z, Sun L, Wang CY et al (2018) Edible insects in China: utilization and prospects. Insec Sci 25(2):184–198

Finke MD (2002) Complete nutrient composition of commercially raised invertebrates used as food for insectivores. Zoo Biol 21(3):269–285

Goodwin T, Srisukh S (1949) The biochemistry of locusts. I. the carotenoids of the integument of two locust species (*Locusta migratoria* migratorioides R. & F. and *Schistocerca gregaria* Forsk.). Biochem J 45(3):263–268

Harrison EH, Kopec RE (2018) Digestion and intestinal absorption of dietary carotenoids and vitamin A. In: Physiology of the gastrointestinal tract. Elsevier, Amsterdam, pp 1133–1151

Holick MF (2004) Sunlight and vitamin D for bone health and prevention of autoimmune diseases, cancers, and cardiovascular disease. Am J Clin Nutr 80(6):1678S–1688S

Huis A, Itterbeeck JV, Klunder H, Mertens E, Halloran A, Muir G et al (2013) Edible insects: future prospects for food and feed security, vol 171. FAO, Rome, p 201

Klamt F, de Oliveira MR, Moreira JCF (2005) Retinol induces permeability transition and cytochrome c release from rat liver mitochondria. Biochim Biophys Acta 1726(1):14–20

Koide S (1998) Chitin-chitosan: properties, benefits and risks. Nutr Res 18(6):1091–1101

Kouřimská L, Adámková A (2016) Nutritional and sensory quality of edible insects. J Food Sci Nutr 4:22–26

Liguori I, Russo G, Curcio F, Bulli G, Aran L, Della-Morte D et al (2018) Oxidative stress, aging, and diseases. Clin Interv Aging 13:757

Mitsuhashi J (2008) Entomophagy: human consumption of insects. In: Encyclopedia of entomology. Springer, Heidelberg, pp 1341–1343

Mohamed EH (2015a) Determination of nutritive value of the edible migratory locust Locusta migratoria, Linnaeus, 1758 (Orthoptera: Acrididae). Int J Adv Pharm, Biol Chem 4:144–148

Mohamed EH (2015b) Fatty acids contents of the edible migratory locust Locusta migratoria, Linnaeus, 1758 (Orthoptera: Acrididae). Int J Adv Pharm, Biol Chem 4:746–750

Muzzarelli R, Terbojevich M, Muzzarelli C, Miliani M, Francescangeli O (2001) Partial depolymerization of chitosan with the aid of papain. Chitin Enzymol 16:405–414

Mwangi MN, Oonincx DG, Stouten T, Veenenbos M, Melse-Boonstra A, Dicke M et al (2018) Insects as sources of iron and zinc in human nutrition. Nutr Res Rev 31(2):248–255

Oonincx D, Van der Poel A (2011) Effects of diet on the chemical composition of migratory locusts (*Locusta migratoria*). Zoo Biol 30(1):9–16

Oonincx D, van Keulen P, Finke M, Baines F, Vermeulen M, Bosch G (2018) Evidence of vitamin D synthesis in insects exposed to UVb light. Sci Rep 8(1):10807

Pandey S, Poonia A (2018) Insects-an innovative source of food. Indian J Nutr Diet 55(1):108

Pennino M, Dierenfeld ES, Behler JL. Retinol, α-tocopherol and proximate nutrient composition of invertebrates used as feed. International Zoo Yearbook Zoological Society of London, London. 1991;30(1):143–149

Schmidt A, Call L-M, Macheiner L, Mayer HK (2019) Determination of vitamin B12 in four edible insect species by immunoaffinity and ultra-high performance liquid chromatography. Food Chem 281:124–129

Smith AD, Warren MJ, Refsum H (2018) Vitamin B12. Advances in food and nutrition research, vol 83. Elsevier, Amsterdam, pp 215–279

Sugihara T, Koda M, Okamoto T, Miyoshi K, Matono T, Oyama K et al (2017) Falsely elevated serum vitamin B12 levels were associated with the severity and prognosis of chronic viral liver disease. Yonago Acta Med 60(1):31

van Huis A (1996) The traditional use of arthropods in sub Saharan Africa. In: Proceedings of the section experimental and applied entomology of The Netherlands Entomological Society. Nederlandse Entomologische Vereniging, Amsterdam

Wang X-D (2012) Lycopene metabolism and its biological significance. Am J Clin Nutr 96(5):1214S–1222S

Xiaoming C, Ying F, Hong Z, Zhiyong C (2010) Review of the nutritive value of edible insects. In: Forest insects as food: humans bite back. FAO, Rome, pp 85–92

Yin W, Liu J, Liu H, Lv B (2017) Nutritional value, food ingredients, chemical and species composition of edible insects in China. Future Foods. https://doi.org/10.5772/intechopen.70085

Zasada M, Budzisz E (2019) Randomized parallel control trial checking the efficacy and impact of two concentrations of retinol in the original formula on the aging skin condition: pilot study. J Cosmet Dermatol. https://doi.org/10.1111/jocd.13040

Zhang Y, Hodgson NW, Trivedi MS, Abdolmaleky HM, Fournier M, Cuenod M et al (2016) Decreased brain levels of vitamin B12 in aging, autism and schizophrenia. PLoS One 11(1):e0146797

Zielińska E, Karaś M, Jakubczyk A (2017) Antioxidant activity of predigested protein obtained from a range of farmed edible insects. Int J Food Sci Technol 52(2):306–312

Chapter 17
Nutrient Composition and Bioactive Components of Mopane Worm (*Gonimbrasia belina*)

Raphael Kwiri, Felix M. Mujuru, and Wishmore Gwala

Abstract Mopane worm *Gonimbrasia (Imbrasia) belina* is of economic and nutritional significance mainly in Southern Africa, where it forms part of people's diet. The mopane worm is found in mopane woodlands characterized by the mopane tree (*Colophospermum mopane*) from where the caterpillar entirely derives its nutrition. Generally, its nutritional value is quite diverse due to the stage of metamorphosis, insect's origin, diet, preparation and processing methods. Mopane worms are nutritionally rich in protein content (approx. 58% dwb and 428.52 mg/g dry weight of the total protein), fat content (approx. 15% dwb of which 38% fatty acids are saturated and 62% are unsaturated), carbohydrates (approx. 8% dwb) and considerable proportions of minerals (approx. 1.335% dwb). Importantly, mopane worm contains significant amounts of fibre commonly chitin and possibly some bioactive compounds and antinutrional factors, though this has not yet established; hence, a study to ascertain their levels is vital. However, looking at the future mopane worm is under threat due to deforestation, human pressure, as populations grow, and erratic weather patterns and participation in the mopane worm trade increase. Additionally, mopane worm provides a favourable environment for microbial survival and growth, thus increasing the possibility of being spoiled and infected by pathogenic microorganisms, such as fungi and their subsequent toxins such as aflatoxins. As such, intervention strategies such as food safety awareness and subsequent HACCP implementation are necessary so as to improve product safety. Legislative aspects and rural community support are critical and a necessity for sustainable production of mopane worm and subsequent utilization of the insect as a potential source of proteins and minerals in food items such as FBFs.

Keywords Bioactive components · Insects · Mopane worm (*G. Belina*) · Nutrient composition

R. Kwiri (✉) · F. M. Mujuru · W. Gwala
Department of Food Processing Technology, School of Industrial Science and Technology,
Harare Institute of Technology, Belvedere, Harare, Zimbabwe
e-mail: rkwiri@hit.ac.zw

© Springer Nature Switzerland AG 2020
A. Adam Mariod (ed.), *African Edible Insects As Alternative Source of Food, Oil, Protein and Bioactive Components*, https://doi.org/10.1007/978-3-030-32952-5_17

Abbreviations

DWB   Dry weight basis
FAO   Food and Agriculture Organization
FBFs   Fortified blended foods
HACCP   Hazard analysis critical control point
MAWF   Ministry of Agriculture, Water and Forestry (Namibia)
NGO   Non-governmental organization
NTFP   Non-timber forest products
UN   United Nations
WHO   World Health Organization
WUR   Wageningen University and Research

17.1   Mopane Worm (*Gonimbrasia belina*) Insect

Raw and processed insects are a significant part of the diet in Africa, especially among the poor (Lautenschläger et al. 2017). Mujuru et al. (2014) confirmed the nutritional and economic significance of mopane worm *Gonimbrasia (Imbrasia) belina* in Southern Africa. According to van Huis (2013) it is the most popular and profitable insect in Africa.

The term mopane (of mopane worm) is derived from the mopane woodlands characterized by the mopane tree (*Colophospermum mopane*) from where the caterpillar exclusively derives its nutrition (Thomas 2013). This is due to the monophagous (immobile) nature of the larvae which have to rely on the mopane tree for oviposition (Hrabar et al. 2009). As reviewed by Hrabar et al. (2009), the factors that are likely to affect the oviposition choice by the female include, among other factors predator avoidance for their offspring, leaf morphology or biomechanical properties, nutrition and distribution of the host. The mopane tree seems to possess these qualities, hence making the perfect host for the mopane worm.

The mopane worm is bivoltine in most areas; that is, two generations are produced each year (the first between November and January, its major outbreak, and the second between March and May with some areas where it is mono or univoltine (Thomas 2013, Hoppe et al. 2009; Madibela et al. 2009). Rainfall has been reported to be a significant factor in the abundance of the mopane worm as it facilitates the laying of eggs by the emperor moth (Thomas 2013).

The life cycle of the mopane worm is reported to have four phases (refer to Fig. 17.1: 1, 2, 3 and 4) in which the first phase the Emperor moth lays the eggs on the leaves of the mopane. In 10 days the eggs hatch and enter into the second phase. The second phase is when the larvae feeds on the leaves and grows in a fully grown worm. The worm has been reported to moult four times before reaching maximum size (Mujuru et al. 2014).

The size of the fully grown worm varies with region, variant and nutrition. Some worms have been reported to reach 10 cm in length. At this stage the worm is

Fig. 17.1 The life cycle of *G. Belina*: (1) eggs (2) larva (3) pupa (4) adult

capable of migrating to other sources for nourishment should the foliage diminish (Kwiri et al. 2014). The third stage is when the mature worms move down the trees into the ground. This is the ideal time to harvest the worms as the nutritional composition is at its peak and it has finished feeding on the leaves. Worms harvested at this stage have the least amount of frass and less bitter due to reduced feed (leaves) in the gut.

Good harvesting practice dictates that a small percentage ideally 10% of the worms be not harvested and allowed to reach the ground to ensure continuity of the life cycle and assure another outbreak in subsequent months. The worm burrows 15 cm into the ground and pupate. They form a hard cocoon and lay dormant for months, which may be up to 7 months, depending on conditions of moisture, temperature and climate. The fourth stage is the mothing where the moth emerges at the onset of summer. The sole purpose of the moth is to mate and lay eggs (Emperor moth) over a 2–3-day period without feeding after which the moth dies (Stack et al. 2003).

17.2   Distribution and Consumption Patterns in Africa

The mopane woodlands stretch from the South-Western parts of Angola extending into the Northern Namibia into the Eastern half of Botswana. Zimbabwe has arguably the vastest expanse stretching into what is termed agricultural regions IV and V. The mopane woodlands also stretch into Southern Zambia, Eastern Mozambique and Northern South Africa as shown in Fig. 17.2.

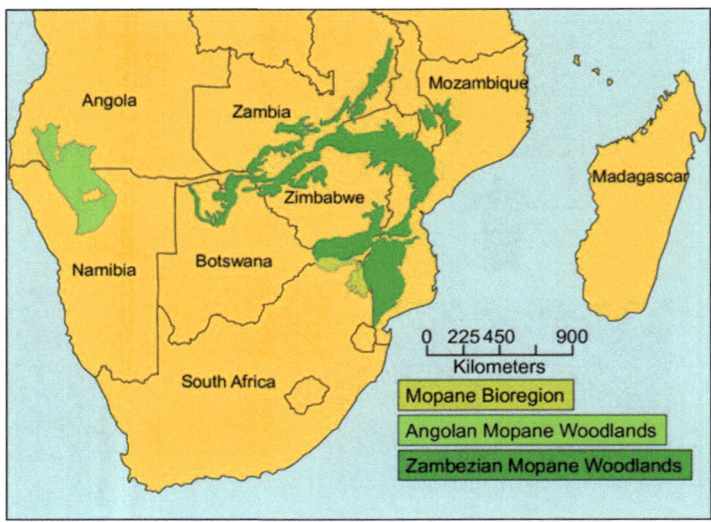

Fig. 17.2 The mopane woodlands of Southern Africa

Characteristic to these woodlands is the low rainfall and drought proneness making farming difficult and non-profitable (Kwiri et al. 2014). The mopane worm becomes a significant commodity for nutritional supplementation as well as income generation in these regions. Hrabar et al. (2009) stated that there is little diversity within the woodlands and the major distinguishing factor is the density of the woodlands and the average height of the mopane tree. Densities ranging from <10 trees/ha in arid north-western Namibia to 481 trees/ha in south-eastern Zimbabwe and 2740 trees/ha in northern South Africa. Heights of between 1.5 and 10 m have been reported showing significant diversity which brings variation in the larvae (Hrabar et al. 2009).

Illgner and Nel (2000) in their report claimed that the effect of globalization has led to the dominance of the western culture which has affected even the diets of most cultures leading to abandonment of traditional foods such as nutritious insects (entomophagy) in favour of processed foods. Entomophagy is becoming an important study area as it promises to offer an alternative to animal protein diets which are beyond reach to most poor communities and as a long term sustainable source of proteins (Manditsera et al. 2018).

In sub-Saharan Africa the mopane worm is a widely studied due to its high protein content and abundance. Kwiri et al. (2014) and Illgner and Nel (2000) reported that the study of entomophagy along with indigenous technical knowledge has gained global recognition with many insects being studied for their nutritional and economic significance with the potential to contribute to food security (Manditsera et al. 2018).

In the case of mopane worms, it is difficult to quantify the tonnage harvested and consumed each season. This is attributed to the fact that most of the trade and consumption of mopane worm occurs in the informal sector where data mining is difficult (Ghazoul 2004). However, some estimations suggest that the annual value of the mopane worm in South Africa's 20,000 km² of mopane woodlands is about £57 m, and Hoppe et al. (2009) estimated an even higher value of US$ 84 m with 40% of that being harvested by rural communities. Illgner and Nel (2000) estimates the cost per kilogram of mopane worm to be between US$ 2.40 and US$4.00 and possibly more. Trade across the borders is also rampant though most of it is confined to the informal sector where accountability is difficult. In a case study of the Zimbabwean scenario, Ghazoul (2004) reported that the marketing of mopane worms is predominantly in the informal sector with a few traders able to participate in the formal market. Ghazoul (2004)' summarizes the harvesting, processing and distribution of mopane worms in Zimbabwe in Fig. 17.3 below.

17.3   Harvesting and Processing

Mopane worm production is quite seasonal and mostly done during the rainy season, that is, from December to January and April to May (Hoppe et al. 2009). Mopane worms are collected from both the ground and from trees, usually the fifth

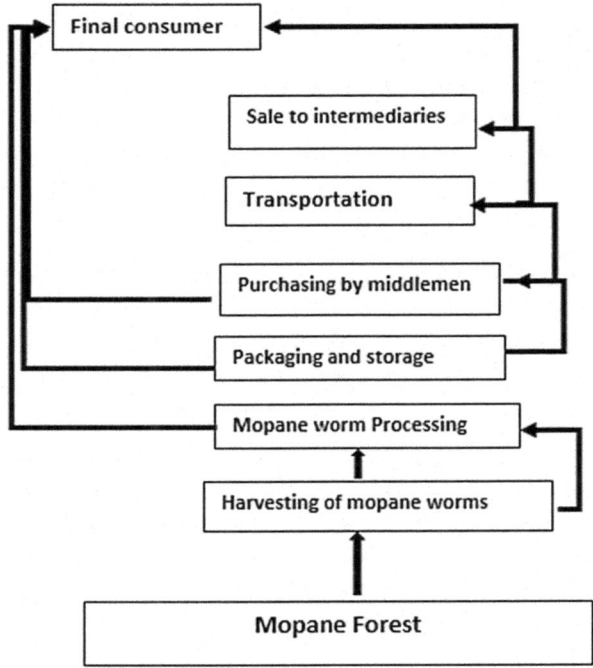

Fig. 17.3 The market chain and distribution of mopane worm in Zimbabwe

instar stage, and the last stage before pupation. Kozanayi and Frost (2002) reported that harvesters prefer collecting mopane worms from shorter trees as they are easily accessible. Harvesters check for waste products below large trees to identify the caterpillar's presence. The rural harvesters generally harvest mature mopane worms when they start crawling down the trees, but during the December and January outbreak, it is not possible as there is not enough worms and the tendency is to collect the worms prematurely. Mature mopane worms ready for picking are identified as being tough and having well-pronounced spikes.

Removal of gut contents of the mopane worm has been shown to be the most difficult and labour-intensive part of processing especially if larvae are collected before they are ready to pupate (van Huis 2013). It has been reported that, when fully grown, larvae have relatively empty guts and their bodies are filled with a "yellow nutritive material". Conventionally, to remove frass from the guts, the larvae are squeezed by placing them between the thumb and forefinger. The pressure during squeezing is adjusted depending on the size of the larvae and the presence of undigested material. This method poses challenges to harvesters in that, the spines of the larvae can puncture the hands and cause bleeding, which can result in contamination of the worms. Taylor (2003) noted that degutting using gloves offered protection from injury as well reduce chances of contamination. In some instances, to reduce damage to fingers, some processors use a bottle as a roller to expel the frass but this has a disadvantage of removing almost all the gut contents, including the desired yellow material. Following the degutting process, mopane worms are then dried. Drying has been shown to reduce mopane worm volume (size) and moisture content (Mujuru et al. 2014). The most common drying method among processors is hot ash drying (the traditional drying used in many parts) owing to the shorter drying time as well as the reduction in size and amount of spikes on the worms. In some instances, the worms may be boiled and then sun dried. However, this method is not very commonly used as the worms will have pronounced spikes and reduced size (Mujuru et al. 2014). The reduced size is attributable to removal of significant amounts of the multilayer water in mopane worm matrices. These changes reduce the market value as consumers perceive them as immature owing to the shrinkage.

17.4 Nutritional Composition

Generally, it has been shown that the nutritional value of edible insects is quite diverse with considerable variations being reported even within the same group of insects depending on the stage of metamorphosis, origin of the insects and their diet (Finke and Oonincx 2014). In addition, changes in nutritional values have been reported owing to the preparation and processing before consumption (e.g. drying, cooking and frying) (van Huis 2013). According to Kourimska and Adamkova (2016), most edible insects contain high amounts of mono- and polyunsaturated fatty acids and can provide sufficient energy and proteins needed in the human diet. The energy value of edible insects depends on their composition, mainly on the fat

Table 17.1 Nutritional composition per dry weight of mopane worms (*Adopted from* Hobane 1994)

Nutrient parameter	% composition
Proteins	58
Carbohydrates	8
Fat	15
Ash	8
Moisture	7
Potassium	0.5
Sodium	0.5
Phosphorus	0.075
Iron	0.01
Calcium	0.25

content (Kourimska and Adamkova 2016). In addition, they are rich sources of minerals such as copper, iron, magnesium, manganese, phosphorus, selenium and zinc.

Gonimbrasia belina (mopane worm) is the larva of the emperor moth (Dithlogo 1996). As shown in Table 17.1 below, mopane worms are nutritionally rich with protein content of around 58% with a fat content of 15% and considerable proportions of minerals (Hobane 1994; Glew et al. 1999; Headings and Rahnema 2002).

17.4.1   Macronutrients

17.4.1.1   Lipids

Lipids consist of triglycerides, where the trihydroxyl alcohol, glycerol, is esterified to three fatty acids (Fennema 1996). The fatty acids could be saturated or unsaturated depending on the number of double bonds in the fatty acid chain (De Man 1999). If no double bonds are present in the fatty acid chain, then the fatty acids are saturated and monounsaturated if one double bond exists whereas they are polyunsaturated if more than one double bond is present (Fennema 1996). The greater the proportion of polyunsaturated fatty acids, the more likely the insect lipids will exist as oil such that they will be liquid at room temperature. Lipids in edible insects have been shown to contain considerable amounts of polyunsaturated essential fatty acids such as linoleic and linolenic acids (Womeni et al. 2009), which the human body cannot synthesize them and should be provided for in the diet (Michaelsen et al. 2009).

Lipids are the most energy-dense macronutrient in food Fennema 1996). In insects, the stage of development has been shown to contribute significantly to the energy value, for example, the larvae or pupae are usually richer in energy compared to adults (Bukkens 1997).

As shown in Table 17.2, mopane worms contain about 15% lipids (Hobane 1994), of which fatty acids constitute 75% of the crude lipid fraction (Glew et al. 1999). About 38% of the fatty acids were saturated and 62% were unsaturated (Rumpold and Schluter 2012). Mopane worm contain a significant proportion of

Table 17.2 Fatty acid composition of mopane worm (*Adopted from* Rumpold and Schluter 2012)

Fatty acid	% composition
Saturated (SFA)	
Myristic (C14:0)	1.15
Palmitic acid (C16:0)	31.9
Stearic acid (C18:0)	4.7
Monounsaturated (MUFAs)	
Palmitoleic (C16:1)	1.8
Oleic acid (C18:1)	34.2
Polyunsaturated (PUFAs)	
Linoleic (C18:2)	6.02
Linolenic (C18:3)	19.6
Arachidonic (C20:4)	0.5
Total unsaturated fatty acids	*62.12*
Total saturated fatty acids	*37.75*

palmitic acid at 32%, oleic acid, a monounsaturated fatty at 34% and linolenic acid, a polyunsaturated fatty acid at 19.6%. As noted by Bukkens (2005), the fatty acid composition of edible insects is a function of the insects' diet during growth and development. The presence of unsaturated fatty acids makes insects-based food products prone to oxidation.

17.4.1.2 Proteins

Proteins represent the main component of the nutrient composition of mopane worms with an average content of 58% on a dry basis in larvae (Glew et al. 1999). However, the amount proteins depends on the processing method used, as highlighted by Bukkens (1997), who noted that dried mopane worms had a relatively higher percentage than dry-roasted ones (57% and 48% respectively). The amino acid content of larvae compares quite well with soybean, with significant proportions of lysine, tryptophan and methionine, which are limiting in maize and legumes respectively (DeFoliart 1989) (Table 17.3). The protein score has been found to be at or above the WHO ideal standard in all essential amino acids (Table 17.4).

Generally, insect protein has low digestibility (Bukkens 1997). Depending of processing methods, protein values and digestibility vary (Dreyer and Wehmeyer 1982). For example, when dried and traditionally prepared, protein digestibility (D) values of about 86%, assimilability (A) of 79% and net protein utilization (NPU) of 68% have been reported for mopane worms. These values compare quite well with conventional protein sources (Bergeron et al. 1988). The relatively low digestibility values could be attributable to the presence of chitin, a nitrogen containing carbohydrate. Chitin is not hydrolyzable in the human gastrointestinal tract due to the absence of chitinase in humans (Bukkens 1997).

Table 17.3 Amino acid content of *G. belina* (*Adopted from* Glew et al. 1999)

Amino acid	Amount (mg/g dry weight)
Tryptophan	5.62
Threonine	27.4
Isoleucine	21.5
Leucine	31.2
Lysine	35.8
Methionine	10
Cysteine	10.4
Phenylalanine	25.5
Tyrosine	30.8
Valine	27.5
Arginine	28.5
Histidine	15
Alanine	25.2
Aspartate	53
Glutamate	60.8
Glycine	22.6
Proline	24.6
Serine	27.1
Total protein content	*482.52*

Table 17.4 Essential amino acid content of *G. belina* compared with the WHO ideal protein (*Adopted from* FAO/WHO/UNU 1985)

Amino acid	% of total amino acids	(% amino acid/ideal) × 100%	WHO ideal
Tryptophan	1.2	109	1.1
Threonine	5.7	168	3.4
Isoleucine	4.5	161	2.8
Leucine	6.5	98	6.6
Lysine	7.4	128	5.8
Methionine + cysteine	4.2	168	2.5
Phenylalanine + tyrosine	11.7	186	6.3
Valine	5.7	163	3.5
Histidine	3.1	111	2.8

17.4.1.3   Minerals

It has been reported that the mopane worm contains a significant amount of important minerals. For instance, the level of iron, a mineral important in prevention of anaemia was present at an average of 300 µg/g dry matter (Table 17.5). This compares well with iron content in beef, which is around 60 µg/g (Bukkens 2005). Stack et al. (2003) attributed the high iron content is attributable to the mopane

Table 17.5 Mineral content
of *G. belina* (mg/g dry
matter) (*Adopted from* Glew
et al. 1999)

Mineral	mg/g dry matter
Calcium	2730
Iron	304
Potassium	15,800
Magnesium	1850
Sodium	18.8
Phosphorus	6340
Zinc	142

worm diet, the leaves they feed. In addition, calcium a mineral important for bone integrity was present at 2730 µg/g dry matter. Of note was also the presence of appreciable amounts of zinc, a mineral important for the normal functioning of the immune system. The amount of zinc in mopane worms was reported to around 140 µg/g dry matter (Glew et al. 1999). Mopane worm did not show significant proportions of sodium but were quite rich in potassium with levels of about 15,800 µg/g dry matter.

17.4.1.4 Fibre Content

Generally, insects contain significant amounts of fibre. The most common form of fibre in insects is chitin, which is the major constituent in the exoskeleton of arthropods (Rumpold and Schluter 2012). Poor digestibility of chitin could result in high fibre content in mopane worms.

17.5 Bioactive Components

Gardiner (2005) indicated that mopane (*Colophospermum mopane*) leaves (on which *G. belina* feed on) are generally not favoured by vertebrate browsers and are used only in drought years, implying that they likely contain plant defence compounds such as phenolic and tannins possibly signifying the presence of some bioactive compounds and antinutritional factors in mopane worm. Likewise, a study carried out in Venetia-Limpopo Nature Reserve on mopane leaf chemistry from different habitats revealed an average of 63.2 g (dry weight basis) of total polyphenolic compounds and 0.59 g (dry weight basis) of condensed tannin–protein ratio (Hrabar et al. 2009). However, studies on bioactive compounds in mopane worm is not available and this have not been studied therefore, a study to ascertain their levels is vital.

17.6   Future Perspectives on Uses of Mopane Worm (*Gonimbrasia belina*) as Food

Interest in mopane worm as a food resource for human is on the rise (Motshegwe et al. 1998; Illgner and Nel 2000; Mpuchane et al. 2000; Ghazoul 2004) as this is the case with people in Africa who eat insects as regular parts of their diets. Mopane worm is the most consumed insect in both rural and urban settlements constituting parts of the traditional diets where it is consumed as snacks or just used as a relish. In most African countries, insects are commonly mixed with, or often consumed as a supplement to predominant diets centred on maize, cassava, sorghum. millet, beans and rice, and form an ingredient to produce other food items (Bukkens 2005). Consumption of the caterpillar is on the increase chiefly because conventional meats such as beef, fish and chicken are unavailable and expensive (FAO/WUR 2013).

Chief among other factors, the African caterpillar could form a foundation for new food products that are based on its substantial nutritive value. It is a great potential source of protein and micronutrients such as minerals that could be utilized to alleviate diet deficiency diseases among most vulnerable groups in society. Kwiri et al. (2014) reported the use of exploiting the nutritional potential of *G. belina* in human diet through fortified blended foods (FBFs) formulations. Thus, making it an alternative substitute for conventional sources of protein, such as soybean, common bean and nuts. In areas where food insecurity is noticeable especially in most African countries, blended food products such as FBFs are typically distributed to the most vulnerable peoples. FBFs are blends of partially precooked and milled cereals, such as soya, beans and pulses, fortified with micronutrients which are largely designed to provide protein and micronutrient supplements in food assistance programmes mostly carried out by organisation such as the UN World Food Programme. Pérez-Expósito and Klein (2009) confirmed the use of insects as a new currently available source of protein and requisite step to alleviate food security problems in many African countries. This was supported by Kinyuru et al. (2011) and FAO/WUR (2013) who supported use of insect as a food ingredient that enhances both food's nutritional quantity and quality. In other examples, insects have been formally used in producing bread containing grubs of the African palm weevil (*Rhynchophorus phoenicis*) that provides major and minor nutrients essential for body growth (Ekpo and Onigbinde 2005), wheat buns enrichment (5% mix) with the termite (*Macrotermes subhyalinus*) (Gahukar 2011a, b), whereas a thin flat bread made from finely ground maize was enriched with ground mealworm *Tenebrio molitor* larvae (Aguilar-Miranda et al. 2002) and termite *Microtermes bellicosus* was used to enrich maize protein (Bukkens 1997).

17.7   Sustainability and Rural Livelihoods Support

The mopane worm, an NTFP, feeds exclusively on the leaves of the mopane tree and causes extensive damage to the forests to the point of being regarded as pests (Thomas 2013). On the other hand, all the highlighted benefits of mopane worm as an important nutritional and economic contributor is under threat of deforestation, human pressure as populations grow, and participation in the mopane worm trade increases. This compounded with erratic weather pattern have contributed to dwindling harvests (Thomas 2013; Manditsera et al. 2018). As such, interventions are necessary for sustainability and to support the rural communities involved in the mopane worm trade.

The mopane tree is unique in that it is a secondary colonizer which is well adapted to arid lands with a shallow and extensively spreading root system. This root system is able to utilize soil water at a matrix lower than grass; hence, its prevalence outcompetes grasses (Ghazoul 2004). The mopane woodlands usually are less dense with a few other tree growing and can be regarded as monospecific. The mopane tree produces copious seed but is difficult to manage as the potency deteriorates rapidly and it is claimed that it may take well over 40 years for a mopane tree to grow to 7 m (Wessels and Potgeiter 2001; Wessels et al. 2001; Ghazoul 2004).

A sustainability intervention study was done in Zimbabwean communities with the objectives of increasing the mopane tree foliage to support higher densities of mopane worms at household and community levels (Ghazoul 2004). Selected trees were coppiced, pruned and pollarded. Observations were made on leaf phenology, changes in leaf size and leaf quality. The findings were that there was no significant difference between treated trees and control trees in all factors considered except for leaf production. It seemed that pollarded and particularly coppiced trees respond better to rainfall in producing secondary leaf flushes. Furthermore, coppiced trees have a capacity to produce new leaves throughout the year and even in the dry season, and are able to retain their leaves for longer. Thus, although coppicing appears to offer no advantage in leaf number or size, an advantage is conferred by virtue of tree vigour and continued productivity (Ghazoul 2004).

Another intervention which many researchers have attempted is the domestication of the mopane worm production at farm level (Ghazoul 2004). The aim was to see if it were possible to grow the mopane independently from the wild. The findings were that within a 3-year period, it is possible to establish and maintain a captive breeding population of G. belina. The advantages of such a system is that fresh mopane worms can be produced at almost any time of year and at much higher quality than before. The challenges encountered included establishing viability of such a venture in the long term, and these lie primarily with the control of viral and bacterial diseases. It may be impossible to entirely eliminate this risk from a captive population that is maintained throughout the year (Ghazoul 2004).

Thomas (2013) looked at sustainability from the view of government intervention in terms of policy formulation. Currently, there exists no sound regulation on harvesting activities of mopane worm and a great danger lies on the long term impact on the environment as more players tap into the mopane worm trade. The Namibian Government is managing forestry resources through various acts and policies administered by the Ministry of Agriculture, Water and Forestry (MAWF). The Forest Act No. 12 of 2001, as amended Forest Act No. 13 of 2005, provides general direction for the management of Namibia's forest resources (Thomas 2013).

In other countries such as Zimbabwe local traditional leadership tries to regulate the harvesting activities by restricting the harvesting to the local population where the outbreaks occur. Only after processing can outside buyers be allowed to buy. This ensures that the local populations benefit from the resource (Ghazoul 2004).

Training is also crucial in ensuring sustainability of the mopane worm trade. Non-governmental organizations (NGOs) and researchers from academic institutions have teamed up to train local communities on good techniques of harvesting and processing to preserve the environment, avoid contamination and ensure consistent quality of the mopane worm. Different processing techniques including evisceration and drying techniques (solar driers) have been tried (Ghazoul 2004)

Efforts to formalize the trade of mopane worn are needed. It has been reported that the mopane worm is a multi-million dollar cross-border trade (Kwiri et al. 2014; Thomas 2013). Formalization may ensure equitable distribution of profits even to the grass roots where rural communities are most vulnerable.

17.8 Food Safety and Quality

Globally, consumer food safety awareness is increasing despite food security concerns in most developing countries. Mopane worm consumption is on the rise despite associated microorganisms that can influence their safety as food. Both insects collected in nature and domesticated may be infected with pathogenic microorganisms, including bacteria, virus, fungi, protozoa and others (Vega and Kaya 2012). Like meat products, mopane worms are rich in nutrients and moisture, providing a favourable environment for microbial survival and growth (Klunder et al. 2012). Basically, in Africa, mopane worm harvesting, processing, packaging and storage practices are largely regarded as poor and most leading causes of their spoilage by microorganisms or fungi (Kwiri et al. 2014) emanating from water of poor quality, insect vectors (such as flies and dipterans) and soil. Traditionally, mopane worms are kept in unclean polypropylene woven bags, plastic or metal buckets and clay pots in which maize or similar products have been stored which in turn possibly increases the risk of end product contamination and spoilage (Allotey et al. 1996; Nyakudya 2004). In a study in Botswana, sun-dried caterpillar's quality deteriorated as the inner flesh changed in colour due to mouldy growth and cavities in the chitinous exoskeleton. The most frequent fungal isolates found were species of *Aspergillus, Penicillium, Fusarium, Cladosporium* and Phycomycetes, of which strains of *Aspergillus, Penicillium* and *Fusarium* are mycotoxin producers. Subsequent quantification of aflatoxins was found varying from 0 to 50 µg per kg of product compared to a maximum safe level set by FAO of 20 µg per kg. This implies that prolonged consumption of contaminated foods possibly pose health risks. However, traditional processing methods, such as boiling, roasting and frying, are often applied to improve sensorial attributes such as taste and palatability, and also have an added advantage of improving food safety.

Global consumer demands for safer and healthier foods have raised concerns over insects handling, processing practices, hygiene and overall food safety. However, FAO (2010b) regarded insects as "health foods" only when collected from forest areas when they are generally clean and free of chemicals as is the case with most African caterpillar. Chemical hazards such as cyanogenetic substances can also be present in *G. belina* as this depends mostly on habitat and plant feed contamination and can only be controlled by selected farming and dietary conditions (Belluco et al. 2013). Blum (1994) reported that the presence of these substances in insects cause inhibition of enzymes such as succinate dehydrogenase and carbonic anhydrase, thereby inhibiting some metabolic pathways for instance oxidative phosphorylation. This is due to the fact that cyanogenetic substances have a high affinity for ferrocytochrome oxidase.

Ingestion of *G. belina* is common in children, probably due to their natural interest. Lee and Hathaway (1998) reported that children who had accidentally ingested caterpillars suffered from symptoms comprising drooling, difficulty swallowing, pain and shortness of breath. FAO (2010a) insisted that ingestion of caterpillars may provoke toxic reactions, even when symptoms suggest an underlying allergic reaction (Okezie et al. 2010). Ingestion of *G. belina* can cause anaphylactic shock. Okezie et al. (2010) reported the case of a 36-year-old female who had two different episodes of anaphylactic shock after *G. belina* consumption (the patient had previously eaten this mopane without reactions), though no allergic reaction test was performed. On the other hand, Kung et al. (2011) described a case of anaphylactic shock in an atopic adolescent who had previously eaten *G. belina*, with mild reactions and both skin prick test and Western blot was performed with positive outcome.

References

Aguilar-Miranda FD, Lopez MG, Escamilla-Santana CA, Barba de la Rosa P (2002) Characteristics of maize flour tortilla supplemented with ground *Tenebrio molitor* larvae. J Agric Food Chem 50:192–195

Allotey J, Mpuchane SF, Gashe BA, Siame GT (1996) Insect pests associated with stored Mophane worm, *Imbrasia belina* Westwood (*Lepidoptera: Saturniidae*) in Botswana. In: Gashe BA, Mpuchane SF (eds) Proceedings of the first multidisciplinary symposium on Phane. NORAD, Gaborone, Botswana

Belluco S, Losasso C, Maggioletti M, Alonzi MC, Paoletti MG, Ricci A (2013) Edible insects in a food safety and nutritional perspective: a critical review. Compr Rev Food Sci Food Saf 12:250–271

Bergeron D, Bushway RJ, Roberts FL, Kornfield I, Okedi J, Bushway AA (1988) The nutrient composition of an insect flour sample from Lake Victoria, Uganda. J Food Compos Anal 1:371–377

Blum M, (1994) The limits of entomophagy: a discretionary gourmand in a world of toxic insects. The Food Insects Newsletter. 7:1. Available from: http://www.food-insects.com/a place to browse.htm. Accessed 6 Oct 2018

Bukkens SGF (1997) The nutritional value of edible insects. Eco Food Nutr 36:287–319

Bukkens SGF (2005) Insects in the human diet: nutritional aspects. In: Paoletti MG (ed) Ecological implications of minilivestock; role of rodents, frogs, snails, and insects for sustainable development. Science Publishers, New Hampshire, pp 545–577

De Man JN (1999) Principles of food chemistry, 3rd edn. Springer-Verlag, Berlin

DeFoliart GR (1989) The human use of insects as food and as animal feed. Bull Entol Soc Am 35:22–35

Dithlogo M (1996) The ecology of Imbrasia belina (Westwood) in North-Eastern Botswana. In: Gashe BF, Mpuchane SF (eds) PHANE, Proc. 1st multi. Disc. Symp. On Phane, 18 June 1996.

Publ. Dept. of Biol., Sci., Univ. of Botswana. Printers: Print. & Publ. Co., Botswana, p 139. ISBN 99912-913-3-4

Dreyer JJ, Wehmeyer AS (1982) On the nutritive value of Mopanie worms. S Afr J Sci 73:33–35

Ekpo KE, Onigbinde AO (2005) Nutritional potentialities of the larvae of Rhynchophorus phoenicis (F.). Pak J Nutr 4:287–290

FAO (2010a) Forest insects as food: humans bite back proceedings of a workshop on Asia-Pacific resources and their potential for development 19–21 February 2008. Chiang Mai, Thailand

FAO (2010b) Development of regional standard for edible crickets and their products. Paper presented at the Joint FAO/WHO meeting Food Standards Programme: FAO/WHO Coordinating Committee for Asia, Bali, Indonesia

FAO/WHO/UNU (1985) Energy and protein requirements. In: Report of a joint FAO/WHO/UNU expert consultation. Food and Agriculture Organization, World Health Organization and the United Nations University, Geneva, Switzerland, p 206

FAO/WUR (2013) Edible insects: future prospects for food and feed security. FAO, Rome

Fennema OR (1996) Food chemistry, 3rd edn. CRC publishers, Boca Raton, Florida

Finke MD, Oonincx D (2014) Insects as food for insectivores. In: Morales-Ramos JA, Rojas MG, Shapiro-Ilan DI (eds) Mass production of beneficial organisms. Invertebrates and entomopathogens. Academic Press, San Diego, CA, USA, pp 583–616

Gahukar RT (2011a) Entomophagy and human food security. Int J Trop Insect Sci 31(3):129–144

Gahukar RT (2011b) Entomophagy and human food security. Int J Trop Insect Sci 31(3):129–144

Gardiner A (2005) Farming Mopane Worms—a household guide. English version. Unpublished manuscript, Harare

Ghazoul J (2004) Mopane woodlands and the mopane worm: enhancing rural livelihoods and resource sustainability, Division of Biology, Imperial College London: DFID Project Reference Number R 7822 Forestry Research Programme (FRP)

Glew RH, Jackson D, Sena L, VanderJagt DJ, Pastuszyn A, Millson M (1999) *Gonimbrasia belina (Lepidoptera: Saturniidae)*, a nutritional food source rich in protein, fatty acids and minerals. Am Entomol 45(4):250–253

Headings ME, Rahnema S (2002) The nutritional value of mopane worms, Gonimbrasia belina (Lepidoptera: Saturniidae) for human consumption. In: Presentation at the ten-minute papers: section B. physiology, biochemistry, toxicology and molecular biology series, 20 November 2002. Ohio State University USA, Ohio

Hobane PA (1994) The urban marketing of the mopane worm: the case of Harare. In: *CASS- NRM Occasional Paper Series*. Harare, Centre for Applied Social Sciences. University of Zimbabwe, Avondale, Harare

Hoppe C, Andersen GS, Jacobsen S, Mølgaard C, Friis H, Sangild PT, Michaelsen KF (2009) The use of whey or skimmed Milk powder in fortified blended foods for vulnerable groups: a literature review. J Nutr 138:145S–161S

Hrabar H, Hattas D, du Toit JT (2009) Intraspecific host preferences of Mopane moths (*Imbrasia belina*) in Mopane (*Colophospermum mopane*) woodland. J Afr Zool 44(2):131–140

Illgner P, Nel E (2000) The geography of edible insects in sub-Saharan Africa: a study of the mopane caterpillar. Geogr J 166:336–351

Kinyuru JN, Kenji GM, Muhoho SN, Ayieko M (2011) Nutritional potential of longhorn grasshopper (RuspoliaDifferens) consumed in Siaya District, Kenya, J. J Agric Sci Technol 12

Klunder HC, Wolkers-Rooijackers J, Korpela JM, Nout MJR (2012) Microbiological aspects of processing and storage of edible insects. Food Contr 26(2):628–631

Kourimska L, Adamkova A (2016) Nutritional and sensory quality of edible insects. NFS J 4:22–26

Kozanayi W, Frost P (2002) Marketing of Mopane worm in southern Zimbabwe. In: Internal report: Mopane woodlands and the Mopane worm: enhancing rural livelihoods and resource sustainability, DFID project no. R7822. Institute of Environmental Studies, University of Zimbabwe, Harare

Kung SJ, Fenemore B, Potter PC (2011) Anaphylaxis to mopane worms (*Imbrasia belina*). Ann Allergy Asthma Immunol 106(6):538–540

Kwiri R, Winini C, Muredzi P, Tongonya J, Gwala W, Mujuru F, Gwala ST (2014) Mopane worm (Gonimbrasia belina) utilisation, a potential source of protein in fortified blended foods in Zimbabwe: a review. Global J Sci Front Res 14(10):1

Lautenschläger T, Neinhuis C, Kikongo E, Henle T, Förster A (2017) Impact of different preparations on the nutritional value of theedible caterpillar *Imbrasia epimethea* from northern Angola. Eur Food Res Technol 243:769–778

Lee JA, Hathaway SC (1998) The challenge of designing valid HACCP plans for raw food commodities. Food Contr 9(2–3):111–117

Madibela OR, Mokwena KK, Nsoso SJ, Thema TF (2009) Chemical composition of Mopane worm sampled at three sites in Botswana and subjected to different processing. Trop Anim Health Prod 41:935–942

Manditsera FA, Lakemond Catriona MM, Fogliano V, Zvidzai CJ, Luning PA (2018) Consumption patterns of edible insects in rural and urban areas of Zimbabwe: taste, nutritional value and availability are key elements for keeping the insect eating habit. Food Secur 8:561–570

Michaelsen KF, Hoppe C, Roos N, Kaestel P, Stougaard M, Lauritzen L, Mølgaard C (2009) Choice of foods and ingredients for moderately malnourished children 6 months to 5 years of age. Food Nutr Bull 30(3):343–404

Motshegwe SM, Holmback J, Yeboah SO (1998) General properties and fatty acid composition of the oil from the Mophane caterpillar, *Imbrasia belina*. J Am Oil Chem Soc 75:725–728

Mpuchane S, Gashe BA, Allotey J, Siame B, Teferra G, Ditlhogo M (2000) Quality deterioration of phane, the edible caterpillar of an emperor moth *Imbrasia belina*. Food Contr 11:453–458

Mujuru F, Kwiri R, Nyambi C, Winini C, Moyo DN (2014) Microbiological quality of Gonimbrasia belina processed under different traditional practices in Gwanda, Zimbabwe. Intern J Curr Micobio Appl Sci 3(9):1089–1094

Nyakudya TT (2004) Determination of the nature and level of contamination of dried larvae of the moth, *Imbrasia belina* (Mopane Worms) by microorganisms. Submitted in partial fulfilment of the requirements for the bachelor of science honours degree in biological sciences. Department of Biological Sciences, Faculty of Science, University of Zimbabwe, Harare

Okezie OA, Kgomotso KK, Letswiti MM (2010) Mopane worm allergy in a 36-year-old woman: a case report. J Med Case Rep 4:42

Pérez-Expósito AB, Klein BP (2009) Impact of fortified blended food aid products on nutritional status of infants and young children in developing countries. Nutr Rev 67(12):706–718

Rumpold BA, Schluter OK (2012) Nutritional composition and safety aspects of edible insects. Mol Nutr Food Res 57:802–823

Stack J, Dorward A, Gondo T, Frost P, Taylor F, Kurebgaseka N (2003) Mopane Worm Utilisation and Rural Livelihoods in Southern Africa. Paper presented at International Conference on Rural Livelihoods, Forests and Biodiversity, 19–23 May, 2003, Bonn, Germany

Taylor FW (2003) Internal report: Mopane woodlands and the Mopane worm: enhancing rural livelihoods and resource sustainability, DFID project no. R7822. Some aspects of innovation and traditional processing, storage and Marketing of Mopane Worms in Botswana. Veld Products Research & Development, Gaborone, Botswana

Thomas B (2013) Sustainable harvesting and trading of mopane worms (*Imbrasia belina*) in northern Namibia: an experience from the Uukwaluudhi area. Inter J Environ Stud 70:494–502

van Huis A (2013) Potential of insects as food and feed in assuring food security. Annu Rev Entomol 58(1):563–583

Vega F, Kaya H (2012) Insect pathology. Academic Press, London

Wessels DCJ, Potgieter MJ (2001) Observation on the survival of Colophospermum mopane seedlings in mopane-veld of the Northern Province, South Africa

Wessels DCJ, Wessels LA, Wessels RD (2001) Observation on the dispersal patterns of Colophospermum fruits from parent tree branches. South African Journal of Batany

Womeni HM, Linder M, Tiencheu B, Mbiapo FT, Villeneuve P, Fanni J, Parmentier M (2009) Oils of insects and larvae consumed in Africa: potential sources of polyunsaturated fatty acids. Ol. Corps Gras Lipides 16(4):230–235

Chapter 18
Nutrient Composition of Desert Locust (*Schistocerca gregaria*)

Abdalbasit Adam Mariod

Abstract In the course of the past few years, there was a new upsurge of interest in insects as food. There are four important locust species in Africa. *Schistocerca gregaria*, the wilderness locust, is probably the most risky of the locust pests due to the capacity of swarms to fly unexpectedly throughout exceptional distances. *Schistocerca g. gregaria*, locusts are popular in African and Arabian diets; however, they also are significant across the world. They are consumed fried, roasted, or boiled. Locusts have acquired attention for their promising sensory properties, nutritionally wealthy composition, and sustainable manufacturing possibilities as a food component. The desert locust energy content was found 179 kcal/100 g with protein content of 14–18 (g/100 g fresh weight).

Keywords *Schistocerca gregaria* · Chemical composition · Nutritional value · Uses

18.1 Introduction

The eating insect resource is an important category of non-wood forest merchandise (NWFPs) gathered from herbal resources (Boulidam 2010). In the course of the past few years, there has been a new upsurge of interest in insects as food. One element that can be accountable is an increasing recognition in the western world that insects are historically and nutritionally important foods for many non-western cultures (DeFoliart 1997).

The four important locust species in Africa are *Schistocerca gregaria* (the desert locust), *Locusta migratoria migratorioides* (the African migratory locust), *Cyrtacanthacris septemfasciata* (the pink locust), and *Locustana pardalina* (the brown locust). The latter three are essential in southern Africa. *Schistocerca g.*

A. A. Mariod (✉)
Indigenous Knowledge and Heritage Centre, Ghibaish College of Science and Technology, Ghibaish, Sudan

© Springer Nature Switzerland AG 2020
A. Adam Mariod (ed.), *African Edible Insects As Alternative Source of Food, Oil, Protein and Bioactive Components*, https://doi.org/10.1007/978-3-030-32952-5_18

gregaria, the feared scourge of the north, is represented with the aid of any other subspecies, *S. g. flaviventris* (Burmeister), in western S. Africa, Namibia, and southern Angola. Despite the fact that flaviventris swarms periodically, it turns into a pest of only quite minor importance (Scholtz and Holm 1985).

Schistocerca gregaria the wilderness locust is probably the most risky of the locust pests due to the capacity of swarms to fly unexpectedly throughout exceptional distances. It has two to five generations consistent with 12 months (Mariod et al. 2017). *Schistocerca g. gregaria*, locusts are popular in African and Arabian diets (Saudi Arabia); however, they are also significant across the world. Famous preparations consist of fried, roasted, or boiled. However, there are over 2000 fit-to-be-eaten insect species that may be eaten at egg, larval, pupal, or grown-up stages, both entire or ground up into food or snack products (Mariod et al. 2017; Van Huis et al. 2013). Locusts are traditionally related to as pests that damage crops, which has influenced studies and purchaser evaluations. In current years, however, through growing interest regarding entomophagy, locusts have acquired attention for their promising sensory properties, nutritionally wealthy composition, and sustainable manufacturing possibilities as a food component (Mariod et al. 2017; Van Huis et al. 2013).

The notorious desert locust, *Schistocerca gregaria gregaria* (Forskal), is extensively dispensed, from northern Burma, Nepal, and Afghanistan throughout southwestern Asia, North and West Africa, besides for a slender band alongside the southern coast of West Africa, to northeastern Congo (Kinshaza) (Zaire) and southern Tanzania (Dirsh 1974). It has been stated as meals across Africa and the center East, and earlier than the approaching of global locust manage packages was extensively advertised across North and West Africa. Commonly, the desert locust can breed in any desert-type region when there may be enough rain. As a pest species, it indicates a few desire for cereals however may be very polyphagous and clearly all crops are at danger. Swarm harm normally outcomes in whole defoliation of the crop and may be devastating over huge regions. Baits, floor and aerial sprays as well as dusts of numerous pesticides are used to govern the hoppers. Barrier spraying with residual insecticides along with dieldrin are used to control the desert locust (https://www.who.int).

Globally, the maximum common insects consumed4 are beetles (Coleoptera) (31%). This isn't unexpected given that the organization includes approximately 40% of all recognized insect species. The intake of caterpillars (Lepidoptera), particularly famous in sub-Saharan Africa, is expected at 18%. Bees, wasps, and ants (Hymenoptera) are available 1/3 at 14% (those bugs are particularly commonplace in Latin America). Following those are grasshoppers, locusts, and crickets (Orthoptera) (13%); cicadas, leafhoppers, planthoppers, scale insects and actual bugs (Hemiptera) (10%); termites (Isoptera) (3%); dragonflies (Odonata) (3%); flies (Diptera) (2%); and other orders (5%). Lepidoptera are fed on nearly totally as caterpillars and Hymenoptera are fed on frequently in their larval or pupal stages. Both adults and larvae of the Coleoptera order are eaten, even as the Orthoptera, Homoptera, Isoptera, and Hemiptera orders are in most cases eaten within the mature stage (Cerritos 2009). Approximately 80 grasshopper species are fed on global, and the huge majority of grasshopper species are edible. Locusts can also arise in swarms, which makes them specifically clean to reap. In Africa, the desert locust, the migratory locust, the red locust, and the brown locust are eaten. However, due to their fame as agricultural pests they will be sprayed with insecticides in governmental manage packages or by farmers. For instance, enormously excessive concentrations of residues of organophosphorus pesticides have been detected in locusts accumulated for food in Kuwait (Saeed et al. 1993).

The well-known eaten insect species are beetles, caterpillars, bees, wasps, ants, grasshoppers, locusts, crickets, cicadas, leaf and plant hoppers, scale insects and true bugs, termites, dragonflies, and flies (FAO 2013).

Islamic literature references to insect eating—including locusts, bees, ants, lice, and termites (El-Mallakh and El-Mallakh 1994). The main references are to locusts, specifically mentioning permission to consume the creatures: It is allowed to eat locusts (Sahih Muslim, 21.4801). Locusts are game of the sea; you may eat them (Sunaan ibn Majah, 4.3222). Locusts are Allah's troops, you may eat them (Sunaan ibn Majah, 4.3219, 3220).

Amar (2003) recommended that consuming positive species of kosher locusts became in large part popular in historical times. The exercise, however, declined among a large part of the Jewish diaspora because of a lack of information about the diverse types of "winged swarming things" stated in the Torah. The way of life became best preserved among Jews of Yemen and in parts of northern Africa. Amar (2003) argued that westernization affects Jews who formerly ate locusts to change their behavior.

One of the greatest current advantages of harvesting plant pests is that it may bring about the decreased use of artificial pesticides. Conversely, there may be a chance that insects may have excessive degrees of pesticides as in the single case of locusts in Kuwait (Saeed et al. 1993). In the Philippines, an outbreak of *Locusta migratoria* was mitigated by paying humans to gather the locusts to sell as food or feed as the authorities were not able to supply pesticides.

18.2 Chemical Composition, Nutritional Value, and Different Uses

Locusts are eaten by most Africans, some Asians, and specifically the Arabs. On their market they appear roasted or grilled in outstanding quantities. When salted, they maintain for some time in storage. They may be used for providing ships, while they may be served as dessert or with coffee. This food is in no way repugnant to observe or by association. It tastes like prawn and is probably more delicately flavored, mainly the females while filled with eggs (FAO 2013).

Ramos Elorduy et al. (1997) analyzed 78 insect species from Oaxaca, Mexico, and determined that caloric content material was within the range of 293–762 kcal consistent with 100 g of dry matter. For instance, the gross energy (which is normally higher than metabolizable energy) of the migratory locust (*Locusta migratoria*) became inside the range of 598–816 kJ in line with 100 g clean weight (recalculated from dry matter), depending on the insect's weight-reduction plan (Oonincx and van der Poel 2011). In Netherlands, the desert locust energy content was found 179 kcal/100 g fresh weight with average protein content of 14–18 (g/100 g fresh weight). As minerals play an essential element in biological processes. The iron content material of locusts (*Locusta migratoria*) varies between 8 and 20 mg per a 100 g of dry weight, depending on their diet (Oonincx et al. 2010). The ability makes use of insects are great. Currently, the usage of insects for the bioconversion of manure and waste has been explored. It might be beneficial to interact industries already producing insects, as an example as pet meals, to promote manufacturing for animal feed and human intake (e.g. mealworm, locusts and crickets) (FAO 2013).

Proximate composition of locust meal consists of a crude protein about 52.3%, oil of 12.0%, CHO approximately 19.0% and 10.0% as ash content material on dry count basis. Desert locust has superb potential as a protein source in broiler diets without causing any physiological sickness as meditated within the hematological evaluation. Phosphatidylcholine, phosphatidylethanolamine, phosphatidylinositols, phosphatidylserine, sphingomyelin, and cardiolipins are present in locust (Mariod et al. 2017).

Fournier et al. (1995) separated six groups of phospholipids from the rectal tissues of the African locust. These phospholipids are phosphatidylcholine, phosphatidylethanolamine, phosphatidylinositols, phosphatidylserine, sphingomyelin, and cardiolipins. Saturated and unsaturated C18 additives have been abundant, as is typically determined in insects. Peculiar-chain (15: zero, 17: zero, 17:1) and long-chain fatty acids were additionally detected. Both C18:$2n-6$ and 20:$4n-6$ were metabolized into prostaglandins and hydroxy-octodecadienoic acids, respectively, by way of the intervention of prostaglandin-endoperoxide synthase. Incorporation of 18:$2n-6$ into phospholipids and neutral lipids or active exchanges of the fatty acid among both classes of lipids pondered different phospholipase, lipase, and transferase activities. Triglycerides seemed as a primary source of fatty acids for locust rectum phospholipids (Fournier et al. 1995).

In tropical nations, insects are frequently fed on complete; however, a few insects, which include grasshoppers and locusts, require the elimination of frame elements (e.g., wings and legs). Relying at the dish, fresh insects can be similarly processed via roasting, frying, or boiling. In many parts of the world, "prepared-to-eat" insects are often bought to local markets after frying or roasting. In such instances, hygienic dealing with is similarly important to prevent the capacity hazard of reinfection and cross-infection. At a family level, fresh insects should be organized hygienically and enough warmness treatment applied to make sure a microbiologically safe meal product. Different simple preservation strategies along with acidifying the insects with vinegar were a success. Another instance is the use of insects for protein enrichment in fermented meals products (FAO 2013). That is a feasible processing alternative with mutual benefits, for the reason that decreased pH in lactic acid-fermented products prevents the boom of probably dangerous microorganisms (Klunder et al. 2012). There has been some achievement in processing and commercializing insects within the Netherlands. Three insect species (yellow mealworm larvae, lesser mealworm larvae, and migratory locusts) may be found in specialized stores in the USA which might be produced and processed especially for human intake. One-day fasting is applied to make sure that the insect has an empty intestine (degutting), and the insect is then freeze-dried complete. This produces a secure product with a quite long shelf life 365 days), if stored correctly in a fab, dry area. Additional benefits of freeze-drying are the upkeep of the insect's nutritional value and the capacity of the product to re-take in water. However, limitations remain: freeze-drying is pricey and regularly results in unwanted oxidation of the lengthy-chained unsaturated fatty acids, lowering the nutritional value of the product and resulting in "off" odors and tastes. A host of other contemporary renovation methods must be explored, along with the software of ultraviolet light and high-pressure technology, in addition to ok packaging techniques. Different important considerations want to be made in deciding on the upkeep technique: the potential to prolong shelf existence (and in flip, include charges), specifically if large quantities of insects need to be processed concurrently; the quantity to which the procedure preserves the nutritional fee of the insects; and the cultural acceptability of the selected renovation/processing technique (FAO 2013). In the Netherlands, agencies that rear insects as pet meals now promote mealworms and locusts for human intake. Kreca is an example of such a company. But mealworms, nonetheless, hold a spot marketplace in the human meal enterprise, and these companies live in particular through the sale of bugs as pet meals (FAO 2013).

Nutritional values are of high importance, but for steady intake a food has to be acceptable from a sensory perspective. Use of insects as food is not common in the Western world, so consumers need to be convinced not only by their nutritional benefits, but also by their tastiness and general sensory appeal (Astrup 2014; Wendin et al. 2017). In cultures where they are eaten, most food insects are considered a delicacy, but they do not all taste the same and different species have the potential to serve different gastronomic functions (Evans et al. 2017). In their book "On Eating Insects," Evans et al. (2017) compiled information on some edible insects and descriptions of their taste/flavor and they reported an intense aroma of cereal, wood, and nuttiness. The flavor has an intense umami and vegetable character, combined with nuttiness and cereal, and notes of Maillard reaction products and relatively low saltiness. They have a crusty, hard, and coarse texture. Books containing recipes for cooking insects have been published (e.g., Van Huis et al. (2014) and Evans et al. (2017)), but there is still a lack of information on how different insects react as ingredients and how a dish tastes (Elhassan et al. 2019).

Locusts are beneficial in terms of their sustainable manufacturing and nutrition. Significantly low amounts of land, water, and feed are needed to rear fit-for-human consumption insects as compared to other farm animals. Moreover, locusts contain excessive quantities of protein, fat, and essential fatty acids that is impacted by diet, age, habitat, and processing method. Future studies investigating how the locusts composition changes at different life stages (nymph versus adults), diets, and seasons would indicate how to optimize locust composition relying on their anticipated quit use. Locust protein extraction after defatting and other processing methods can enhance the protein content material, increase water and oil-conserving capability for insoluble protein fractions, and decrease the dark brown color of complete locusts (Clarkson et al. 2018).

Locusts are used globally as human food in Africa, Southern USA, and Asia, both in rural and urban areas. Business farming of locusts is growing for food and feed in South East Asia and rice field grasshoppers are harvested for food in Japan, China, and Korea. In Africa, the desert locust (*Schistocerca gregaria*), the migratory locust (*Locusta migratoria*), the red locust (*Nomadacris septemfasciata*), and the brown locust (*Locustana pardalina*) are commonly eaten. They are an important food source, as are other insects, adding proteins and fats to the daily weight loss plan, especially in times of food crisis. But, in lots of African, center eastern and Asian countries, locusts are considered a delicacy and eaten in abundance. They may be additionally served on skewers in some Chinese food markets (Mohamed 2015).

References

Amar Z (2003) The eating of locusts in Jewish tradition after the Talmudic period. Torah u-Madda J 11:186–202

Astrup PJ (2014) Disgusting or delicious—utilization of bee larvae as ingredient and consumer acceptance of the resulting food. PhD thesis, Copenhagen University, Copenhagen, Denmark

Boulidam S (2010) Forest insects as food: humans bite back, proceedings of a workshop on Asia-Pacific resources and their potential for development. FAO Regional Office for Asia and the Pacific, Thailand, Bangkok, pp 131–140

Cerritos, R. 2009. Insects as food: an ecological, social and economical approach CAB reviews: perspectives in agriculture, veterinary science, nutrition and natural resources Wallingford CAB International, 4(27): 1–10

Clarkson C, Mirosa M, Birch J (2018) Potential of extracted *Locusta migratoria* protein fractions as value-added ingredients. Insects 9(20):1–12. https://doi.org/10.3390/insects9010020

DeFoliart G (1997) An overview of the role of edible insects in preserving biodiversity. Ecol Food Nutr 36:109–132

Dirsh VM (1974) Genus Schistocerca (Acridomorpha, Insecta). Ser. Entomologica, vol 10. W. Junk, The Hague. (Overview)

Elhassan M, Wendin K, Olsson V, Langton M (2019) Quality aspects of insects as food—nutritional, sensory, and related concepts. Foods 8(95):1–14

El-Mallakh OS, El-Mallakh RS (1994) Insects of the Qur'an (Koran). Am Entomol 40:82–84

Evans J, Flore R, Frøst MB (2017) On eating insects essays, stories and recipes. Phaidon, London, UK

FAO (2013) FORESTRY PAPER 171. In: van Huis A, Itterbeeck JV, Klunder H, Mertens E, Halloran A, Muir G, Vantomme P (eds) Edible insects: future prospects for food and feed security. Food and Agriculture Organization of the United Nations, Rome

Fournier BR, Wolff RL, Nogaro M, Radallah D, Darret D, Larrue J, Girardie A (1995) Fatty acid composition of phospholipids and metabolism in rectal tissues of the African locust. Comp Biochem Physiol B Biochem Mol Biol 111(3):361–370. https://www.who.int/water_sanitation_health/resources/vector357to384.pdf

Klunder HC, Wolkers-Rooijackers J, Korpela JM, Nout MJR (2012) Microbiological aspects of processing and storage of edible insects. Food Control 26:628–631

Mariod AA, Saeed Mirghani ME, Hussein I (2017) Chapter 44. *Schistocerca gregaria* (desert locust) and *Locusta migratoria* (migratory locust). In: Unconventional oilseeds and oil sources. Academic Press, Cambridge, MA, USA, pp 293–297. ISBN 978-0-12-809435-8

Mohamed EHA (2015) Determination of nutritive value of the edible migratory locust *Locusta migratoria*, Linnaeus, 1758 (Orthoptera: Acrididae). Int J Adv Pharm Biol Chem 4(1):144–148

Oonincx DGAB, van der Poel AFB (2011) Effects of diet on the chemical composition of migratory locusts (Locusta migratoria). Zoo Biol 30:9–16

Oonincx DGAB, van Itterbeeck J, Heetkamp MJW, van den Brand H, van Loon J, van Huis A (2010) An exploration on greenhouse gas and ammonia production by insect species suitable for animal or human consumption. PLoS One 5(12):e14445

Ramos Elorduy J, Pino JM, Prado EE, Perez MA, Otero JL, de Guevara OL (1997) Nutritional value of edible insects from the state of Oaxaca, Mexico. J Food Compos Anal 10:142–157

Saeed T, Dagga FA, Saraf M (1993) Analysis of residual pesticides present in edible locusts captured in Kuwait. Arab Gulf J Sci Res 11(1):1–5

Scholtz CH, Holm E (1985) Insects of Southern Africa. Butterworths, Durban, p 502

Van Huis A, van Itterbeek J, Klunder H, Mertens E, Halloran A, Muir A, Vantomme P (2013) Edible insects: future prospects for food and feed security. Food and Agriculture Organization of the United Nations, Rome, Italy. ISBN 978-92-5-107596-8

Van Huis A, Van Gurp H, Dicke M (2014) The insect cookbook—food for a sustainable planet; Colombia. University Press, New York, NY, USA

Wendin K, Norman C, Forsberg S, Langton M, Davidsson F, Josell Å, Prim M, Berg J (2017) Eat them or not? Insects as a culinary delicacy. In: Mikkelsen B, Ofei KT, Olsen Tvedebrink TD, Quinto Romano A, Sudzina F (eds) In Proceedings of the 10th International Conference on Culinary Arts and Sciences, Copenhagen, Denmark, 6–7 July 2017. Aalborg University, Copenhagen, Denmark, pp 100–106. ISBN 978-87-970462-0-3

Chapter 19
Nutritional Value of Brood and Adult Workers of the Asia Honeybee Species *Apis cerana* and *Apis dorsata*

Sampat Ghosh, Bajaree Chuttong, Michael Burgett, Victor Benno Meyer-Rochow, and Chuleui Jung

Abstract We assessed nutrients composition of the Asian honeybee *Apis cerana* and giant honeybee *Apis dorsata*. Abundances of individual amino acids and thus their total amounts were found to be considerably higher in *A. cerana* pupae than those of *A. dorsata* pupae. The immature developmental stages of honey bees contained higher amounts of fat than the adults. MUFA proportions were found to be higher in adults than pupae of both *A. dorsata* and *A. cerana*. In contrast, *A. cerana* larvae contained slightly more SFA than MUFA. Mineral contents were lower in both *A. dorsata* and *A. cerana* in comparison to previously reported European honeybee *A. mellifera*, except phosphorus. From our previous as well as present results, we provide evidence that honey bee brood would be an excellent nutritional source

S. Ghosh
Agriculture Science and Technology Research Institute, Andong National University, Andong, GB, Republic of Korea

Department of Life Science, Sardar Patel University, Balaghat, MP, India

B. Chuttong
Science and Technology Research Institute, Chiang Mai University, Chiang Mai, Thailand

M. Burgett
Department of Horticulture, Oregon State University, Corvallis, OR, USA

V. B. Meyer-Rochow
Agriculture Science and Technology Research Institute, Andong National University, Andong, GB, Republic of Korea

Department of Ecology and Genetics, Oulu University, Oulu, Finland

C. Jung (✉)
Agriculture Science and Technology Research Institute, Andong National University, Andong, GB, Republic of Korea

Department of Plant Medicals, Andong National University, Andong, GB, Republic of Korea
e-mail: cjung@andong.ac.kr

© Springer Nature Switzerland AG 2020
A. Adam Mariod (ed.), *African Edible Insects As Alternative Source of Food, Oil, Protein and Bioactive Components*, https://doi.org/10.1007/978-3-030-32952-5_19

265

which also advocate the conservation of honeybee species and enhance the livelihood of small and medium scale bee keepers.

Keywords Bee keeping · Brood · Conservation · Sustainable food · Livelihood

19.1 Introduction

How to feed the global population in the future is a much debated issue. Although there are nowadays fewer cases of outright starvation than 50 years ago, global food security has not very significantly improved since Meyer-Rochow (1975) in the journal "Search" of the Australia and New Zealand Association for the Advancement of Science suggested that the use of insects as food and feed could ease the problem of food shortages in the world. Sure enough, insects have been getting more attention as an alternative food resource as their value as suppliers of nutrients became more widely accepted and it was realized that insects were consumed as a traditional food item by members of many different populations worldwide. It has been estimated that probably 2 billion people include at least sometimes insects in their regular diet and that the number of edible insects amounts to at least 2000 (Mitsuhashi 2008; Jongema 2015). Moreover, stable carbon isotope analyses have demonstrated that South African *Australopithecus* was not only frugivorous but also consumed significant amounts of C4 foods, which presumably included insects (termites) in addition to grasses and sedges (Sponheimer et al. 2005). This demonstrates the long history of human entomophagy.

Even today, various species of honey bees as well as wasps are traditionally components of the diets of various ethnicities, especially in East and Southeast Asia and Africa. To mention just one example: honey bee brood and the broods of wasps and hornets are considered delicacies by the ethnic populations of the Sabah region of Malaysia (Chung et al. 2002). People consume the brood, primarily larvae and pupae, either raw or processed, boiled in porridge or rice like fried eggs or use the honey comb together with the brood to squeeze out and extract the honey (Chung et al. 2002). In Thailand use of the brood as food involves the bee species *Apis cerana*, the dwarf honey bee *Apis florea* and the giant honey bee *Apis dorsata* (Johnson 2010). People harvest the colonies not only to obtain the honey but also to collect the brood for consumption (Wongsiri et al. 1997).

However, irrespective of the species, the traditions of eating insects are declining in many developing countries while, ironically, the acceptance of insects as a food item is increasing in developed, industrialized countries. Daniel Ambühl of Switzerland, for example, strongly advocates the use of honey bees and in particular drones and their larvae as food for humans (Ambühl 2016). The decreasing willingness to consume insects in developing countries, on the other hand, is thought to be due to a host of reasons. Not wanting to be regarded as backward and a desire to be seen by Westerners as "civilized" could be one reason as pointed out by Meyer-Rochow (1975). The ever increasing amount and variety of imported food stuffs from foreign countries and changing agricultural practices could be other reasons. A lack of an awareness of the nutritive value of food insects and the misconception that they are dirty, unsafe, or unsavory could further have led to their decreasing popularity.

Honey bees more than any other species of insect have been "partners" of human civilizations and have assisted humans in many ways. Bees are the principal "horse power" when it comes to pollination and as a consequence of that food production. Honey has been used as a sweetener as well as a medicinal compound (the word medicine stems from the fermented honey comb known as "mead": Taylor 1975) by almost all the communities of the world since ancient time and has been mentioned in writings such as the Ayurveda, the Bible and the Quran. Beekeeping primarily to obtain honey and wax as well as other hive products is thought to have begun around 3000 years ago in the region of Tel Rehov in Israel. Almost a 1000 years later bee-keeping was known to be practiced in Korea, Japan and China (Jung 2014; Kohsaka et al. 2017).

The European bee *Apis mellifera* currently receives the most attention worldwide in the commercial beekeeping sector when compared with other honey bee species. Prior to the introduction of *A. mellifera* in the twentieth century, the Asian honey bee *A. cerana* had been domesticated in the Indian sub-continent as well as China, Thailand, South Korea and Japan (Tej et al. 2017; Ai et al. 2012; Kohsaka et al. 2017). In the South Indian state of Tamil Nadu state there are 50,000 beekeepers involved these days with the rearing of *A. cerana* (Tej et al. 2017). In a survey carried out in Nagaland (North East India) on traditional beekeeping Singh (2014) has been able to show that traditional beekeepers primarily utilize *A. cerana* and sometimes the stingless bee species *Tetragonula iridipennis* in their cultures (Singh 2014). Out of a total of interviewed bee keepers, 65.3% were interested in rearing bees in traditional indigenously manufactured boxes (Singh 2014). There is a recent concern regarding the decline of *A. cerana* populations in many parts of Asia (Theisen-Jones and Bienefeld 2016).

Numerous scientific reports have been published describing honey's nutritional value and its usefulness as therapeutic material based on the honey's antimicrobial and antioxidant properties (Khalil et al. 2010; Mandal and Mandal 2011; Kwakman and Zaat 2012; Vallianou et al. 2014). However, despite the knowledge that honey bees and wasps themselves and not just their brood can be eaten, they have not receive much attention as a potential source of nutrition. Scientific studies on the nutritional properties of the honey bees have focused entirely on *A. mellifera* (Finke 2005; Ghosh et al. 2016). Although the practice of consuming honeybee brood exists, there is limited information on the nutritional value available for species other than *A. mellifera*. This lack of background knowledge often limits the possibilities of introducing honey bee brood as a food item. In our previous study we argued that in addition to hive products like honey, wax, bee venom, and royal jelly, bee brood could be another hive product at least for medium scale beekeeper. In the present study we assess the nutritional value of the Asian honey bee species *A. cerana* and the giant honey bee *A. dorsata*.

19.2   Material and Methods

19.2.1   Sample Collection

Apis cerana and *Apis dorsata* colonies were collected from the Chiang Mai province in northern Thailand in 2017 and 2018. *Apis dorsata* pupae were collected from two different locations, namely, Thaweephol Condo in metropolitan Chiang Mai (TC) and the Mae Jo University campus (MJ). The brood combs and adult worker bees were taken to the Chiang Mai University Post Harvest Research and Technology facility for sample processing. Bee pupae were individually removed from the combs with fine tweezers and then stored at −20 ° C. Wings of the adults were removed prior to freezing. Both adult worker bees and pupae were freeze-dried, ground to powder and prepared as dry matter (DM) for further analysis. All the solvents and chemicals used in the study were of analytical grade.

19.2.2   Amino Acid Analysis

Amino acid composition was determined by a Sykam Amino Acid analyser S433 (Sykam GmbH, Germany) equipped with LCA K07/Li (PEEK—column 4.6 × 150 mm) and photometer detector (570 nm and 440 nm) following the standard method of AOAC (1990). The ground samples were hydrolyzed in 6 N HCl for 24 h at 110 °C under nitrogen atmosphere and then concentrated with rota-evaporator (Ghosh et al. 2016). The concentrated samples were reconstituted with sample dilution buffer supplied by the manufacturer (0.12 N, pH 2.20). The hydrolyzed samples were analyzed for amino acid composition.

19.2.3   Fatty Acid Analysis

Fatty acid composition was analyzed by gas chromatography equipped with flame ionization detector (GC-14B, Shimadzu, Tokyo, Japan), following the standard method of the Korean Food Standard Codex (2010). The samples were derivatized into fatty methyl esters (FAMEs) following the method described by of Lepage and Roy (1986). Identification and quantification of FAMEs were accomplished by comparing the retention times of peaks with those of pure standards purchased from Sigma and analyzed under the same conditions. The results were expressed as a percentage of individual fatty acids in the lipid, that is, sum of total fatty acid fraction.

19.2.4 Minerals Analyses

Minerals were analyzed following the standard method of the Korean Food Standard Codex (2010). The dried powder samples were digested with nitric and hydrochloric acid (1:3) at 200 °C for 30 min. Each sample was filtered using Whatman filter paper (0.45 μm) and stored in washed glass vials before analysis (Ghosh et al. 2016). Minerals were analyzed by inductively coupled plasma-optical emission spectrophotometer (ICP-OES 720 series Agilent).

19.3 Results and Discussion

Table 19.1 represents the comparative account of amino acid compositions of *A. dorsata* and *A. cerana* pupae. Seventeen proteinergic amino acids were identified, and their amounts were determined. Abundances of individual amino acids and thus their total amounts were found to be considerably higher in *A. cerana* pupae than those of *A. dorsata* pupae and *A. mellifera* (Ghosh et al. 2016). The kinds of amino acids present were not different from what had been reported in earlier

Table 19.1 Amino acid composition of *Apis dorsata* and *A. cerana* pupae (g/100 g dry weight basis)

	Apis dorsata pupae	*Apis cerana* pupae	*Apis mellifera* pupae[a]
Valine	2.190	3.084	2.4
Isoleucine	1.667	2.446	2.3
Leucine	3.260	4.376	3.2
Lysine	2.222	2.966	3.0
Tyrosine	1.267	1.894	2.0
Threonine	1.681	2.163	1.9
Phenylalanine	1.544	2.076	0.2
Histidine	0.976	1.271	1.1
Methionine	1.392	1.705	–
Arginine	1.905	2.522	2.3
Aspartic acid	5.171	6.333	3.5
Serine	1.910	2.359	2.0
Glutamic acid	4.310	5.301	8.4
Glycine	2.878	3.698	2.5
Alanine	3.316	4.925	2.9
Cysteine	0.547	0.715	0.4
Proline	2.699	3.380	–
Other	–	–	2.8
Total	38.935	51.214	40.9

[a]Ghosh et al. (2016)

reports on the nutritional value of honey bees, but with regard to relative abundances differences were noted. For example, leucine was the most common and predominant among the essential amino acids followed by valine in the pupae of *A. dorsata* and *A. cerana*. However, in previous reports on *A. mellifera* (Finke 2005; Ghosh et al. 2016) and *A. florea*, irrespective of their developmental stages, glutamic acid was the most dominating amino acid followed by aspartic acid. The reverse was found true for *A. dorsata* and *A. cerana*. From a nutritional point of view among the essential amino acids lysine receives the most attention as it is the limiting amino acid in cereal based diets, which are most popular in Asia. Despite some of the differences, it is clear that all of the analyzed honey bee species contained substantial amounts of lysine.

Fatty acid compositions are shown in Table 19.2. In general, the immature developmental stages of honey bees contain higher amounts of fat than the adults.

Table 19.2 Fatty acid composition (mg/100 g dry weight basis) of different honey bee species of different developmental stages

Fatty acid	*A. cerana* (adult)	*A. cerana* (pupae)	*A. cerana* (larvae)	*A. dorsata* (adult)	*A. dorsata* (pupae) TC[a]	*A. dorsata* (pupae) MJ[a]
Lauric acid (C12:0)	20	30	10	10	30	30
Tridecanoic acid (C13:0)	10	10	0	10	0	10
Myristic acid (C14:0)	80	190	240	30	200	150
Palmitic acid (C16:0)	770	1990	2320	450	2060	1830
Stearic acid (C18:0)	510	670	490	450	730	790
Arachidic acid (C20:0)	40	40	20	30	30	40
SFA (subtotal)	*1430*	*2930*	*3080*	*980*	*3050*	*2850*
Palmitoleic acid (C16:1)	50	30	20	40	20	30
Oleic acid (C18:1n-9,Cis)	2440	3160	2850	1910	2950	3090
Cis-11-Eicosenoic acid (C20:1)	190	150	90	130	110	110
MUFA (subtotal)	*2680*	*3340*	*2960*	*2080*	*3080*	*3230*
Linoleic acid (C18:2n−6, Cis)	110	60	30	70	50	50
Linolenic acid (C18:3n−3)	10	10	10	0	0	0
PUFA (subtotal)	*120*	*70*	*40*	*70*	*50*	*50*

[a]TC and MU are the sampling locations, Thaweephol Condo in metropolitan Chiang Mai (TC) and the Mae Jo University campus (MJ)

Presumably the adults' foraging activity and other daily routine work requires faster energy sources in carbohydrates other than just fat. This result is in agreement with those in which the chemical composition of different species of honey bees, including the *A. mellifera* (Finke 2005; Ghosh et al. 2016) was examined. Based on their degree of saturation, fats can be classified as members of three groups, namely saturated fatty acids (SFA), monounsaturated fatty acids (MUFA) and polyunsaturated fatty acids (PUFA). Among the SFAs palmitic acid is the most abundant one followed by stearic acid. From a nutritional point of view saturated fatty acids are generally not desirable because of their linkage with atherosclerotic disorders. However, the only saturated fatty acid that is credited with playing a positive role with regard to the removal of "bad" cholesterol in the human body, according to Bonanome and Grundy (1988), is stearic acid.

Table 19.3 represents the mineral content of different developmental stages of *A. dorsata* and *A. cerana*. All of the minerals were found to be higher as an individual matured, so that adults had a higher mineral content than pupal and larval instars. With the exception of phosphorus all the minerals were less abundant in both *A. dorsata* and *A. cerana* than in *A. mellifera* (Ghosh et al. 2016). The reduced content of sodium increases the potential of the nutritional value of these species.

In our previous papers we have discussed the nutritional benefits of each component (Ghosh et al. 2016, 2017). A comparison between the various honey bee species demonstrated that *A. cerana* pupae can be considered to contain the highest amount of protein. MUFA proportions were found to be higher in adults rather than pupae of both *A. dorsata* and *A. cerana*. However, *A. cerana* larvae contained slightly higher SFA than MUFA. From our previous works as well as this paper, we provide evidence that honey bee brood would be an excellent nutritional source. People traditionally harvest honey bee colonies from the wild for honey extraction in many regions. What we suggest is that this should include the use of honey bee brood as a nutritional product as its consumption could enhance the health and improve the livelihood of the people.

Table 19.3 Mineral content (mg/100 g dry weight basis) of different honey bee species of different developmental stages

	A. cerana (adult)	A. cerana (pupae)	A. cerana (larvae)	A. dorsata (adult)	A. dorsata (pupae) TC[a]	A. dorsata (pupae) MJ[a]
Calcium	91.1	62.9	63.1	99.2	68.9	78.5
Potassium	1538.8	1153.2	823.1	1573.9	1136.6	1254.3
Sodium	77.1	44.4	37.2	67.8	48.6	53.9
Magnesium	148.8	104.3	86.6	144.7	103.4	113.3
Phosphorus	1283.9	931.5	715.6	1187.9	905.0	972.3
Iron	11.1	7.1	5.9	8.9	5.8	7.6
Copper	1.9	1.2	1.0	1.6	1.1	1.2
Zinc	12.9	7.7	7.3	9.5	6.4	7.4
Manganese	0.2	0.18	1.1	0.2	0.1	0.1

[a]TC and MU are the sampling locations, Thaweephol Condo in metropolitan Chiang Mai (TC) and the Mae Jo University campus (MJ)

Acknowledgments We wish to acknowledge the support from the Priority Research Centre's Program of the National Research Foundation of Korea (NRF) funded by the Ministry of Education, Science and Technology (NRF-2018R1A6A1A03024862).

References

Ai H, Yan X, Han R (2012) Occurrence and prevalence of seven bee viruses in *Apis mellifera* and *Apis cerana* apiaries in China. J Invertebr Pathol 109(1):160–164

Ambühl D (2016) "Beezza" das Bienenkochbuch/the honey bee cook book. Skyfood, Zürich

AOAC (1990) Official methods of analysis, 15th edn. Association of Official Analytical Chemists, Washington DC, USA

Bonanome A, Grundy SM (1988) Effect of dietary stearic acid on plasma cholesterol and lipoprotein levels. N Engl J Med 318(19):1244–1248

Chung AYC, Khen CV, Unchi S, Binti M (2002) Edible insects and entomophagy is Sabah, Malaysia. Malay Nat J 56(2):131–144

Finke MD (2005) Nutrient composition of bee brood and its potential as human food. Ecol Food Nutr 44(4):257–270

Ghosh S, Jung C, Meyer-Rochow VB (2016) Nutritional values and chemical composition of larvae, pupae and adults of worker honey bee, *Apis mellifera ligustica* as a sustainable food source. J Asia Pac Entomol 19(2):487–495

Ghosh S, Lee S-M, Jung C, Meyer-Rochow VB (2017) Nutritional composition of five commercial edible insects in South Korea. J Asia Pac Entomol 20(2):686–694

Johnson DV (2010) The contribution of edible forest insects to human nutrition and to forest management. In: Durst PB, Johnson DV, Leslie RN, Shono K (eds) Forest insects as food: humans bite back. Proceedings of a workshop on Asia-Pacific resources and their potential for development.19-21 February 2008, Chiang Mai. United Nation Food and Agriculture Organization, Bangkok, Thailand

Jongema Y (2015) List of edible insects of the world. Wageningen University, Wageningen. http://tinyurl.com/mestm6p

Jung C (2014) A note on the early publication of beekeeping of western honeybee, Apis mellifera in Korea: Yangbong Yoji (Abriss Bienenzucht) by P. Canisius Kugelgen. Korean J Apic 29:73–77

Khalil MI, Sulaiman SA, Boukrea L (2010) Antioxidant properties of honey and its role in preventing health disorder. Open Nutraceuticals J. 3:6–16

Kohsaka R, Park MS, Uchiyama Y (2017) Beekeeping and honey production in Japan and South Korea: past and present. J Ethn Foods 4:72–79

Korean Food Standard Codex (2010) Ministry of Food and Drug safety. Republic of Korea

Kwakman PHS, Zaat SAJ (2012) Antibacterial components of honey. IUBMB Life 64:48–55

Lepage G, Roy CC (1986) Direct transesterification of classes of lipids in a one-step reaction. J Lipid Res 27(1):114–120

Mandal MD, Mandal S (2011) Honey: its medicinal property and antibacterial activity. Asian Pac J Trop Biomed 1(2):154–160

Meyer-Rochow VB (1975) Can insects help to ease the problem of world food shortage? Search 6:261–262

Mitsuhashi J (2008) Sekai Konchu Shoko Taizen. YasakaShobo, Tokyo

Singh AK (2014) Traditional beekeeping shows great promise for endangered indigenous bee *Apiscerana*. Indian J Tradit Know 13(3):582–588

Sponheimer M, Lee-Thorp J, de Ruiter D, Cordon D, Cordon J, Baugh AT, Thackeray F (2005) Hominins, sedges, and termites: new carbon isotope data from the Sterkfontein valley and Kruger national Park. J Hum Evol 48(3):301–312

Taylor RJ (1975) Butterflies in my stomach. Woodbridge Press, Santa Barbara

Tej MK, Aruna R, Mishra G, Srinivasan MR (2017) Beekeeping in India. In: Omkar (ed) Industrial entomology. Springer Nature, Singapore, pp 35–66

Theisen-Jones H, Bienefeld K (2016) The Asian honey bee (*Apiscerana*) is significantly in decline. Bee World 93(4):90–97

Vallianou NG, Gounari P, Skourtis A, Panagos J, Kazazis C (2014) Honey and its anti-inflammatory, anti-bacterial and anti-oxidant properties. Gend Med 2. https://doi.org/10.4172/2327-5146.1000132

Wongsiri S, Lekprayoon C, Thapa R, Thirakupt K, Rinderer TE, Sylvester HA, Oldroyd BP, Booncham U (1997) Comparative biology of *Apis andreniformi s*and Apis florea in Thailand. Bee World 78(1):23–35

Chapter 20
Nutrient Composition of Mealworm (*Tenebrio molitor*)

Abdalbasit Adam Mariod

Abstract The mealworm (*Tenebrio molitor* L.) is perhaps the greatest creepy beetles feeding on stored items. Grown-up hatchling weighs about 0.2 g and is 25–35 mm long. The grown-up larvae are utilized as human sustenance in certain countries of the world. The pupa is a free-living, 12–18 mm long and of velvety white color. Mealworms begin to lay eggs 4–17 days after intercourse. The live mealworm is made out of 20% protein, 13% fat, 2% fiber, and 62% moisture, while the dried mealworm is made out of 53% protein, 28% fat, 6% fiber, and 5% moisture. Mealworms are anything but difficult to breed and encourage, and have a profitable protein profile. Consequently, they are delivered modernly as feed for pets and zoo animals, including winged animals, reptiles, little well evolved animals, batrachians, and fish. They are normally encouraged live, yet they are likewise sold canned, dried, or in powder structure.

Keywords Mealworm (*Tenebrio molitor*) · Nutrient composition · Nutritional value · Uses

20.1 Introduction

The mealworm (*Tenebrio molitor* L.) is probably the greatest insect (ca. 15 mm long) invading put-away stored food items (Fig. 20.1). It searches on plant items and makes harm their complete mass and nutritive worth. The mealworm eats put-away sustenance yet additionally taints it with exuviates, waste products, and dead insects. *Tenebrio molitor* lays ovoid, prolonged eggs, secured with some sticky substance with which it joins eggs to the substrate. Little (around 3 mm long), whitish larvae hatch from the mealworm's eggs. Following a couple of days, they turn yellowish and produce a hard, chitinous exoskeleton. Grown-up larva weighs about 0.2 g and

A. A. Mariod (✉)
Indigenous Knowledge and Heritage Centre, Ghibaish College of Science and Technology, Ghibaish, Sudan

© Springer Nature Switzerland AG 2020
A. Adam Mariod (ed.), *African Edible Insects As Alternative Source of Food, Oil, Protein and Bioactive Components*, https://doi.org/10.1007/978-3-030-32952-5_20

Fig. 20.1 The adult mealworm (*Tenebrio molitor* L.)

Fig. 20.2 Mealworm (*Tenebrio molitor*) larva

is 25–35 mm long. Furthermore, larvae are utilized as human sustenance in certain countries of the world. The pupa is a free-living animal, 12–18 mm long and of velvety white color. Mealworms begin to lay eggs 4–17 days after sexual intercourse. A solitary female may lay up to 500 eggs. The ideal brooding temperature is 25–27 °C, at which the embryonic advancement keeps going 4–6 days (Siemianowska et al. 2013; Mariod et al. (2017), Ghaly and Alkoaik 2009).

The yellow mealworm (*Tenebrio molitor* L., Coleoptera: Tenebrionidae) is an eatable insect, disseminated worldwide and a reasonable hotspot for industrial scale production (Van Huis et al. 2013). Consumable insects are recommended as an increasingly economical wellspring of animal protein. Mealworm bugs are indigenous to Europe and are presently disseminated around the world. Mealworms experience four life stages: egg, hatching, pupa, and grown-up. Larvae commonly measure about 2.5 cm (Fig. 20.2) or more, though grown-ups are by and large somewhere in the range of 1.25 and 1.8 cm long (Bryning et al. 2005).

20.2 Chemical Composition, Nutritional Value, and Different Uses

The composition of live mealworm is made out of 20% protein, 13% fat, 2% fiber, and 62% moisture, while the dried mealworm is made out of 53% protein, 28% fat, 6% fiber, and 5% moisture. The mealworm lipid has an acid value estimation of 7.6 (mg KOH/g) and iodine estimation of 96 (gI/100 g), saponification estimation of 162 (mg KOH/g) with low peroxide estimation of 0.27 (meq/kg) (Zheng et al. 2013). Fresh larva contained 56% of water, 18% of complete protein, 22% of absolute fat, and 1.55% of ash. High substance of minerals were found in the larvae: magnesium (87.5 mg/100 g), zinc (4.2 mg/100 g), iron (3.8 mg/100 g), copper (0.78 mg/100 g), and manganese (0.44 mg/100 g). The extent of $n - 6/n - 3$ unsaturated fats was profitable and added up to 6.76. Larvae powder contained twice higher substance of protein, fat, ash remains and minerals. Larva of mealworm is a significant source of supplements in sums more beneficial for human life form than conventional meat food. The powdered mealworm larvae is a high-grade item to be connected as an enhancement to traditional dishes (Siemianowska et al. 2013). Lysine was the most plentiful basic amino acid in the insect larval suppers (*Tenebrio molitor*), though glutamic acid was the most plenteous dispensable one. As in *Tenebrio molitor* supper, glutamic insect larval supper was a decent wellspring of methionine and threonine. The *Tenebrio molitor* meal demonstrated higher lysine, methionine, and threonine substance (De Marcoa et al. 2015). Mealworms have high measures of unsaturated fats, predominantly linoleic and oleic acids and the immersed unsaturated fatty acid is palmitic acid (Lenaerts et al. 2018; Paul et al. 2017). The gross energy of the mealworm was 206 kcal/100 g new weight (recalculated from dry issue), contingent upon the insect's diet. The scope of protein substance of mealworm was 14–25 g/100 g fresh weight). Vitamin B12 occurs just in foods of animal origin and is found in mealworm larvae as 0.47 μg per 100 g (Finke 2002).

Siemianowska et al. (2013) detailed that the new larvae of *Tenebrio molitor* contained 56% of water, 18% of all out protein, 22% of all out fat and 1.55% of ash. High substance of minerals were found in the larvae: magnesium (87.5 mg/100 g), zinc (4.2 mg/100 g), iron (3.8 mg/100 g), copper (0.78 mg/100 g), and manganese (0.44 mg/100 g). The extent of $n - 6/n - 3$ unsaturated fats was beneficial and added up to 6.76.

Mealworm larvae are monetarily utilized for animal feed and in certain nations, for example, Asia and Africa, for human nourishment in light of their high fat, protein and mineral substance (Rumpold and Schlüter 2013). In the Netherlands and China, they are utilized as food ingredients and progressively are viewed as an option in contrast to meat (Jeon et al. 2016). Protein is a noteworthy segment of palatable insects, containing among 30% and 65% of the complete dry issue (Rumpold and Schlüter 2013; Van Huis et al. 2013). For enhancing food or feed items with insect-based ingredients, broad learning of protein properties is required.

Customarily, the fat (e.g., oil) of some insect species is utilized widely for browning meat and other sustenance items (Van Huis et al. 2013). Glutamate dehydrogenase was sanitized 4300-overlap from the yellow mealworm fat body by gel filtration. The filtered enzyme had a maximal explicit activity of 2.1 µkat/mg protein when measured toward glutamate synthesis with NADH as a coenzyme (Teller 1988).

Mealworms are anything but difficult to breed and feed, and have an important protein profile. Hence, they are created modernly as feed for pets and zoo creatures, including birds, reptiles, little well evolved animals, batrachians and fish. They are normally sustained live, yet they are additionally sold canned, dried, or in powder structure (Veldkamp et al. 2012). Mealworms are regularly fed live; however, canned and dried larvae are industrially accessible. In feeding trials, larvae have been dried at 50 °C for 24 h, dried at 100 °C for 200 min; dried in the sun for 2 days or boiled in water for 3 min and afterward oven-dried at 60–100 °C (Aguilar-Miranda et al. 2002).

20.3   Processing of Mealworms (*Tenebrio molitor*)

Kröncke et al. (2018) explored the impact of various drying advancements and procedure parameters on the stability of nutrients and the conceivable event of mealworms lipid oxidation. These authors planned to get a capacity stable item, without decreasing the quality by excessive stress during drying, and to locate an option in contrast to customary freeze drying. Drying strategies changed nutritional components. Microwave drying, fluidized bed drying and drying with vacuum decreased the protein solvency. Freeze-dried mealworms showed highest oxidation than the other dried mealworms. Kröncke et al. (2018) study demonstrated that drying with a vacuum oven and microwave drying can be an option in contrast to traditional freeze drying for mealworms (Kröncke et al. 2018).

Lenaerts et al. (2018) examined whether microwave drying could be a proper alternative to drying microwave drying for mealworms. Two drying techniques were used to investigate the proximate composition, vitamin B12 content, unsaturated fat profile, oxidation status, and color parameters of mealworms. Besides, the influence of the use of vacuum during microwave drying was examined. The different drying advancements brought about little differences in the proximate composition, while the vitamin B12 substance was just diminished by microwave drying. The fat part of freeze-dried mealworms demonstrated a higher oxidation status than the fat of microwave dried mealworms. Utilization of a vacuum during the microwave drying procedure did not appear to offer favorable circumstances (Lenaerts et al. 2018). Freeze drying is the commercial approach to dry mealworms on an industrial scale to balance them out after harvesting. It ensures a decent item quality because of the evacuation of oxygen and low preparing temperatures (EFSA 2015). This system depends on sublimation and is dominatingly utilized for excellent items where it is essential to protect surface, nutritive value, fragrance, or color, yet it is likewise a vitality requesting and costly strategy in view of long processing occasions joined with the utilization of vacuum and low temperatures (Khalloufi et al. 2000; Sandulachi and Tatarov 2012).

Siemianowska et al. (2013) decided the healthy benefit of larvae of mealworm (*Tenebrio molitor* L.). They utilized a 3-month-old mealworm larva 25–30 mm long. Larvae were boiled for 3 min and next dried in 60 °C. Contents of water, ash, minerals, protein, fat, and fatty acid profile have been studied. Larvae powder contained twice higher substance of protein, fat, ash, and minerals. Larva of mealworm is an important source of nutrients in amounts more gainful for human than customary meat food. Powdered larva is a high-grade item to be considered as an enhancement to conventional meals.

References

Aguilar-Miranda ED, Lopez MG, Escamilla-Santana C, De La Rosa BAP (2002) Characteristics of maize flour tortilla supplemented with ground *Tenebrio molitor* larvae. J Agric Food Chem 50:192–195. https://doi.org/10.1021/jf010691y

Bryning GP, Chambers J, Wakefield ME (2005) Identification of a sex pheromone from male yellow mealworm beetles, *Tenebrio molitor*. J Chem Ecol 31:2721–2730

De Marcoa M, Martínez S, Hernandez F, Madrid J, Gai F, Rotolo L, Belforti M, Bergeroa D, Katz H, Dabboud S, Kovitvadhi A, Zoccarato I, Gascoc L, Schiavone A (2015) Nutritional value of two insect larval meals (*Tenebrio molitor* and *Hermetia illucens*) for broiler chickens: apparent nutrient digestibility, apparent ileal amino acid digestibility and apparent metabolizable energy. Anim Feed Sci Technol 209:211–218

EFSA Scientific Committee (2015) Risk profile related to production and consumption of insects as food and feed. EFSA J 13:5–60

Finke MD (2002) Complete nutrient composition of commercially raised invertebrates used as food for insectivores. Zoo Biol 21:269–285

Ghaly AE, Alkoaik FN (2009) The yellow mealworm as a novel source of protein. Am J Agric Biol Sci 4:319–331

Jeon Y-H, Son Y-J, Kim S-H, Yun E-Y, Kang H-J, Hwang I-K (2016) Physiochemical properties and oxidative stabilities of mealworm (*Tenebrio molitor*) oils under different roasting conditions. Food Sci Biotechnol 25:105–110

Khalloufi S, Giasson J, Ratti C (2000) Water activity of freeze dried mushrooms and berries. Can Agric Eng 42:51–56

Kröncke N, Böschen V, Woyzichovski J et al (2018) Comparison of suitable drying processes for mealworms (*Tenebrio molitor*). Innov Food Sci Emerg Technol 50:20–25

Lenaerts S, Van der Borght M, Callens A, Van Campenhout L (2018) Suitability of microwave drying for mealworms (*Tenebrio molitor*) as alternative to freeze drying: impact on nutritional quality and colour. Food Chem 254:129–136

Mariod AA, Saeed Mirghani ME, Hussein I (2017) Chapter 44-Schistocerca gregaria (DesertLocust) and Locusta migratoria (Migratory Locust). In: Unconventional oilseeds and oil sources. Academic, Cambridge, MA, USA, pp 293–297. ISBN 978-0-12-809435-8

Paul A, Frederich M, Megido RC, Alabi T, Malik P, Uyttenbroeck R, Francis F, Blecker C, Haubruge E, Lognay G, Danthine S (2017) Insect fatty acids: a comparison of lipids from three Orthopterans and *Tenebrio molitor* L. larvae. J Asia Pac Entomol 20:337–340

Rumpold BA, Schlüter OK (2013) Potential and challenges of insects as an innovative source for food and feed production. Innov Food Sci Emerg Technol 17:1–11

Sandulachi EI, Tatarov PG (2012) Water activity concept and its role in strawberries food. Chem. J. Mold 7:103–115

Siemianowska E, Kosewska A, Aljewicz M, Skibniewska KA, Polak-Juszczak L, Jarocki A, Jędras M (2013) Larvae of mealworm (*Tenebrio molitor L.*) as European novel food. Agric Sci 4(6):287–291

Teller JK (1988) Purification and some properties of glutamate dehydrogenase from the mealworm fat body. Insect Biochem 18(1):101–106

Van Huis A, Van Itterbeeck J, Klunder H, Mertens E, Halloran A, Muir G, Vantomme P (2013) Edible insects: future prospects for food and feed security. FAO Forestry Paper, Rome, p 171

Veldkamp T, van Duinkerken G, van Huis A, Lakemond CMM, Ottevanger E, Bosch G, van Boekel MAJS (2012) Insects as a sustainable feed ingredient in pig and poultry diets—a feasibility study. Rapport 638. Wageningen Livestock Research, Wageningen

Zheng L, Hou Y, Li W, Yang S, Li Q, Yu Z (2013) Exploring the potential of grease from yellow mealworm beetle (*Tenebrio molitor*) as a novel biodiesel feedstock. Appl Energy 101:618–621

Chapter 21
Nutrient Composition of Termites

Oladejo Thomas Adepoju

Abstract Edible insects are generally abundant, nutrient-dense, and economically valuable. They constitute an important part of the daily diet of a large population worldwide either as a snack or as a meal. Commonly consumed insects include winged adult termites (*Macrotermes bellicosus/Macrotermes notalensis*), adult crickets (*Brachytrypes* spp.), adult short-horned grasshoppers (*Cytacanthacris aeruginosus unicolor*), scarab beetle larvae (*Oryctes boas*), and larvae of butterfly and moth (*Anaphe* spp.). Research findings have revealed that insects often contain more protein, fat, and carbohydrates than equal amounts of beef or fish, and a higher energy value than soybeans, maize, beef, fish, lentils, or other beans. The quantity and quality of proteins, lipids, vitamins, minerals, and calories in edible insect caterpillars are comparable to those of beef, fish, lamb, pork, chicken, milk, and eggs; and they are many times higher in protein and fat than the plants upon which they feed.

Termites, classified as social insects with colonies, are widely consumed in many parts of the world, especially Africa. They are usually classified as *Isoptera* and are mainly found in the tropical areas and subtropical climates in large and diverse group consisting of over 2600 species worldwide. Termite utilization was recorded in 29 countries across three continents, with Africa being the continent with the highest number of records, followed by the USA and Asia. Termites have been used as food in sub-Saharan Africa, Asia, Australia, and Latin America. They are relished as a delicacy as part of a meal or simply eaten as a snack. Dried termites have been used by some mothers as sprinkle in baby porridge, and also converted into other unrecognizable forms in muffins and crackers. Termites can provide food security in many developing and developed countries of the world, as they contain essential nutrients, which are often lacking in the diets of many people in these countries.

Keywords Edible insects · Termites · Adult winged termites · Termite consumption · Termite · Mounds

O. T. Adepoju (✉)
Department of Human Nutrition, University of Ibadan, Ibadan, Nigeria

© Springer Nature Switzerland AG 2020 281
A. Adam Mariod (ed.), *African Edible Insects As Alternative Source of Food, Oil, Protein and Bioactive Components*, https://doi.org/10.1007/978-3-030-32952-5_21

21.1   Introduction

There is increase in the utilization of insects as a sustainable and secure source of animal-based food for the human diet over the years (Dossey 2010). Several ethnic groups in about one hundred and thirty (130) countries of the world make use of insects as an essential component of their diet. Also, there is an increasing interest in insect-based food products in the United States of America (USA) in recent years (Dossey et al. 2016).

Termites, as social insects with colonies, are divided into "castes" which include the workers, soldiers, winged reproductives, a queen, and a king. They are usually classified as *Isoptera* and are popularly but erroneously called white ants. They are mainly found in the tropical areas (Umeh 2000) and subtropical climates, in a large and diverse group consisting of over 2600 species worldwide. Africa is the richest continent in termite diversity with over 660 species (Eggleton 2000). The utilization of termites was recorded in 29 countries across three continents, with Africa being the continent with the highest number of records of nineteen countries, and the USA and Asia having five countries each.

Termite availability is seasonal, especially during the rainy or wet seasons between the months of April and July, and October to December (Ayieko et al. 2010a). They are the most widely accepted insects as food in Nigeria (Fasoranti and Ajiboye 1993). They are source of food with high economic and social importance, being readily accessible to the rural poor; and important source of food that may contribute to improvement in dietary intake of humans, particularly the people who suffer from protein malnutrition (Jongema 2014).

21.2   Economic Importance of Termites

Termites have profound economic importance both on the ecosystem and human activities. They are important in the tropics, where they are considered a nonconventional food with great economic and social importance (Van Huis and Vantomme 2014). They are serious pests of wood and wood products (Horn 1989). They feed on dead plant materials such as wood, leaf litter, or soil; and about 10% of the 4000 odd species (about 2600 taxonomically known) have been identified as economically important as pests that can cause serious structural damage to buildings, crops or plantation forests. Most of the species that damage crops, trees, and rangeland belong to the family Termitidae, which consists of four subfamilies: *Macrotermitinae*, *Nasutitermitinae*, *Termitinae*, and *Apicotermitinae*. Over 90% of the termite damage in agriculture, forestry, and urban settings are attributable to members of the *Macrotermitinae* (Mitchell 2002), which build large mounds (Fig. 21.1) that form spectacular feature of the African landscape (Malaisse 1978).

Termites are known to feed on materials that are rich in lignin and carbohydrate, especially cellulose (Waterhouse 1991). Their diet range from living vegetation such as trees, grasses, and roots to dead wood, decaying or rotten plant materials,

Fig. 21.1 Various termite mounds as recognized by Farmers. (**a**) Big termite mounds. (**b**) Small termite mounds. (**c**) Wood. (**d**) Cap-shaped termite mounds. (Source: Loko et al. (2017) *in Farmers' perception of termites in agriculture production and their indigenous utilization in Northwest Benin. Journal of Ethnobiology and Ethnomedicine*)

while some species feed on dried animal dung, humus, or soil rich in organic matter (Braitiiwaite et al. 1988; Waterhouse 1991). They can also fix nitrogen by gut symbionts (Schaefer and Whitford 1981), and mix soil matter, thereby increasing the soil fertility (Umeh 2003). However, the affinity of termites for cellulose materials makes them deleterious to many plant species, including timber, and agricultural products such as maize, sorghum, yam, rice, peanut, Bambara groundnut, and millet. Lands containing termite mounds are reported to be fertile, and studies revealed that termite mounds are used as fertilizer by some farmers in Ethiopia, Uganda, Zambia, Zimbabwe, Tanzania, Niger, and Sierra Leone because the soil of mounds is usually rich in minerals like calcium, magnesium, potassium, sodium, and available phosphorus.

Soil from termite mounds is also used for granary construction, because the treated termite mound clay silos demonstrate great potential for reducing temperature fluctuations and maintaining stored grain quality. The mounds are also used as medicines for the treatment of various human diseases (Ekpo et al. 2009). The insects are commonly used in entomotherapeutic practices and traditional popular medicine (Raubenheimer and Rothman 2012), and in the treatment of various diseases such as influenza, asthma, bronchitis, whooping cough, sinusitis, tonsillitis, and hoarseness (Alves and Alves 2010).

21.3 Consumption of Termites as Food

Termites have been used as food in sub-Saharan Africa, Asia, Australia, and Latin America. Consumption of edible insects has been reported in Nigeria (Adepoju and Omotayo 2014), among which the termites had the highest mean frequency (Cloutier 2015). Although termite harvest begins with the onset of the rains and the swarming of the winged termites, studies have shown that some termites could be induced to emerge even during the dry seasons, making them available throughout the year. Over ten (10) different insect species are consumed by the people of Benue State in Nigeria, with *M. natalensis* having the highest mean frequency of consumption (Agbidye et al. 2009).

Termites are relished as a delicacy by many households in Western Kenya. They are consumed as part of a meal or as a complete meal with tapioca, bread, roasted corn, or simply eaten as a snack. Dried termites are usually grinded into powder by some mothers for use as a sprinkle in baby porridge (Bergeron et al. 1988). They are also eaten raw directly after collection from the emergence hole (Ayieko et al. 2010b). In Nigeria, termites are usually fried or roasted and are usually consumed as a snack. The insect has also been converted into other unrecognizable forms in muffins and crackers (Ayieko et al. 2010b). Different kinds of edible insects are listed in Table 21.1 below.

Table 21.1 Termite species used as food or feed

Species	Use Feed	Use Food	Country(ies)
Hodotermitidae			
Hodotermes mossambicus		X	Botswana
Microhodotermes viator		X	South Africa
Kalotermitidae			
Kalotermes flavicollis	X	X	Brazil, Thailand
Rhinotermitidae			
Coptotermes formosanus		X	China
Reticulitermes flavipes		X	Thailand
Reticulitermes tibialis		X	Mexico
Termitidae			
Cubitermes atrox		X	Indonesia
Labiotermes labralis		X	Columbia
Macrotermes acrocephalus		X	China
Macrotermes annandalei		X	China
Macrotermes barneyi		X	China
Macrotermes bellicosus	X	X	Central African Republic, Congo, Democratic Republic of the Congo, Nigeria, Angola, Zambia, Kenya, Guinea, Senegal, Tanzania, Uganda
Macrotermes falciger		X	Zimbabwe, South Africa

(continued)

Table 21.1 (continued)

Species	Use		Country(ies)
	Feed	Food	
Macrotermes gabonensis		X	Congo
Macrotermes herus	X		Tanzania
Macrotermes lilljeborgi	X		Cameroon, Guinea
Macrotermes michaelseni	X		Malawi
Macrotermes muelleri		X	Congo, Cameroon, Guinea
Macrotermes natalensis		X	Central African Republic, Zimbabwe, Congo, Nigeria
Macrotermes subhyalinus	X	X	Angola, Zambia, Kenya
Macrotermes vitrialatus		X	Zambia
Microcerotermes dubius		X	Malaysia
Nasutitermes ephratae		X	Venezuela
Odontotermes badius		X	South Africa, Zambia
Odontotermes capensis		X	South Africa
Odontotermes kibarensis		X	Uganda

Source: de Figueirêdo et al. (2015)

21.4   Harvesting and Processing of Termites

Winged reproductive termites are usually collected in some parts of Africa, especially in Zambia, the Central African Republic, Angola, the Democratic Republic (DR) of Congo (Malaise 2005), and Nigeria at the time of nuptial flights, when adults emerge in large numbers from the termitaries subsequent to the maiden rains. Their harvesting is mostly seasonal. In Ghana and Nigeria, *Macrotermes bellicosus* is available for harvesting only in June and July (Anankware et al. 2016),in the rainy months between March and May in East Cameroon and Kenya (Muafor et al. 2014), as well as during the long raining season from September to December in Kenya.

The most popular and easy method of collecting termite in Africa is by placing a basin of water right under a light source in the evening during their swarming period. As light is reflected on the water, they are attracted and trapped on the water surface (Chung 2008). In most western and eastern African countries, termites are collected during the rainy seasons as they emerge from holes on the ground. In the DR of Congo, a basket is put upside down over an emergence hole of the mound. Alternatively, a dome-shaped framework of sticks is built up, or elephant grass is covered with banana leaves or a blanket, to cover part of the emergence hole near which a receptacle is placed to collect flying termites. Continuous beating and drumming on the ground around the hill trigger certain termite species to emerge to extract soldiers from the mound. Women and children push grass blades or parts of tree pods or the bark into the shafts of a termite mound or prepare smoke from charcoal from certain trees and blow it into the opening. Soldiers stripped into a container are then collected. Sometimes nests are dug up to collect queens (Van-Huis 2003).

Processing and preparation of the insect mostly include dewinging followed by frying or toasting in its own oil, with salt occasionally added (Kinyuru et al. 2010, 2013). In most parts of Nigeria, termites are prepared by washing, salting to taste, with mild frying in its own fat, or roasted and eaten as snack or sold in open market. They can also be killed by drowning, boiling or roasting for a few minutes, followed by sun-drying (Silow 1983). In Cote d'Ivoire, termites prepared for human consumption are either dried or fresh and prepared through grilling, baking, frying, seasoning, or roasting (Niaba et al. 2012), and then sold in the open market. Termites have a nutty flavor when prepared and its delicious taste makes it a good meal for all groups.

21.5   Winged Termite

There are many species of edible winged termites such as *Macrotermes nigeriensis*, *Macrotermes notalensis*, *Macrotermes subhyalinus*, and *Macrotermes bellicosus* which are harvested during the swarming season (Mbah and Elekima 2007). *Macrotermes nigeriensis* is an eusocial insect of the order *Isoptera* and family *Termitidae*. The commonly available and consumed among the termites are the reproductive individuals, which are also called alates. They are the fully developed adult stage of the termites.

Fair variation exists among the winged adults, probably due to geographic locations or developmental stages. However, they are all simply termed "termites," "winged termites," or "Macrotermes species." Locally, they are known as *Aku*, *Khinyea*, and *Esusun* by the Ibo, Hausa, and Yoruba speaking tribes of Nigeria, respectively (Ntukuyoh et al. 2012). Winged termites (*Macrotermes* species) often emerge with the first rains at the ends of the dry season, and are usually differentiated by the morphological features, and the time and season when they emerge. The winged reproductive adult termites are available during the onset of rainy season, during which they fly off in large number from their nest for their "nuptial flight," which ends with a pair of male and female alates isolating themselves from others. Then, their wings fall off and they retire to a suitable spot where they establish a new colony where they become the potential king and queen of the new colony (Allotey and Mpuchane 2003).

During the flight which occurs mainly at early hours of the night, the alates are harvested by placing a bowl of water under sources of illumination to which they are attracted. The alates trapped in the water are washed and roasted with or without salt. Once roasted, the wings are removed by sifting the roasted insects or by rubbing them between the palms of the hands. The finished product is either consumed at home by the members of the family, especially the young ones; or sold in markets in the Western and Eastern parts of Nigeria as a snack (Banjo et al. 2006). However, some consumers prefer them raw. Roasted termites can also be sun-dried for future use. Figure 21.2(a) shows the picture of winged termites in their nuptial flight while Fig. 21.2(b) shows the picture of roasted winged termites.

Fig. 21.2 (**a**) Winged termites (*Macrotermes bellicosus*). (**b**) Roasted winged termites (*Macrotermes bellicosus*). (Source: Adepoju and Ajayi (2016))

21.6   Nutritional Importance of Termites

The nutritional benefit of insects varies as a result of the wide range of edible species available. It has been shown that nutritional and medicinal values could also differ within the same species, based on the metamorphic stage of the insect, the habitat, and their diet. They are a readily available source of protein, lipids, certain essential vitamins and minerals. The protein content in insects is high, equivalent to that of fish and meat which is similar to the one found in a human body, making it easier to be utilized by the body compared to plant protein (Bukkens 1997). The energy content of insects is on average compared to that of meat (on a fresh weight basis) except for pork because of its particularly high-fat content (Sirimungkararat et al. 2010).

Dried winged termites have been reported to be a good source of dietary protein, fat, and micronutrients (Banjo et al. 2006; Adepoju and Omotayo 2014). Adepoju and Omotayo (2014) reported termites to be low in anti-nutrients and suggested its possible inclusion in formulating adequate, nutrient-dense complementary foods with nutraceutical benefits.

Termites have been reported to be rich in proteins, fats, vitamins, and many essential minerals, and thus provide food security for poor households. They constitute a food source of great nutritional value with high protein and essential amino acids, especially tryptophan, which is generally limited in the food insects (De Foliart 2005). They are also rich in minerals and essential fatty acids (Booth 1998). Termites are good source of heme iron, and their inclusion in the daily diet could improve iron status and help prevent anemia in developing countries.

Termites' central role as edible insects in Africa owes this prestigious status to its rich fat (44.82–47.31 g/100 g) and protein contents (33.51–39.74 g/100 g) (Bukkens 1997; Ekpo et al. 2009; Igwe et al. 2011; Kinyuru et al. 2013; Adepoju and Omotayo 2014). *Macrotermes bellicosus* alates have been reported to contain 31.8 g crude protein, 16.4 g crude fat, 3.8 g ash, 43.0 g carbohydrates, 361.13 mg potassium, 98.4 mg sodium, 227.5 mg calcium, 24.33 mg magnesium, 361.3 mg phosphorus, 2.07 mg iron, 15.03 mg zinc and produced 450.7 kcal of energy/100 g of roasted insects (Adepoju and Ajayi 2016). They have been found to be good sources of vitamins, with contents of 2.89 μg/100 g for vitamin A; 0.12–1.98 mg vitamin B2 and 2.83–3.41 mg vitamin C/100 g sample (Banjo et al. 2006; Adepoju and Ajayi 2016). Mbah and Elekima (2007) found alates to contain 21 mg calcium, 1.36 mg phosphorus, 27 mg iron, and 0.15 mg magnesium /100 g of termites. Banjo et al. (2006) also found that *M. bellicosus* contains 43.3% carbohydrate, while Mbah and Elekima (2007) reported the oil content of the same termite species as 28.37%.

Igwe et al. (2011) determined the proximate, minerals, vitamins and fatty acids composition of termites and reported 20.94% proteins, 20.74% carbohydrates, 34.23% lipids, 3360.00 mg potassium, 1120.00 mg sodium, 9.56 mg iron, and 0.97 mg zinc/ kg sample; 17.76 mg ascorbic acid, 2.74 mg niacin, and 1.56 mg riboflavin/100 g sample of insect. Identified fatty acids were oleic acid (52.45%), palmitic acid (31.39%), and linoleic acid (7.57%).

Studies have reported the use of different species of termites as a source of enrichment of both human food and animal feeds. The use of *Macrotermes bellicosus* in enriching complementary foods formulated with maize and sorghum resulted in significant reduction in moisture content with significant increase in crude protein, fat, ash, total carbohydrates, and mineral and gross energy content of complementary foods (Banjo et al. 2006). Winged reproductive *Macrotermes nigeriensis*, naturally rich in oil, and has a higher value of essential amino acids such as lysine, methionine, and histidine than soybean meal has been used with soybean (*Glycine max*) meals blend as solitary source of protein in the diet of *Heterobranches bidorsalis* fingerlings. This resulted in marked increases in body protein and lipid levels of the fingerlings as the inclusion level of *Macrotermes nigeriensis* increased (Solomon et al. 2007).

The insects have also been converted into other unrecognizable forms in muffins and crackers (Ayieko et al. 2010b).Termites can therefore provide food security in many poor African countries as they contain essential nutrients, which are often lacking in the diets of people in those countries (Ayieko 2007).

21.7 Medicinal Value of Termites

Termite and its products form an important part of traditional medicine. In several traditional African medicinal systems, the paste of termite and mound is boiled and topically applied to prevent infection of external wounds and ingested to treat internal hemorrhages (Hoare 2007). Also, termites can be used as medical device to insert a drug substance subcutaneously, by injecting it under the skin through biting into the area of the patient's skin coated with the substance (de Figueirêdo et al. 2015). Some termite species are used as an alternative treatment for physiological and spiritual problems. For instance, *M. nigeriensis* is used in Nigeria in the treatment of wounds and sickness of pregnant women and as a charm for spiritual protection problems (de Figueirêdo et al. 2015). Also, *M. bellicosus* is used in Somalia to suture wounds.

References

Adepoju OT, Ajayi K (2016) Assessment of quality and safety of winged termites (*Macrotermes bellicosus*) enriched locally formulated complementary foods. J Food Res 6(5):117–130

Adepoju OT, Omotayo OA (2014) Nutrient composition and potential contribution of winged termite (*Marcrotermes bellicosus* Smeathman) to micronutrient intake of consumers in Nigeria. British J App Sci Technol 4(7):1149–1158

Agbidye FS, Ofuya TI, Akindele SO (2009) Some edible insect species consumed by the people of Benue State, Nigeria. Pak J Nutr 8(7):946–950

Allotey J, Mpuchane SF (2003) Utilization of useful insects as food source. Afr J Food Agric Nutr Dev 2:160–168

Alves RRN, Alves HN (2010) The faunal drugstore: Animal-based remedies used in traditional medicines in Latin America. J Ethnobiol Ethnomed 7:1–43

Anankware J, Osekre E, Obeng-Ofori D, Khamala C (2016) Identification and classification of common edible insects in Ghana. Int J Entomol Res 1(5):33–39

Ayieko M, Oriaro V, Nyambuga I (2010a) Processed products of termites and lake flies: improving entomophagy for food security within the Lake Victoria Region. Afri J Food Agric Nut Dev 10:2085–2098

Ayieko MA (2007) Nutritional value of selected species of reproductive Isoptera and Ephemeroptera within the ASAL of Lake Victoria basin. Discov Innov 19(2):126–130

Ayieko MA, Ndong'a FO, Tamale A (2010b) Climate change and the abundance of edible insects in the Lake Victoria Region. J Cell Anim Biol 7:112–118

Banjo AD, Lawal OA, Songonuga EA (2006) The nutritional value of fourteen species of edible insects in southwestern Nigeria. Afr. J. Biotechnol 5:298–301

Bergeron D, Rodney JB, Franklin LR, Irv K, John O, Alfred AB (1988) The nutrient composition of an insect flour sample from Lake Victoria, *Uganda*. J Food Compos Anal 11:371–377

Booth RG (1998) A review of the species resembling *Chilocorus nigrita* (*Coleoptera: coccinellidae*): Potential agents for biological control. Bull Entomol Res 88:361–367

Braitiiwaite RW, Miller L, Wood JT (1988) The structure of termite communities of the Australian tropics. Australian J Ecol 13:375–391

Bukkens SGF (1997) The nutritional value of edible insects. Ecol Food Nutr 36:287–319

Chung AYC (2008) An overview of edible insects and entomophagy in Borneo. In Forest insects as food: humans bite. FAO Regional Office for Asia and the Pacific, Bangkok, pp 1–9

Cloutier J (2015) Edible insects in Africa: an introduction to finding, using and eating insects, 1st edn. Agromisa Foundation and CTA, WageningenDigigrafi, Veenendaal, Netherlands. ISBN Agromisa: 978-90-8573-146-7, ISBN CTA: 978-92-9081-577-8

de Figueirêdo RECR, Alexandre V, Iamara SP, Alves RRN (2015) Edible and medicinal termites: a global overview. J Ethnobiol Ethnomed 11(29):1–7

De Foliart GR (2005) Overview of role of edible insects in preserving biodiversity. In: Paoletti MG (ed) Ecological implications of minilivestock. Enfield NH Science Pubpp, New Hampshire, pp 123–140

Dossey A, Morales-Ramos J, Guadalupe R (2016) Insects as sustainable food ingredients: production, processing and food applications, 1st edn. Academic Press, Cambridge, MA, USA

Dossey AT (2010) Insects and their chemical weaponry: new potential for drug discovery. Nat Prod Rep 27(12):1737–1757

Eggleton P (2000) Global patterns of termite diversity. In: Abe T, Bignell DE, Higashi M (eds) Termites: evolution, sociality, symbioses, ecology. Kluwer Academic Publishers, Dordrecht, The Netherlands, pp 25–51

Ekpo KE, Onigbinde AO, Asia IO (2009) Pharmaceutical potentials of the oils of some popular insects consumed in Southern Nigeria. Afri J Pharm Pharm 3:51–57

Fasoranti JO, Ajiboye DO (1993) Some edible insects of Kwara State. Nigeria. Am Entomol 39(11):3–6

Hoare AL (2007) The use of non-timber forest products in the Congo Basin: constraints and opportunities, vol 1. The Rainforest Foundation, London, pp 1–57

Horn DJ (1989) Ecological approach to pest management, vol 285. Elsvier, London

Igwe CU, Ujowundu CO, Nwaogu LA, Okwu GN (2011) Chemical analysis of an edible African termite, *Macrotermes nigeriensis*; a potential antidote to food security problem. Biochem Anal Biochem 1:1–4

Jongema Y (2014) List of edible insect species of the world. http://www.wageningenur.nl/en/ Expertise Services/Chairgroups/Plant Sciences/Laboratory-of Entomology/Edible insects/ Worldwidespecies-list.htm

Kinyuru JN, Kenji GM, Njoroge SM, Ayieko M (2010) Effect of processing methods on the in vitro protein digestibility and vitamin content of edible winged termite (*Macrotermes subhyalinus*) and grasshopper (*Ruspolia differens*). Food Bioprocess Technol 3:778–782

Kinyuru JN, Konyole SO, Roos N, Onyango CA, Owino VO, Owuor BO, Estambale BB, Friis H, Aagaard-Hansen J, Kenji GM (2013) Nutrient composition of four species of winged termites consumed in western Kenya. J Food Compos Anal 30:120–124

Loko LE, Orobiyi A, Agre P, Dansi A, Tamò M, Roisin Y (2017) Farmers' perception of termites in agriculture production and their indigenous utilization in Northwest Benin. J Ethnobiol Ethnomed 13(64):1–12

Malaise F (2005) Human consumption of Lepidoptera, termites, Orthoptera and ants in Africa. In: Ecological implications of Minilivestock: potential of insects, rodents, frogs and snails. Scince Publishers, Inc, Hauppauge, New York, pp 175–230

Malaisse F (1978) High termitaria. In: Werger MJA (ed) Biogeography and ecology of Southern Africa. Dr. W. Junk, The Hague, pp 1279–1300

Mbah CE, Elekima GOV (2007) Nutrient composition of some terrestrial insects in Ahmadu Bello University, Samaru, Zaria Nigeria. Sci World J 2(2):17–20. https://doi.org/10.4314/swj. v2i2.51728

Mitchell JD (2002) Termites as pests of crops, forestry, rangeland and structures in southern Africa and their control. Sociobiology 40:47–69

Muafor FJ, Levang P, Le Gall P (2014) A crispy delicacy: Augosoma beetle as alternative source of protein in East Cameroon. Int J Biodiver 2014:214071, 7 pages. https://doi.org/10.1155/2014/214071

Niaba KV, Atchibri LO, Gbassi KG, Beugre AG, Adou M, Anon AB (2012) Consumption survey of edible winged termites in Cote d'ivoire. Int J Agri Food Sci 2:149–155

Ntukuyoh AI, Udiong DS, Ikpe E, Akpakpan AE (2012) Evaluation of Nutritional Value of Termites (*Macrotermes bellicosus*): soldiers, workers, and queen in the Niger Delta Region of Nigeria. Int J Food Nutr Saf 1(2):60–65

Raubenheimer D, Rothman J (2012) Nutritional ecology of entomophagy in humans and other primates. Annu Rev Entomol 58:141–160

Schaefer DA, Whitford WG (1981) Nutrient cycling by the subterranean termite *Gnathamitermes tuhiformans* in a Chihuahuan desert ecosystem. Oecologia 48:277–283

Silow CA (1983) Notes on Ngangela and Nkoyaethnozoology. Ants and termites Etnol Stud, vol 36. Göteborgs etnografiska museum, Göteborg, p 177

Sirimungkararat S, Saksirirat W, Nopparat T, Natongkham A (2010) Edible products from eri and mulberry silkworms in Thailand. In: Durst PB, Johnson DV, Leslie RN, Shono K (eds) Forest insects as food: humans bite back. FAO, Bangkok, Thailand, pp 189–200

Solomon SG, Tiamiyu LO, Sadiku SOE (2007) Wing reproductive termite (Macrotermes nigeriensis)-soybean (Glycine max) meals blend as dietary protein source in the practical diets of Heterobranchus bidorsalis fingerlings. Pak J Nutr 6:267–270

Umeh VC (2000) Advances in the control of termite pests of some tropical crops using naturally occurring pesticides. In: Paper presented at 14th Africa Association of Insect Scientists and the 9th Crop Protection society of Ethiopia joint conference, June 4–8 2001. EthiopianAgricultural Research organization, Addis Ababa

Umeh VC (2003) An appraisal of farmers' termite control methods in some traditional cropping system of West Africa. Nigerian J Entom 20:25–39

Van Huis A, Vantomme P (2014) Conference report: insects to feed the world. Food Chain 4:184–193

Van-Huis H (2003) Insects as food in sub-Saharan Africa. Insect Sci. Appl 23(3):163–185

Waterhouse DP (1991) Insects and humans in Australia. In: *The insects of Australia,* CSIRO, Division of Entomology. University Press, Melbourne, pp 221–225

Chapter 22
Termites in the Human Diet: An Investigation into Their Nutritional Profile

Sampat Ghosh, Daniel Getahun Debelo, Wonhoon Lee,
V. Benno Meyer-Rochow, Chuleui Jung, and Aman Dekebo

Abstract Many different entomophagous communities of the world consume termites particularly in time of insect's swarming. We analysed the nutritional composition of the termites that are being used as food and found that protein and fatty acid contents differed between adult and nymphal stages. All the tested amino acids satisfied the level of a nearly ideal protein pattern. Monounsaturated fat predominated among the categories of fatty acids. Calcium and iron contents were found to be relatively high and thus helpful in mitigating some widespread deficiencies

S. Ghosh
Agriculture Science and Technology Research Institute, Andong National University, Andong, Gyeongbuk, Republic of Korea

Department of Life Sciences, Sardar Patel University, Balaghat, Madhya Pradesh, India

D. G. Debelo
Program of Applied Biology, Adama Science and Technology University, Adama, Ethiopia

W. Lee
Department of Plant Medicine and Institute of Agriculture & Life Science, Gyoungsang National University, Jinju, Gyeongnam, Republic of Korea

V. B. Meyer-Rochow
Department of Plant Medicals, Andong National University, Andong, Gyeongbuk, Republic of Korea

Department of Genetics and Physiology, Oulu University, Oulu, Finland

C. Jung (✉)
Agriculture Science and Technology Research Institute, Andong National University, Andong, Gyeongbuk, Republic of Korea

Department of Plant Medicals, Andong National University, Andong, Gyeongbuk, Republic of Korea
e-mail: cjung@andong.ac.kr

A. Dekebo
Program of Applied Chemistry, Adama Science and Technology University, Adama, Ethiopia

© Springer Nature Switzerland AG 2020
A. Adam Mariod (ed.), *African Edible Insects As Alternative Source of Food, Oil, Protein and Bioactive Components*, https://doi.org/10.1007/978-3-030-32952-5_22

prevalent. In the context of rapid population increase and unanticipated climate change the preservation of traditional food is clearly of importance.

Keywords Protein · Amino acids · Fatty acids · Minerals · Iron · Calcium

22.1 Introduction

Using insects as a food source has a long tradition in human societies (Bequaert 1921; Bergier 1941; Bodenheimer 1951; Van Huis et al. 2013; Kim and Jung 2013; Nadeau et al. 2014; Ghosh et al. 2017, 2018). However, it was a paper by Meyer-Rochow (1975) that first suggested that edible insects represented a resource to combat global food shortages. Besides their various ecological roles termites in particular have received attention in tropical countries as a food source and panacea since prehistoric times. Termites have been involved in treatments "of an almost equally wide spectrum of diseases and disorders than bees" (Meyer-Rochow 2017), including malnutrition (Costa-Neto 2005). Isotope studies have demonstrated an association between termites and early human societies that found its reflection even in the ware patterns of stone tools evidencing termite collecting by humans (Sponheimer et al. 2005).

Termite consumption in southern Africa was taken note of more than 100 years ago (Livingstone 1857) and even today termites make up an important dietary component of the inhabitants, for example, of Zambia (Silow 1983 cited from van Huis 2017) and elsewhere in sub-Saharan Africa (van Huis 2017). An ethnographic account of the San people of the Kalahari also revealed their acceptance of termites as food (Nonaka 1996) and a study focusing on rural households in Limpopo (South Africa) showed that in 93% of them insects such as termites, grasshoppers and flying ants were consumed (Twine et al. 2003). Termites are also known to be accepted as a delicacy in northeastern and southern parts of India (Chakravorty et al. 2013, 2016; Wilsanand 2005) as well as numerous other places in the world (de Figueiredo et al. 2015).

As part of a survey of termite diversity in Ethiopia, it was noticed, mainly in the conversation, that the local inhabitants of the Benishangul Gumuz zone of western Ethiopia might consume the termite *Macrotermes* sp. (Debelo and Degaga 2015). An ethnozoological use of termites, mainly however for treating illnesses, has also been reported from the Metema Woreda region located nearby Benishangul Gumuz (Kendie et al. 2018). The Benishangul Gumuz region is the habitat of many different ethnic communities like Berta, Amhara, Oromo, Shinasha, and Agaw-Awi. Swarming of termites often indicates pending rainfall. The pedologic activity of termites in this region of erratic rainfall (700–1100 mm/year) significantly increases water retention in the soil, which in turn helps plants to grow even under conditions of low precipitation.

Sandy and loamy soils possess a moderate potential for the production of food, which is why almost every year the local people in order to meet their nutritional requirements gather wild produce from the surrounding areas. Variations with regard to the extent termites are consumed exist among the different communities, which is in total agreement with scientific studies from numerous other countries. Thus, it would not be an overstatement that besides being essential for maintaining structural and functional integrity in the agro-environment, termites also have a considerable potential as a food source. Regrettably, tribal people are often reluctant to tell whether or not they consume termites and other insects, because they feel "entomophagy" can make them seem backward and primitive and poses a hindrance in being accepted as "civilized" in the eyes of westerners—an observation made by Meyer-Rochow as early as 1975.

In the Volta region and the greater Accra in Ghana, 95% and 56.7% of the interviewed, respectively, reported that they consumed termites (Akutse et al. 2012). Inhabitants of Uganda out of 14 termite species consumed 10; the other four being neglected mainly because of their small size or difficulties in finding alates (Nyeko and Olubayo 2005). In general, humans seem to prefer reproductive castes, which is why in most cases winged specimens, that is, alates, are consumed. However, the latter are also easier to collect than wingless castes, although an accidental consumption of soldiers is also known (Nyeko and Olubayo 2005). Incidentally, Makiritare Amerindians of Venezuela have been reported to consume preferentially soldiers rather than alates of the termite *Syntermes* sp. (Paoletti et al. 2003). Redford and Dorea (1984) analysed nutritional value of different termite species from central part of Brazil. They pointed out that the people do not only go for the nutritional value of the specie but availability and easy access also plays crucial role. Preparation for human consumption involves mainly roasting. However, children often eat fresh termites raw while collecting them. Similarly, the raw consumption of queens was reported from Ghana by Atu (1993) and Akutse et al. (2012). People fond of eating termites adopted different methods to preserve them for longer periods, presumably to use them as food in times of non-availability. In Uganda banana leaves are used to wrap dried termites for the purpose of preservation (Van Huis 2017). Even food products made of or using ground termites are not uncommon. In Sudan, people prepare cakes made out of termite flour and in Uganda a kind of sauce known as "ekipooli" is prepared in which termites are essential ingredients (Bennett et al. 1965). The aim of our study was to examine the nutritional potential of edible termites in Ethiopia and to explore ways to improve the livelihoods of the local residents possibly with enhancing nutritional security.

22.2 Materials and Methods

22.2.1 Sample Preparation

Edible termites were collected from termite mounds in the Adama Science and Technology University campus, located in the central part of Ethiopia as well as different parts of the country. Samples were placed in a freeze-box and transferred to the laboratory. The samples were separated on the basis of developmental stage and castes like adults, nymphs, workers, soldiers, and reproductives. The samples were freeze-dried and processed for further analyses. Prior to mineral analyses the samples were defatted by Soxhlet extraction method. Following the fat extraction the rest defatted samples were analysed for mineral estimation.

22.2.2   Species Identification

Field collected specimens were further identified based on the morphological and genetic characteristics. Total DNA was extracted with a DNeasy Blood &Tissue kit (QIAGEN, Inc., Dusseldorf, Germany) following the manufacturer's protocol. Two primers, LCO1490 (5′-GGT CAA CAA ATC ATA AAG ATA TTG G-3′) and HCO2198 (5′-TAAACT TCA GGG TGA CCA AAA AAT CA-3′), for the mitochondrial *COI* gene were used (Folmer et al. 1994). The *COI* gene corresponding to the "DNA Barcode" region (Herbert et al. 2003) was amplified by the polymerase chain reaction (PCR) using AccuPower PCR PreMix (Bioneer, Daejeon, Korea). Sequencing was carried out by BIONICS (Seoul, Korea). All sequences were generated in both directions. *COI* sequences generated in this study were aligned and underwent a neighbour joining analysis based on *COI* database of the National Centre for Biotechnology Information (NCBI) (http://www.ncbi.nlm.nih.gov). A neighbour joining (NJ) tree was constructed to check the genetic relationship between haplotypes using MEGA 5.2 (Tamura et al. 2011).

22.2.3   Amino Acid Analysis

Amino acid composition was determined using a Sykam Amino Acid analyser S433 (Sykam GmbH, Eresing, Germany) following a standard routine (AOAC 1990). The ground samples were hydrolysed in 6 N HCl for 24 h at 110 °C under a nitrogen atmosphere and concentrated in a rotary evaporator. The concentrated material was reconstituted with dilution buffer provided by the manufacturer (0.12 N citrate buffer, pH 2.20). The hydrolysed samples were then used for the determination of the amino acid composition. The amino acid score was calculated based on the ideal protein pattern provided by the consultation of FAO/WHO/UNU (2007). If 1 g of protein contains equal or more than recommended amount of individual essential amino acids, the score should be 100% or higher.

22.2.4   Fatty Acid Composition Analysis

Constituent fatty acids were identified by using the gas chromatography-flame ionization detector (GC-14B, Shimadzu, Tokyo, Japan), equipped with an SP-2560 column, following the recommended method (Korean Food Standard Codex 2010). The samples were derivatised into fatty acid methyl esters (FAMEs). Identification and quantification of FAMEs was achieved by comparing peak retention times with those of pure standards purchased from Sigma (Yongin, Republic of Korea) processed under the same conditions.

22.2.5   Mineral Analyses

Minerals were analysed following standard procedures (Korean Food Standard Codex 2010). The samples, consisting of dried powder, were digested for 30 min with nitric and hydrochloric acid (1:3) at 200 °C. Each sample was then filtered using Whatman filter paper (0.45 μm) and stored in washed glass vials before analyses could commence. Mineral content was then determined with the help of an inductively coupled plasma-optical emission spectrophotometer (ICP-OES 720 series; Agilent; Santa Clara, CA, USA).

22.3   Results and Discussion

The termites consumed by the local communities in Ethiopia were identified as species of the genus *Macrotermes*. Three different species, namely, *M. falciger*, *M. herus* and *M. jeanneli*, were present based on the genetic analysis (Figs. 22.1 and 22.2).

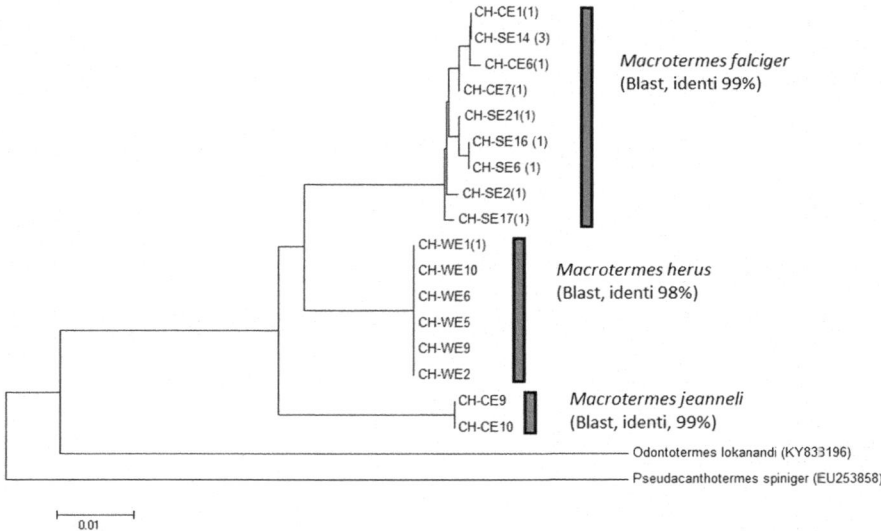

Fig. 22.1 A neighbour joining analysis of termites collected from termite mounds from the three regions of Ethiopia; CE (Central Ethiopia), SE (South Ethiopia), WE (Western Ethiopia) with other termites, *Odontotermes lokanandi* and *Pseudacanthotermes spiniger*, as outgroups. Numbers represent the sampling locations in the campus [CH does not carry any technical meaning, this is just for laboratory working purpose]

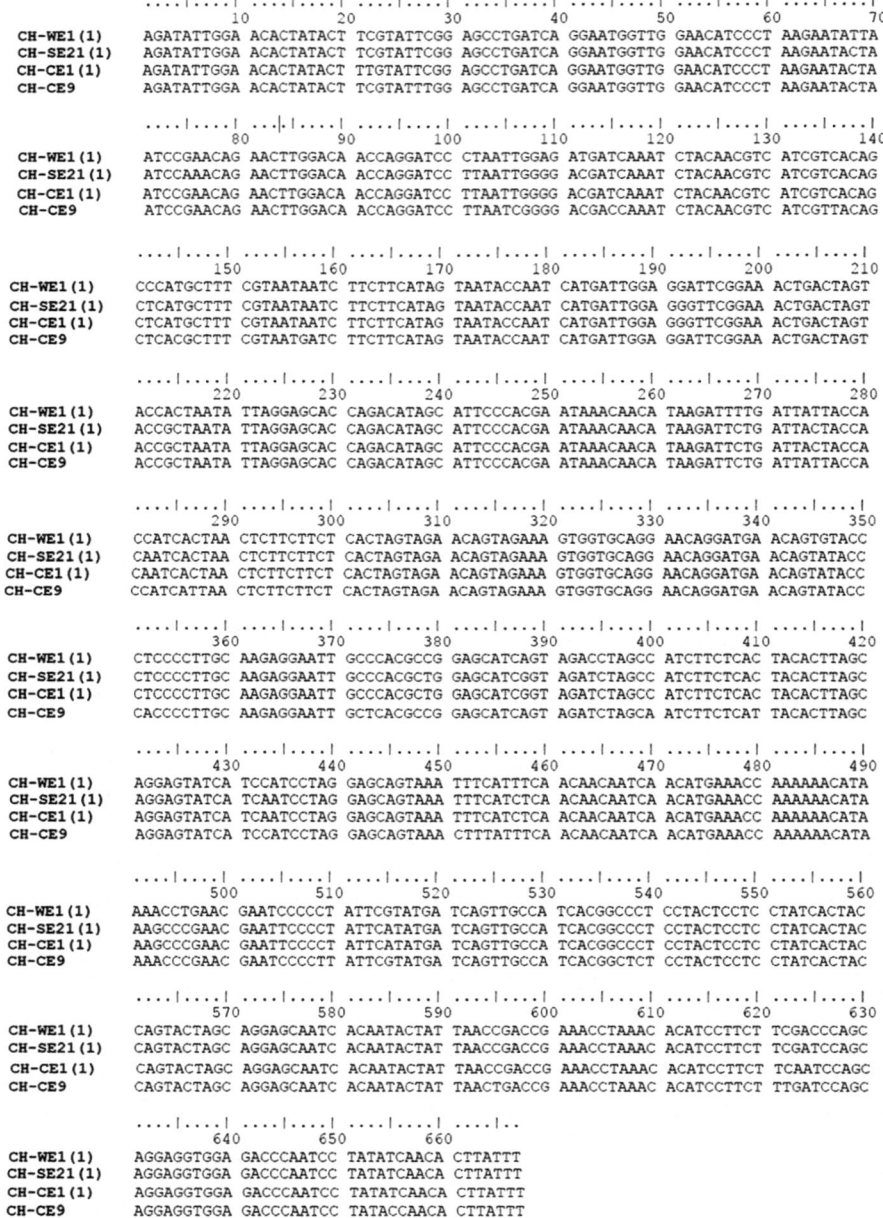

Fig. 22.2 Representative *COI* sequences of termites collected from termite mounds from the three regions of Ethiopia; CE (Central Ethiopia), SE (South Ethiopia), and WE (Western Ethiopia)

The termites collected from the central and south Ethiopia mostly belonged to *M. falciger* (99% Blast identical) with a few being *M. jeanneli* (99% Blast identical). All of the termites collected from mounds located in western Ethiopia were found to be *M. herus* (98% Blast identical). Benishangul Gumuz region is located in the western part of Ethiopia and thus the termite species consumed there would be most likely *M. herus* species.

Protein receives the most attention in order to understand the nutritional value of food, and amino acid profiles are essential for a better understanding of the quality of protein. Table 22.1 represents the amino acid content of *Macrotermes* (Table 22.1). Altogether 17 amino acids including 8 essential and 9 non-essential amino acids were determined. Tryptophan was not detected presumably because of acid digestion of the protein and methionine and cysteine were also not reliably estimated owing to the limitation of the process (Pickering and Newton 1990) and the fact that these acids appear to be present in insects in only very small quantities (Ghosh et al. 2016). The proportion of essential to non-essential amino acids was approximately 4:6. A similar result was reported for the edible termite, *Odontotermes* sp. from India (Chakravorty et al. 2016). Among the essential amino acids, leucine dominated, followed by another solely ketogenic amino acid, lysine (Table 22.1). Lysine is the limiting amino acid in cereals in particular those based on wheat, rice, cassava, and maize, which happen to be the most prevalent cereal constituents in the developing and underdeveloped regions of the world. The high level of lysine in

Table 22.1 Amino acid composition (mean ± SD) of adults and nymphs of *Macrotermes* termite (g/100 g DM and % of total amino acids)

	Adult		Nymph	
	g	%	g	%
Valine	2.32 ± 0.035	7.5	1.65 ± 0.014	7.9
Isoleucine	1.47 ± 0.014	4.8	1.02 ± 0.007	4.9
Leucine	2.54 ± 0.021	8.2	1.69 ± 0.007	8.1
Lysine	2.22 ± 0.354	7.2	1.37 ± 0.014	6.5
Threonine	1.06 ± 0.014	3.4	1.06 ± 0.049	5.1
Phenylalanine	1.56 ± 0.014	5.1	1.11 ± 0.007	5.3
Methionine	0.18 ± 0.000	0.6	0.09 ± 0.007	0.4
Histidine	1.02 ± 0.007	3.3	0.64 ± 0.000	3.1
Tyrosine	2.39 ± 0.014	7.8	1.88 ± 0.007	9.0
Arginine	1.78 ± 0.233	5.8	0.97 ± 0.071	4.6
Aspartic acid	1.94 ± 0.007	6.3	1.36 ± 0.21	6.5
Glutamic acid	3.94 ± 0.007	12.8	2.62 ± 0.014	12.5
Serine	1.67 ± 0.057	5.4	1.11 ± 0.014	5.3
Glycine	1.54 ± 0.007	5.0	0.86 ± 0.035	4.1
Alanine	2.22 ± 0.007	7.2	1.38 ± 0.042	6.6
Cysteine	0.83 ± 0.028	2.7	0.63 ± 0.000	3.0
Proline	2.15 ± 0.014	7.0	1.54 ± 0.007	7.3
Total amino acids	*30.83*		*20.98*	

the protein part of *Macrotermes* sp. could be expected to benefit human health especially in the Ethiopian region, where cassava is the most popular staple, and where *injera*, a sourdough flatbread made of teff flour is the national dish. The share of the indispensable amino acids in the total protein fraction satisfies the ideal protein pattern (Table 22.2).

The fatty acid content of *Macrotermes* termites is shown in Table 22.3. These termites possess higher MUFA regardless of developmental stages (Table 22.4), a finding in agreement with previous reports on lipid compositions of the nymphs, workers, or soldiers of several other edible termite species (Kinyuru et al. 2013; Paoletti et al. 2003; Oyarzun et al. 1996). However, the percentage of the SFAs in our nymphs (41.4%) was somewhat lower than that of 49.5% reported for *Macrotermes bellicosus* in Kenya (Kinyuru et al. 2013) and 52.9% in *Odontotermes* sp. from India (Chakravorty et al. 2016). Oleic acid (a MUFA) was the most abundant

Table 22.2 Chemical score of the essential amino acids compared with the ideal protein pattern suggested by the FAO/WHO/UNU 2007 joint consultation

Amino acids	FAO/WHO/UNU (2007) mg/g protein	Chemical score (%)	
		Adult	Nymph
Valine	39	193.1	201.5
Lysine	45	160.0	145.1
Isoleucine	30	159.0	162.0
Leucine	59	139.7	136.6
Phenylalanine + tyrosine	38	337.1	375.0
Threonine	23	149.6	219.6
Histidine	15	220.7	203.3
Methionine + cysteine	22	148.6	140.5

Table 22.3 Fatty acid profile (mg/100 g dry matter) of adults and nymphs of *Macrotermes* termites

	Adult	Nymph
Lauric acid (C12:0)	ND[a]	19.35
Myristic acid (C14:0)	ND[a]	142.78
Palmitic acid (C16:0)	58.11	7218.89
Heptadecanoic acid (C17:0)	24.64	107.32
Stearic acid (C18:0)	98.56	1767.57
Palmitoleic acid (C16:1)	ND[a]	492.46
Oleic acid (C18:1n−9) cis	397.06	10328.78
Elaidic acid (C18:1n−9) trans	4.30	ND[a]
Cis-11-Eicosenoic acid (C20:1)	ND[a]	153.37
Linoleic acid (C18:2n−6) cis	233.72	2128.30
SFA	181.31	9255.02
MUFA	401.36	10974.61
PUFA	233.72	2128.30
Total fatty acids	816.39	22357.93

[a]ND means not detected

Table 22.4 A comparative inventory of the fatty acid distribution of the termite lipids

Termite species	Condition	stage	SFA	MUFA	PUFA	Total fat	Reference
			Relative proportion (%)			g/100 g DM	
Macrotermes subhyalinus	DWB	Alate	35.1	52.8	12.2	44.82	Kinyuru et al. 2013
Pseudacanthotermes militaris	DWB	Alate	32.2	56.1	11.7	46.59	Kinyuru et al. 2013
Macrotermes bellicosus	DWB	Alate	49.5	44.6	5.9	47.03	Kinyuru et al. 2013
Pseudocanthotermes spiniger	DWB	Alate	35.8	52.9	11.3	47.31	Kinyuru et al. 2013
Odontotermes sp.	DWB	Alate	52.9	44.5	2.6	50.93	Chakravorty et al. 2016
Syntermes sp.	TA	Soldier	22.8	42.3	34.5	1.20[a]	Paoletti et al. 2003
Syntermes sp.	H	Soldier	18.6	46.1	34.9	1.36[a]	Paoletti et al. 2003
Nasutitermes sp.	WB	Soldier	33.5	36.6	29.9	11.23	Oyarzun et al. 1996
Nasutitermes sp.	WB	Worker	31.8	53.1	15.1	2.21	Oyarzun et al. 1996
Nasutitermes sp.	WB	Alate	38.6	52.9	8.6	40.23	Oyarzun et al. 1996
Macrotermes bellicosus	WB	Alate	49.0	17.9	33.1	36.12[a]	Ekpo and Onigbinde 2007
Macrotermes sp.	DWB	Adult	22.2	49.2	28.6	0.82	Present study
Macrotermes sp.	WB	Nymph	41.4	49.1	9.5	22.36	Present study

DWB Dewinged whole body, *TA* Thorax and Abdomen, *H* Head, *WB* Whole body
[a]Calculated from total fatty acids

among fatty acid of *Macrotermes* sp., a finding that holds true for most of other edible termite species with the exception of alates of *M. bellicosus* from Nigeria (Ekpo and Onigbinde 2007).

MUFAs contribute to the diet in many parts of the world. Most remarkably they represent at least one-third of the fatty acid consumption in the Mediterranean region. Olive oil, which is composed of mainly the MUFA oleic acid (70–80%) is the most widely used culinary fat in the Mediterranean region (Sales-Campos et al. 2013), a region famous for the low incidence of coronary heart disease (Keys 1970; deLorgeril and Salen 2006). From a nutritional point of view, oleic acid demonstrates a series of beneficial effects on human health. Besides its wound healing properties, research suggests that oleic acid exerts a beneficial effect on the autoimmune system and counteracts inflammatory diseases (Cardoso et al. 2004; Kremer et al. 1990; Sales-Campos et al. 2013). The presence of olive oil in the diet leads to improved immune responses associated with the elimination of bacteria and fungi.

Moreover, having the highest in vitro inhibition of prolyl endopeptidase (PEP), which is critically involved in the formation of amyloid plaque in the brain, oleic acid has been suggested to have a potential role to play in decreasing brain-related disorders such as dementia and Alzheimer's disease (Aluko 2012). Attenuation of the store-operated Ca^{2+} entry (SOCE) process by oleic acid could reduce Ca^{2+} influx into cells and as a result diminish tumorous cell proliferations, suggesting a function of oleic acid in the therapeutic management of colorectal cancer (Aluko 2012).

The second most abundant category of fatty acids after MUFAs in the oil of *Macrotermes* sp. was the SFAs. Foods or oils containing a higher proportion of SFAs are considered undesirable, because of their suspected connection with atherosclerotic maladies. The most abundant SFA in *Macrotermes* sp. adults was palmitic acid followed by stearic acid, a situation also seen in other species of termites such as *Odontotermes* sp., *M. subhyalinus*, *M. bellicosus*, *P. militaris*, *P. spiniger*, *Nasutitermes* sp. (Chakravorty et al. 2016; Kinyuru et al. 2013; Oyarzun et al. 1996). However, in nymphs, stearic acid was higher than palmitic acid as reported for *Syntermes* sp. soldiers (Paoletti et al. 2003) indicating different roles of fatty acids in the termite's development. The least abundant oil category determined in our *Macrotermes* sp. was that of the PUFAs, which contained almost entirely $n - 6$ linoleic acid. This result differs from that of some other reports: *Syntermes* sp. thorax and abdomen and *M. bellicosus* oils (from Nigeria) contained high amounts of PUFAs (30–35%) (Paoletti et al. 2003; Ekpo and Onigbinde 2007). However, the total fat content in termites varies depending on the caste of a termite (and possibly season as well). Winged alates, that is, the reproductive castes, contain much more fat than workers and soldiers (Oyarzun et al. 1996), which could be related to the labour-intensive activities these castes perform as part of their roles in the social framework of these insects. Perhaps due to their higher fat content, queen termites are used as a fattening agent for livestock (Kendie et al. 2018). Based on calculations of total fatty acids, it became obvious that the soldier caste contained less fat than the reproductive caste.

The mineral content of *Macrotermes* termite is shown in Table 22.5. In comparison with other foods of animal origin, *Macrotermes* termites were found to be exceptionally rich in iron. The high value could be due to the defatting procedure of

Table 22.5 Mineral composition of *Macrotermes* sp.

Minerals	mg/100 g defatted matter
Calcium	379.7
Magnesium	119.0
Sodium	89.5
Potassium	365.4
Phosphorus	490.6
Iron	204.8
Copper	5.5
Zinc	17.4
Manganese	57.2

the sample prior to analysis (i.e. after exraction of fat), facilitating iron detection. Consequences of iron deficiency through inadequate supplies of iron to cells of the body requiring this element are manifold, anaemia being the most serious (Kotze et al. 2009). A recent epidemiological study revealed that about 31.8% of pregnant women in the Asossa zone of Benishangul Gumuz are anaemic and 54% were diagnosed as moderately anaemic (Abay et al. 2017). Even the prevalence of prenatal anaemia is high in the region (Abay et al. 2017). Not only is the termite's iron content high, termites also contain relatively high amounts of zinc and copper. Zinc is an essential component of a large number (at least 300) of enzymes and copper is an important trace element in humans. The important role of copper in redox reactions and as a scavenger of free radicals (Linder and Hazegh-Azam 1996) depends on copper's ability of easily accepting and donating electrons by shifting between the cuprous (Cu^+) and cupric (Cu^{2+}). Furthermore, copper in concert with zinc functions as a structural component in the antioxidant enzyme superoxide dismutase (Turnland 2006).

Absorption of calcium is relatively inefficient, especially in the presence of phytate and oxalate which significantly hinder calcium absorption (Rafferty and Heaney 2008; Weaver et al. 1997). Vitamin D also influences calcium absorption (Christakos et al. 2011). The sources of dietary calcium are predominantly milk, dairy products, animal meats and oil sardines. Oil sardines are virtually absent in the region and meat consumption is not at all regular: 62.2% of the local residents consume meat only once a week (Abay et al. 2017). Calcium plays a vital role in the formation of bone and teeth, is involved in neuromuscular actions, many enzyme-mediated processes, excretion and blood clotting. The most vulnerable section of the population is post-menopausal women who often suffer from osteoporosis due to calcium deficiency. Given good bioavailability, minerals present in termites could be expected to mitigate the risks of calcium, zinc, and iron deficiency disorders.

From a nutritional point of view, termites can supplement the nutritional status of those in the local population who suffer from food deficits most. With a population that doubled during the last decade, the total population of the Benishangul Gumuz region is now about one million according to the Central Statistics Agency of Ethiopia (Population projection of Ethiopia for all regions at Wereda from 2014 to 2017 by Federal Democratic Republic of Ethiopia Central Statistical Agency). The livelihood of almost 95% of the population depends on subsistence agriculture (Flatie et al. 2009) and the situation has improved little given the small number of livestock and presence of epizootic diseases that ever so often reduce any improvements made (Benishangul Rehabilitation and Development Association 2000).

While foods of animal origin are generally costly and beyond the buying capacity for poor local folk, termite harvests are priceless. However, termites alone cannot solve malnutrition, but in concert with other approaches to fight nutritional inadequacies they can make a contribution. Obviously their nutritional qualities had led to the traditional acquisition of termite-eating by local folk before western attitudes caused them to scale down their customary uses of termites. In order to explore the community-based acceptance of termites as a food source, an in-depth ethnographic study needs to be conducted and a critical assessment as to the risk of overexploiting the termite resource has to be prepared. Ideally the farming of termites, that is, rearing them under controlled conditions, would be the way forward, but that remains the challenge of the future.

Acknowledgements In addition to the informants in the field we wish to acknowledge the support received from the Priority Research Centre's Program of the National Research Foundation of Korea (NRF) funded by the Ministry of Education, Science and Technology (NRF-2018R1A6A1A03024862).

References

Abay A, Yalew HW, Tariku A, Gebeye E (2017) Determinants of prenatal anemia in Ethiopia. Arch Public Health 75:51. https://doi.org/10.1186/s13690-017-0215-7

Akutse KS, Owusu EO, Afreh-Nuamah K (2012) Perception of farmers' management strategies for termites control in Ghana. J Appl Biosci 49:3394–3405

Aluko RE (2012) Functional foods and nutraceuticals. Springer, New York

AOAC (1990) Official methods of analysis, 15th edn. Association of Official Analytical Chemists, Washington DC

Atu UG (1993) Cultural practices for the control of termite (Isoptera) damage to yams and cassava in South-Eastern Nigeria. J Int Pest Manag 39(4):446–462. https://doi.org/10.1080/09670879309371841

Benishangul Rehabilitation and Development Association (2000) Nutritional Survey in Assosa Zone. Assosa

Bennett FJ, Mugalula-Mukiibi AA, Lutwama JSW, Nansubugaet G (1965) An inventory of the Kiganda foods. Uganda J 29:45–53

Bequaert J (1921) Insects as food: how they have augmented the food supply of mankind in early and recent years. Nat Hist J 21:191–200

Bergier E (1941) Peuples entomophages et insects comestibles: étude sur les moeurs de l'homme et de l'insecte. Imprimerie Rullière Frères, Avignon

Bodenheimer FS (1951) Insects as human food. W. Junk, The Haugue

Cardoso CR, Souza MA, Ferro EA Jr, Favoreto S, Pena JD (2004) Influence of topical administration of n-3 and n-6 essential and n-9 nonessential fatty acids on the healing of cutaneous wound. Wound Repair Regen 12(2):235–243. https://doi.org/10.1111/j.1067-1927.2004.012216.x

Chakravorty J, Ghosh S, Megu K, Jung C, Meyer-Rochow VB (2016) Nutritional and anti-nutritional composition of Oecophylla smaragdina (hymenoptera: Formicidae) and *Odontotermes* sp. (Isoptera: Termitidae): two preferred edible insects of Arunachal Pradesh. J Asia Pac Entomol 19(3):711–720. https://doi.org/10.1016/j.aspen.2016.07.001

Chakravorty J, Ghosh S, Meyer-Rochow VB (2013) Comparative survey of entomophagy and entomotherapeutic practices in six tribes of eastern Arunachal Pradesh (India). J Ethnobiol Ethnomed 9:50. https://doi.org/10.1186/1746-4269-9-50

Christakos S, Dhawan P, Porta A, Mady LJ, Seth T (2011) Vitamin D and intestinal calcium absorption. Mol Cell Endocrinol 347(1–2):25–29. https://doi.org/10.1016/j.mce.2011.05.038

Costa-Neto EM (2005) Entomotherapy, or the medicinal use of insects. J Ethnobiol 25(1):93–114. https://doi.org/10.2993/0278-0771(2005)25[93:EOTMUO]2.0.CO;2

de Figueiredo RECR, Vasconcellos A, Policarpo IS, Alves RRN (2015) Edible and medicinal termites: a global review. J Ethnobiol Ethnomed 11:29. https://doi.org/10.1186/s13002-015-0016-4

Debelo DG, Degaga EG (2015) Farmers' knowledge, perceptions and management practices of termites in the central rift valley of Ethiopia. Afr J Agric Res 10(36):3625–3635. https://doi.org/10.5897/AJAR2014.9283.

deLorgeril M, Salen P (2006) The Mediterranean diet in secondary prevention of coronary heart disease. Clin Invest Med 29(3):154–158

Ekpo KE, Onigbinde AO (2007) Characterization of lipids in winged reproductive of the termite *Macrotermesbellicosus*. Pak J Nutr 6(3):247–251

FAO/WHO/UNU (2007) Protein and amino acid requirements in human nutrition. Report of a joint WHO/FAO/UNU Expert Consultation. WHO Technical Report Series No. 935

Flatie T, Gedif T, Asres K, Gebre-Mariam T (2009) Ethnomedical survey of berta ethnic group Assosa zone, Benishangul-Gumuz regional state, mid-West Ethiopia. J Ethnobiol Ethnomed 5:14. https://doi.org/10.1186/1746-4269-5-14

Folmer O, Black M, Hoeh W, Lutz R, Vrijenhoek R (1994) DNA primers for amplification of mitochondrial cytochrome c oxidase subunit 1 from diverse metazoan invertebrates. Mol Mar Biol Biotechnol 3(5):294–299

Ghosh S, Jung C, Meyer-Rochow VB (2016) Nutritional value and chemical composition of larvae, pupae, and adults of worker honey bee, Apis mellifera *ligustica* as a sustainable food source. J Asia Pac Entomol 19:487–495

Ghosh S, Lee SM, Jung C, Meyer-Rochow VB (2017) Nutritional composition of five commercial edible insects in South Korea. J Asia Pac Entomol 20(2):686–694. https://doi.org/10.1016/j.aspen.2017.04.003

Ghosh S, Meyer-Rochow VB, Jung C (2018) Importance of neglected traditional food to ensure health and Well-being. Int J Food Nutr Sci 8(1):555729. https://doi.org/10.19080/NFSIJ.2018.08.555729

Hebert PD, Cywinska A, Ball SL, deWaard JR (2003) Biological identifications through DNA barcodes. Proc Biol Sci 270:313–321. https://doi.org/10.1098/rspb.2002.2218

Kendie FA, Mekuriaw SA, Dagnew MA (2018) Ethnozoological study of traditional medicinal appreciation of animals and their products among the indigenous people of MetemaWoreda, North-Western Ethiopia. J Ethnobiol Ethnomed 14:37. https://doi.org/10.1186/s13002-018-0234-7

Keys A (1970) Coronary heart disease in seven countries. Circulation 14 (Suppl.):S1–211

Kim HS, Jung C (2013) Nutritional characteristics of edible insects as potential food materials. Kor J Apic 28:1–8

Kinyuru JN, Konyole SO, Roos N, Onyango CA, Owino VO, Owuor BO, Estambale BB, Friis H, Aagaard-Hansen J, Kenji GM (2013) Nutrient composition of four species of winged termites consumed in Kenya. J Food Compos Anal 30(2):120–124

Korean Food Standard Codex (2010) Ministry of Food and Drug Safety (Republic of Korea)

Kotze MJ, van Velden DP, van Ransburg SJ, Erasmus R (2009) Pathogenic mechanisms underlying iron deficiency and iron overload: new insights for clinical application. Electron J Int Fed Clin Chem Lab Med 20(2):108–123

Kremer JM, Lawrence DA, Jubiz W, DiGiacomo R, Rynes R, Bartholomew LE, Sherman M (1990) Dietary fish oil and olive oil supplementation in patients with rheumatoid arthritis. Clinical and immunologic effects. Arthritis Rheumatol 33(6):810–820

Linder MC, Hazegh-Azam M (1996) Copper biochemistry and molecular biology. Am J Clin Nutr 63(5):797S–811S

Livingstone D (1857) Missionary travels and researches in South Africa. John Murray, Albemarte St., London

Meyer-Rochow VB (1975) Can insects help to ease the problem of world food shortage? Search 6(7):261–262

Meyer-Rochow VB (2017) Therapeutic arthropods and other, largely terrestrial, folk-medicinally important invertebrates: a comparative survey and review. J Ethnobiol Ethnomed 13:9. https://doi.org/10.1186/s13002-017-0136-0

Nadeau L, Nadeau I, Franklin F, Dunkel F (2014) The potential for entomophagy to address undernutrition. Ecol Food Nutr. https://doi.org/10.1080/3670244.2014.930032

Nonaka K (1996) Ethnoentomology of the central Kalahari san. Afr Stud Monogr 22(Supplementary issue):29–46. https://doi.org/10.14989/68378

Nyeko P, Olubayo FM (2005) Participatory assessment of Farmers' experiences of termite problems in agroforestry in Tororo district, Uganda. Agriculture Research and Extension Network (AgREN) Network Paper No. 143. ISBN 0850037441

Oyarzun SE, Crawshaw GJ, Valdes EV (1996) Nutrition of the Tamandua: I. nutrient composition of termites (*Nasutitermes* spp.) and stomach contents from wild Tamanduas (Tamandua tetradactyla). Zoo Biol 15(5):509–524. https://doi.org/10.1002/(SICI)1098-2361(1996)15:5<509::AID-ZOO7>3.0.CO;2-F

Paoletti MG, Buscardo E, Vanderjagt DJ, Pastuszyn A, Pizzoferrato L, Huang Y-S, Chuang L-T, Glew RH, Millson M, Cerda H (2003) Nutrient content of termites (*Syntermes* soldiers) consumed by Makiritare Amerindians of the alto Orinoco of Venezuela. Ecol Food Nutr 42(2):177–191. https://doi.org/10.1080/036702403902-2255

Pickering MV, Newton P (1990) Amino acid hydrolysis: old problems, new solutions. LC/GC 8(10):778–781

Rafferty K, Heaney RP (2008) Nutrient effects on the calcium economy: emphasizing the potassium controversy. J Nutr 138(1):166S–171S

Redford KH, Dorea JG (1984) The nutritional value of invertebrates with emphasis on ants and termites as food for mammals. J Zool 203:385–395

Sales-Campos H, de Souza PR, Peghini BC, da Silva JS, Cardoso CR (2013) An overview of the modulatory effects of oleic acid in health and disease. Mini Rev Med Chem 13(2):1–10. https://doi.org/10.2174/1389557511313020003

Silow CA (1983) Notes on Ngangela and Nkoya ethnozoology. Ants and termites. Entologiska Studier

Sponheimer M, Lee-Thorp J, de Ruiter D, Codron D, Codron J, Baugh AT, Thackeray F (2005) Hominins, sedges, and termites: new carbon isotope data from the Sterkfontein valley and Kruger national park. J Hum Evol 48(3):301–312

Tamura K, Peterson D, Peterson N, Stecher G, Nei M, Kumar S (2011) MEGA5: molecular evolutionary genetics analysis using maximum likelihood, evolutionary distance, and maximum parsimony methods. Mol Biol Evol 28(10):2731–2739. https://doi.org/10.1093/molbev/msr121

Turnland JR (2006) Copper. In: Shils ME, Shike M, Ross AC, Caballero B, Cousins RJ (eds) Modern nutrition in health and disease, 10th edn. Lippincott Williams and Wilkins, Philadelphia, PA, pp 286–299

Twine WD, Moshe T, Netshiluvhi TR, Siphugu V (2003) Consumption and direct-use values of savanna bio-resources used by rural households in Mametja, a semi-arid area of Limpopo province, South Africa. South Afr J Sci 99:467–473

Van Huis A (2017) Cultural significance of termites in sub-Saharan Africa. J Ethnobiol Ethnomed 13:8. https://doi.org/10.1186/s13002-017-0137-z

Van Huis A, Itterbeek JV, Klunder H, Mertens E, Halloran A, Muir G, Vantomme P (2013) Edible Insects: Future Prospects for Food and Feed Security. FAO Forestry Paper 171

Weaver CM, Heaney RP, Nickel KP, Packard PI (1997) Calcium bioavailability from high oxalate vegetables: Chinese vegetables, sweet potatoes and rhubarb. J Food Sci 62(3):524–525. https://doi.org/10.1111/j.1365-2621.1997.tb04421.x

Wilsanand V (2005) Utilization of termite, *Odontotermes formosanus* by tribes of South India in medicine and food. Explorer 4(2):121–125

Index

A

AccuPower PCR PreMix, 296
Acheta domesticus, 96
Acridians, 170
Acridians consumption, 179
Adown stream camera system, 126
African conventional foods, 177
African Edible Acridians (Grasshoppers)
  aluminum, 181
  baking, 187
  barium, 181
  boron, 181
  calcium content, 183
  carbohydrates, 179
  chromium, 184
  cobalt, 183
  copper, 184
  dietary fiber, 180
  diversity, 170
  domestic markets, 170
  eating habits, 171
  energy intake, 180
  environmental pollution, 189
  heavy metals, 189
  iron, 185
  lead, 187
  lipids, 177–178
  magnesium, 185–186
  manganese, 185
  nutrients, 170
  phosphorus, 186
  potassium, 184
  proteins, 171–177
  sodium, 186
  Southern Africa, 171
  species, 172–176
  vitamins, 179
  vutritive value, 182
  water levels, 189
  zinc, 187
African edible insects
  antibacterial and anticancer drugs, 97
  beetle grubs, 97
  caterpillars, 96, 97
  composition of, 99
  entomophagy, 97
  food security, 96
  insect orders, 98
  nutritional composition, 97, 98
  termites, 96, 97
African insect consumption, 60
African palm weevil, 38, 63
African palm weevil larva, 46
African protein malnutrition programs, 177
Agonoscelis pubescens, 150, 156, 157
AgriProtein, 32, 37
Air separation, 127
Albeit critical difference, 117
Antimicrobial peptides (AMPs), 107
Apis cerana, 266, 267, 269, 271
Apis dorsata, 266, 268, 269, 271
Aquatic insects, 118
Artificial neural network classifiers
  (ANN), 130
Asia Honeybee species
  amino acid analysis, 268
  amino acid compositions, 269
  commercial beekeeping sector, 267
  fatty acid analysis, 268
  fatty acid composition, 270

Asia Honeybee species (*cont.*)
    food insects, 267
    food production, 267
    food resource, 266
    global population, 266
    livelihood, 271
    mineral content, 271
    minerals analyses, 269
    nutritional benefits, 271
    sample collection, 268
    wasps, 266
Aspire Food Group, 38
Aspongopus viduatus, *see* Watermelon bug
Aspongopus viduatus (Pentatomidae), 160
Australian Centre for International
        Agricultural Research (ACIAR), 38
Australopithecus, 266
Automation
    cost-effective farming systems, 124
    FAO technical consultation, 124
    industrial cultivation, 124
Average daily gain (ADG), 56

B
Bacterial endospores, 72
Bee keeping, 267
Bioactive components, 214, 250
Biodiesel, 165
Biological macromolecules, 232
Black soldier fly (BSF)
    adult and larvae, 196
    allergies, 206
    benefit, 196, 204
    biology, 198
    bioreactors, 203
    BSFL, 197
    characteristics, 196
    diets, 203
    edible and nutritious, 204
    eggs/larvae, 198
    fatty acid profiles, 200
    feedstock, 201
    insecticides, 206
    insects, 199
    livestock, 202–204
    macronutrients, 198
    and microbes, 205
    mineral content, 202
    MUFA, 203
    nutritional value, 199
    populations, 201
    protein digestibility, 199
    protein levels, 201
    PUFA, 200
    sensory tests, 204
    structure, 197
    supplementation, 203
    vitamins, 201
Black soldier fly larvae (BSFL), 197
Boiled cricket (*Brachytrupes* spp.), 45
Brain-related disorders, 302
Brood, 266–268, 271
Business farming, 262

C
Cannibalism, 220
Carbohydrates, 179, 233
Carotenoids, 237
Caterpillars (*Cirina forda*), 96
Cecropin, 97
Central Statistics Agency of Ethiopia, 303
Chinese and Indian conventional drugs, 227
Chinese medicine, 227
Chitin, 120, 199, 206, 235, 236
Codex Alimentarius, 143
Coleoptera, 258
Commercial approach, 278
Commercialization, 21
Consuming insects, 59
Consumption
    African countries, 21, 30
    catching insect, 31
    Central African Republic, 22
    climate, 20
    cultivation and utilization, 21
    customer-centric strategy, 32
    definition, 19
    demand and low supply, 34
    disease, 32
    domestic consumption, 30
    doom-spelling factors, 21
    edible insects, 22, 27
    entomophagy, 34
    entotherapy/zootherapy, 20
    equilibrium, 33
    FAO, 21
    fishmeal, 36
    funding agencies, 31
    Ghana, 28
    harvest and preservation, 29
    insect business, 32
    insect species, 24–26
    large-scale production, 31
    linoleic and linolenic acids, 21
    livestock, 22
    malnutrition, 22

market, 20, 29
meat, 20
meat protein, 36
mopane caterpillar, 21
Nigeria, 26, 28–29
nutritional, medicinal values, 21
population, 20
preparation forms, 31
processing, 33
tropical rainforest, 20
Contemporary renovation methods, 261
Controlled-diet study, 201
Conventional African sources, 177
Cossus, 55
Creutzfeld–Jakob disease, 205
Cricket life cycle
adult, 216
courtship process, 217
digestive system, 216
eggs, 215
feeding, 220
housing, 219
human consumption, 219
laboratory experiments, 218
nutrient and bioactive components, 217–219
nutritional analysis, 219
nymph, 215
reproduction, 216–217
Cricket production, 222–223
Cricket protein, 217
Cricket rearing production system, 9
Crickets (Gryllidae)
animal protein, 213
biology, 214
consumption, 222
human consumption, 213
mating stock, 221
processing, 221
production, 214
Criquet puant, 4

D
DEAE-cellulose, 226
Defatting, 134–138
Democratic Republic of Congo (DRC), 26,
101, 118, 285
Desert locust, see Schistocerca gregaria
Diesel fuel viscosity, 165
Dietary fibre, 180
Dietary recommendations, 200
Diptera, 258
DNA Barcode region, 296
Docosahaexaenoic acid (DHA), 106

Dried mealworm larvae, 136
Drying, 133–134
Drying strategies, 278

E
Eating edible insects, 57
Economic important
edible insect
African edible bush cricket, 6
mopane caterpillar, 4–5
shea caterpillar, 6
Edible Grasshoppers, 177
Edible insect lobbying groups, 205
Edible insects, 3, 9, 10, 54–57, 60, 119,
284, 288
African (see African edible insects)
Anglo-Saxon countries, 144
animal-source food, 95
Australia and New Island, 145
Belgium, 145
bioactive compounds, 107, 108
consumer acceptance, 40
EU Parliament, 145
FDA, 144
Finland, 146
as foods, 100–104
Germany, 146
Great Britain, 147
health benefits, 117
import and sales, 144
Kingdom of Denmark, 146
legislative framework, 148
livestock, 21
magical insects, 43–47
marketing, 42
medicinal Services, 42–43
microbiology, 61
Netherlands, 146
Non-Western Countries, 147–148
North-American country, 144
Norway, 147
as oil, 106, 107
organized market, 39
packaging, 39
processing, 41
as protein, 104, 105
scientific names, 77–79
storage, 39
sub-Saharan Africa, 44
Switzerland, 147
technology and research, 39
USA, 144
Eicosapentaenoic acid, 106

Elobied Agricultural Studies, 160
Energy, 234
Entomophagy, 53–55, 60, 226, 232, 244, 295
    allergic reaction, 57
    beetles, 54
    benefits, 55–57
    consumption of insects, 54
    edible insects, 56
    energy and protein, 54
    food and feed outputs, 54
    insect species, 54
    spoilage/wastage, 57
    toxicity, 57
Essential amino acids (EAAs), 156
Essential fatty acids (EFA), 36
EU Food Safety Authority (EFSA), 143
European Food Safety Authority, 84

F
FAO/WHO Codex Coordinating Committee, 142
Farming insects, 60
    crickets, 9
    edible insect, 7
    palm weevil, 7–9
    simple rearing methods, 7
Fasopro, 38
Fat, 233
Fatty acid composition, 118, 152
Fatty acid methyl esters (FAMEs), 268, 296
Federal Agency for the safety of the food
        chain (FASFC), 143
Federal Democratic Republic of Ethiopia
        Central Statistical Agency, 303
Fermentation, 71
Fiber content, 120–121
Fibre, 233
Fish-oil production, 36
Food and Agricultural Organization (FAO),
        21, 232
Food and Agriculture Organization of the
        United Nations (UN), 95
Food-borne infections, 67
Food-borne pathogens, 64, 74
Food Hygiene Regulations of the European
        Union, 85
Food nutrients, 20
Food safety, 60, 67, 72
Food safety regulations
    allergenic structures, 88
    allergens, 84
    Codex Alimentarius, 85
    contaminants, 84
    EFSA, 85

European Union regulation, 85
    grasshoppers, 84
    health hazards, 84
    human consumption, 85
    insect-eating, 83
    metamorphosis, 84
    microbial safety, 85–87
    molds and mycotoxins, 87
    parasites, 88
    physical hazards, 90
    protein source, 83
    technologies, 84
    toxic, 89
Food security, 3, 5, 10, 11
Foodstuff processing, 67
Force balancing, 128
Fortified blended foods (FBFs), 251
Four fibroin proteins of weaver ants
        (WAF1–4), 226, 228
Fourier transform infrared spectroscopy
        (FTIR), 166
Freeze-dried mealworms, 278
Functional model, 125

G
Gas chromatography (GC), 154
Genuine bugs, 160
Globe Customs Organization, 144
Glutamate dehydrogenase, 278
Glutamic acid, 171
Glycoprotein, 226
Grain proteins, 118
Grasshopper sale, 46
Grasshoppers (Acridids), 96
Greenhouse gases (GHG), 20–21
Gross domestic product (GDP), 20
Gross energy, 259

H
Halal gelatin, 166
Harvesting, 217
    and post-harvest management, 221
    process, 221
Hatchings, 198
Hazard Analysis, 143
Health foods, 253
Healthy benefit, 279
Heat treatment, 70
Heavy metals, 89
Hemijana variegata, 102
Hemiptera, 258
Hermetia illucens, 67

Heterobranches bidorsalis, 288
House crickets, 223
Hymenoptera, 258

I
Image processing, 131
Imitative magic, 44
Inductively coupled plasma-optical emission
        spectrophotometer (ICP-OES), 269
Insect as feed, 6–7
Insect-based feed, 142
Insect-based food or products, 223
Insect-based foods, 100
Insect breeding, 127
Insect derived food products, 89
Insect farming, 66
Insect farming and processing, 87
Insect farming systems, 67
Insects, animal feed
    defatting, 134
    drying, 133–134
    European countries, 133
    insect-processing industry, 133
    mealworm larvae, 133
    physicochemical properties, 137
    press process, 136
Insects as food
    dryland farming and pastoral rangeland
            systems, 2
    nutrition, 9–10
    pest insects, 3–4
    processing, 10
    sub-Saharan Africa, 1–3
Insect-specific pathogenic microorganisms, 85
International Development Research Centre,
        Canada (IDRC), 38
Intrinsic flora, 61
Isoptera, 258, 282

K
Kembaar, 43
Kitchen hygiene, 73

L
Labial/larval salivary glands, 228
Larval condition, 131
Legal frameworks, 141
Legislation
    EU Commission, 142
    FAoLEx, 141
    food and feed production, 142

foodstuffs, 143
organic waste, 141
risk assessments and containment, 142
small-scale production and trade, 143
species and quantities, 142
traditional dietary intake, 143
Lepidoptera, 258
Linoleic acid, 178
Lipids, 247, 248
Livestock, 21
Lobbying, 142
Locust (Fara), 45
Locusta migratoria
    bioactive ingredients
        antioxidant peptides, 237
        body, 236
        carotenoids, 237
        chitin, 235, 236
        vitamin A, 236
        vitamin B$_{12}$, 236
        vitamin D, 237
    concentrations, 238
    entomophagy, 232
    nutrients
        carbohydrates, 233
        diet, 235
        dry matter, 235
        energy, 234
        fat, 233
        fibre, 233
        minerals, 233
        protein, 232
        vitamins, 234
Lysine, 177, 299

M
Macronutrients
    bioactive components, 250
    fibre content, 250
    lipids, 247, 248
    minerals, 249, 250
    proteins, 248, 249
Macrotermes, 297, 299
Macrotermes bellicosus, 101, 285, 288
Macrotermes nigeriensis, 286, 288
Macrotermes subhyalinus, 101
Marketing
    application, 41
    geography, 41
    product type, 40
Mealworm (*Tenebrio molitor*)
    larva, 276
    life stages, 276

Mealworm (*Tenebrio molitor*) (*cont.*)
    nutrient composition, 277–278
    nutritional value, 277–278
    processing, 278–279
    stored food items, 275
    uses, 277–278
Meat consumption, 36
Medihoney™, 43
Melon bug oil (MBO), 161, 162
    antibacterial activity, 165
    deodorization temperatures, 165
    kinematic viscosity, 165
    laboratory scale, 163
    metabolism, 166
    oxidative balance, 164
    phosphorus, 164
    quality and stability, 165–166
    SKO, 164
    viscosity values, 165
Melon bugs, 160
Microbial safety, 85–87
Microbiological analyses, 61
Microbiological criteria, 74
Microbiological safety parameters,
        75–76
Microbiome, 61–63, 66, 67
Micronutrient insufficiency, 119
Micronutrients, 119
Microorganisms/fungi, 253
Migratory locust, *see Locusta migratoria*
Milling edible insects, 42
Minerals, 119, 233, 249, 250
Mini-livestock, 142
Ministry of Agriculture, Water and Forestry
        (MAWF), 252
Molds and mycotoxins, 87
Monounsaturated fatty acids (MUFA), 271
Mopane Worm (*Gonimbrasia belina*)
    distribution and consumption patterns,
        Africa, 243–245
    food resource, 251
    food safety and quality, 253, 254
    good harvesting practice, 243
    harvesting and processing, 245, 246
    insect, 242, 243
    life cycle, *G. Belina*, 243
    nutritional and economic significance, 242
    nutritional composition, 243, 246, 247
    phases, 242
    rainfall, 242
    sustainability and rural livelihoods support,
        251–253
Mounds, 282, 283, 285, 289
Mpaylaar, 43

N
N-acetyl-D-glucosamine, 235
National Centre for Biotechnology
        Information (NCBI), 296
Next Generation Nutrition (NGN), 32
Nigeria animal protein consumption, 35
Non-African edible insect species, 65–66
Non-governmental organizations (NGOs), 253
Non-wood forest merchandise (NWFPs), 257
Nsenene, 6
Nutrient
    and bioactive components, 214
    composition, 248
Nutrient Value Score of crickets, 116
Nutritional assessment
    dietary energy, 117
    proteins and amino acids, 117
Nutritional composition, 97, 98
Nutritional supplementation, 244
Nutritional values, 177, 261
Nutrition-related questions, 198
Nutritious food, 96
Nwaigu, 44

O
Odonata, 258
Odontotermes lokanandi, 297
Oecophylla smaragdina
    nutrition and bioactive components
        amino acid, 226
        fibroins, 226
        glycoprotein, 226
        lectins, 226
        lipids, 226
    O. longinoda, 225
    uses
        Chinese medicine, 227
        entomophagy, 226
        harvest, 227
        herbal fibers, 228
        herbal habitats, 227
        larvae and pupae, 227
        markets, 227
        meals and conventional remedy, 227
        silk, 228
        WAF1–4, 228
Oil aliquots, 155
Oleic acid, 300
Omega-3fatty acids, 106
Optical measuring system, 130
Optimum housing, 219
Organic compounds, 180
Orthoptera, 22, 258

Oryctes monoceros, 102
Ovipositor, 215

P

Palm weevil, 41
Palm weevil delicacy, 100
Palm weevil larvae, 118
Parasites, 88
Pesticides, 90
Phospholipids, 260
Physical hazards, 90
Pierisin, 97
Polyandry, 216
Polymerase chain reaction (PCR), 296
Polyrhachis vicina, 226
Polyunsaturated fatty acids (PUFAs), 106, 271
Population growth, 95
Poultry feed, 36, 198
Primitive foodstuffs, 60
Prolyl endopeptidase (PEP), 302
Protein-containing foods, 88
Protein solubility, 134
Protein source, 116
Proteins, 54, 232, 248, 249, 299
Pseudacanthotermes spiniger, 297

Q

Quality Insect for Feed and Food in Nigeria
(QUIFFAN), 33

R

Rearing system, 125
Relative humidity (RH), 215
Renovation/processing technique, 261
Rhynchophorus phoenicis, 66, 101, 102
Roller mils, 137
Root system, 252

S

Saturated fatty acids (SFA), 106, 200, 271
Scanning calorimetry, 166
Schistocerca gregaria
    advantages, 259
    African and Arabian diets, 258
    business farming, 262
    Cyrtacanthacris septemfasciata, 257
    desert-type region, 258
    gross energy, 259
    iron content, 260
    Locusta migratoria migratorioides, 257

Locustana pardalina, 257
    nutritional values, 261
    organophosphorus pesticides, 259
    phospholipids, 260
    prepared-to-eat, 260
    processing and commercializing, 260
    quantities, 259
    renovation/processing technique, 261
    uses, 260, 261
    winged swarming things, 259
Sensory quality
    consumable species, 116
    cultivating conditions, 116
    dietary benefit, 116
    nutritional benefit, 116
    protein sources, 116
Shea tree caterpillar (*Cirina butyrospermi*), 97
Shorter development period, 220
Silk, 228
Simple rearing methods, 7
Small-scale bioreactors, 198
Small-scale production, 141
Sodium dodecyl sulfate–polyacrylamine gel
        electrophoresis (SDS-PAGE),
        156, 166
Sorghum bug
    biochemistry and nutrition, 150
    edible oil, 151–156
    oil content, 151
    palm weevil, 149
    physiological conditions, 150
Sorghum bug oil (SBO), 151
    antioxidant interest, 154
    biodiesel, 156
    in deep-frying, 155
    fatty acid composition, 155
    fatty acids, 152
    frying, 154
    kinematic viscosity values, 156
    lipid classes, 153
    oxidative stability, 155
    sterols, 153
    tocopherol content, 152
    tocopherols, 153
    transesterification, 156
Soybean (*Glycine max*), 288
Sphenarium purpurascens, 56
Stearic acid, 302
Steroids, 178
Sunflower kernel oil (SKO), 164
Supply–demand gap, 35
Swine feed, 199
Swiss Federal Institute of Aquatic Science and
        Technology, 197

T
Tenebrio molitor, 124, 130, 133
Termites, 101, 102, 106
    agro-environment, 295
    amino acid, 296, 299, 300
    amino acid composition, 299
    animal-based food, 282
    calcium, 303
    communities, 295
    consumption, 294
    diversity, 294
    ecological roles, 294
    economic importance, 282, 283
    ethnic groups, 282
    fatty acid composition, 296
    fatty acid content, 300
    fatty acid distribution, 301
    fatty acid profile, 300
    food consumption, 283–285
    food products, 295
    food source, 294
    harvesting and processing, 285, 286
    human consumption, 295
    insect-based food products, 282
    iron deficiency, 303
    isotope studies, 294
    medicinal value, 289
    mineral composition, 302
    mineral content, 302
    minerals, 297
    nutritional importance, 287–289
    pedologic activity, 294
    protein, 299
    protein malnutrition, 282
    representative *COI* sequences, 298
    sample preparation, 295
    sandy and loamy soils possess, 294
    species identification, 296
    stearic acid, 302
    Syntermes sp., 295
    vitamin D, 303
    white ants, 282
    winged, 286
    zinc functions, 303
Tetragonula iridipennis, 267
Toxicity, 89
Traditional animal source foodstuffs, 28
Traditional entomophagy, 74

Traditional harvesting methods, 8
Traditional insect processing, 69
Traditional preserving methods, 73
Transesterification, 156, 165
Triacyglycerol (TAG), 152
Tryptophan, 299

U
University of Applied Sciences
        Bremerhaven, 126
Unsaturated fatty acids, 106
Usta terpsichore, 101

V
Vitamin A, 120, 236
Vitamin B_{12}, 236
Vitamin D, 237
Vitamin E, 120
Vitamins, 120

W
Wageningen University and Research Centre
        (WUR), 21
Wasps, 266
Waste management, 196–198, 201, 207
Watermelon (*Citrullus vulgaris*), 160
Watermelon bug
    biochemistry and nutrition, 160
    oil extracts, 163
    oil fatty acid composition, 161–162
    tocopherol and sterol content, 162
Watershed algorithm, 131
Weaver ants, *see Oecophylla smaragdina*
White ants, 282
Winged termite, 46, 47
World Food Program (WFP), 20
World Health Organization (WHO),
        104, 232

Z
Zig-zag airseparation, 127
Zig-zag separator, 128
Zinc insufficiency, 120
Zoonotic infections, 85

Printed by Printforce, the Netherlands